Biomass for Energy Application

Biomass for Energy Application

Editor

David Herak

MDPI • Basel • Beijing • Wuhan • Barcelona • Belgrade • Manchester • Tokyo • Cluj • Tianjin

Editor
David Herak
Czech University of Life Sciences Prague
Czech Republic

Editorial Office
MDPI
St. Alban-Anlage 66
4052 Basel, Switzerland

This is a reprint of articles from the Special Issue published online in the open access journal *Energies* (ISSN 1996-1073) (available at: https://www.mdpi.com/journal/energies/special_issues/biomass_energy_application).

For citation purposes, cite each article independently as indicated on the article page online and as indicated below:

LastName, A.A.; LastName, B.B.; LastName, C.C. Article Title. *Journal Name* **Year**, *Volume Number*, Page Range.

ISBN 978-3-0365-0268-7 (Hbk)
ISBN 978-3-0365-0269-4 (PDF)

© 2021 by the authors. Articles in this book are Open Access and distributed under the Creative Commons Attribution (CC BY) license, which allows users to download, copy and build upon published articles, as long as the author and publisher are properly credited, which ensures maximum dissemination and a wider impact of our publications.

The book as a whole is distributed by MDPI under the terms and conditions of the Creative Commons license CC BY-NC-ND.

Contents

About the Editor vii

Preface to "Biomass for Energy Application" ix

Giovanni Ferrari, Andrea Pezzuolo, Abdul-Sattar Nizami and Francesco Marinello
Bibliometric Analysis of Trends in Biomass for Bioenergy Research
Reprinted from: *Energies* 2020, 13, 3714, doi:10.3390/en13143714 1

Martin Lisý, Hana Lisá, David Jecha, Marek Baláš and Peter Križan
Characteristic Properties of Alternative Biomass Fuels
Reprinted from: *Energies* 2020, 13, 1448, doi:10.3390/en13061448 23

Saaida Khlifi, Marzouk Lajili, Saoussen Belghith, Salah Mezlini, Fouzi Tabet and Mejdi Jeguirim
Briquettes Production from Olive Mill Waste under Optimal Temperature and Pressure Conditions: Physico-Chemical and Mechanical Characterizations
Reprinted from: *Energies* 2020, 13, 1214, doi:10.3390/en13051214 41

Cimen Demirel, Gürkan Alp Kağan Gürdil, Abraham Kabutey and David Herak
Effects of Forces, Particle Sizes, and Moisture Contents on Mechanical Behaviour of Densified Briquettes from Ground Sunflower Stalks and Hazelnut Husks
Reprinted from: *Energies* 2020, 13, 2542, doi:10.3390/en13102542 55

Anna Brunerová, Hynek Roubík, Milan Brožek, Agus Haryanto, Udin Hasanudin, Dewi Agustina Iryani and David Herák
Valorization of Bio-Briquette Fuel by Using Spent Coffee Ground as an External Additive
Reprinted from: *Energies* 2020, 13, 54, doi:10.3390/en13010054 75

Moritz von Cossel, Anja Mangold, Yasir Iqbal and Iris Lewandowski
Methane Yield Potential of Miscanthus (*Miscanthus* × *giganteus* (Greef et Deuter)) Established under Maize (*Zea mays* L.)
Reprinted from: *Energies* 2019, 12, 4680, doi:10.3390/en12244680 91

Subhashis Das, Rajnish Kaur Calay, Ranjana Chowdhury, Kaustav Nath and Fasil Ejigu Eregno
Product Inhibition of Biological Hydrogen Production in Batch Reactors
Reprinted from: *Energies* 2020, 13, 1318, doi:10.3390/en13061318 109

Li Ji, Pengfei Li, Fuhou Lei, Xianliang Song, Jianxin Jiang and Kun Wang
Coproduction of Furfural, Phenolated Lignin and Fermentable Sugars from Bamboo with One-Pot Fractionation Using Phenol-Acidic 1,4-Dioxane
Reprinted from: *Energies* 2020, 13, 5294, doi:10.3390/en13205294 123

Hengli Zhang, Chunjiang Yu, Zhongyang Luo and Yu'an Li
Investigation of Ash Deposition Dynamic Process in an Industrial Biomass CFB Boiler Burning High-Alkali and Low-Chlorine Fuel
Reprinted from: *Energies* 2020, 13, 1092, doi:10.3390/en13051092 141

José Juan Alvarado Flores, José Guadalupe Rutiaga Quiñones, María Liliana Ávalos Rodríguez, Jorge Víctor Alcaraz Vera, Jaime Espino Valencia, Santiago José Guevara Martínez, Francisco Márquez Montesino and Antonio Alfaro Rosas
Thermal Degradation Kinetics and FT-IR Analysis on the Pyrolysis of *Pinus pseudostrobus*, *Pinus leiophylla* and *Pinus montezumae* as Forest Waste in Western Mexico
Reprinted from: *Energies* **2020**, *13*, 969, doi:10.3390/en13040969 153

Małgorzata Wzorek
Evaluating the Potential for Combustion of Biofuels in Grate Furnaces
Reprinted from: *Energies* **2020**, *13*, 1951, doi:10.3390/en13081951 179

Cheng Li, Xiaochen Yue, Jun Yang, Yafeng Yang, Haiping Gu and Wanxi Peng
Catalytic Fast Pyrolysis of Forestry Wood Waste for Bio-Energy Recovery Using Nano-Catalysts
Reprinted from: *Energies* **2019**, *12*, 3972, doi:10.3390/en12203972 195

Bartosz Matyjewicz, Kacper Świechowski, Jacek A. Koziel and Andrzej Białowiec
Proof-of-Concept of High-Pressure Torrefaction for Improvement of Pelletized Biomass Fuel Properties and Process Cost Reduction
Reprinted from: *Energies* **2020**, *13*, 4790, doi:10.3390/en13184790 207

About the Editor

David Herak, Ph.D. (1978), is currently the Vice Dean for International Relations and a professor at the Department of Mechanical Engineering, Faculty of Engineering, Czech University of Life Sciences, Prague. He specializes in the application of agricultural technology in tropical countries and energy efficiency for processing agricultural products. He has been involved in several scientific projects focused on the processing of tropical oil seeds worldwide (Ethiopia, Indonesia, Malaysia, Cambodia, Myanmar, Turkey and Ukraine). In addition, he is the coordinator of summer schools in Indonesia, the Philipinnes and Turkey. He also served as a guest professor at several foreign universities. Professor Herák is the author of 161 publications in scientific databases, the author of two patents, a member of the European Physical Society, a member of The Union of Czech Mathematicians and Physicians, as well as the editor and editorial board member of several high-level scientific journals.

Preface to "Biomass for Energy Application"

Considering new worldwide regulations related to the minimization of fossil fuel utilization to eliminate the negative impacts of global warming and prevent the exhaustion of resources, the utilization of biomass for energy applications has become one of the most common forms of renewable energy. The development of the utilization of renewable resources has highlighted a number of other tasks and constraints linked to the nature of renewable resources, including treatment, processing, conversion, and applied technologies, and thus, there is large number of critical views on this issue. Thus, this book, compiled from the new rigorously peer-reviewed research studies, tries to help the reader look at the utilization of biomass for energy applications from a different perspective. I hope that the research results published in this book will help us at least a little to keep the Earth in a sustainable form forever.

David Herak
Editor

Review

Bibliometric Analysis of Trends in Biomass for Bioenergy Research

Giovanni Ferrari [1], Andrea Pezzuolo [1,*], Abdul-Sattar Nizami [2] and Francesco Marinello [1]

1. Department of Land, Environment, Agriculture and Forestry, University of Padua, 35020 Legnaro, Italy; giovanni.ferrari.7@studenti.unipd.it (G.F.); francesco.marinello@unipd.it (F.M.)
2. Sustainable Development Study Centre, Government College University, Lahore 54000, Pakistan; nizami_pk@yahoo.com
* Correspondence: andrea.pezzuolo@unipd.it

Received: 29 June 2020; Accepted: 15 July 2020; Published: 19 July 2020

Abstract: This paper aims to provide a bibliometric analysis of publication trends on the themes of biomass and bioenergy worldwide. A wide range of studies have been performed in the field of the usage of biomass for energy production, in order to contribute to the green transition from fossil fuels to renewable energies. Over the past 20 years (from 2000 to 2019), approximately 10,000 articles have been published in the "Agricultural and Biological Sciences" field on this theme, covering all stages of production—from the harvesting of crops to the particular type of energy produced. Articles were obtained from the SCOPUS database and examined with a text mining tool in order to analyze publication trends over the last two decades. Publications per year in the bioenergy theme have grown from 91 in 2000 to 773 in 2019. In particular the analyses showed how environmental aspects have increased their importance (from 7.3% to 11.8%), along with studies related to crop conditions (from 10.4% to 18.6%). Regarding the use of energy produced, growing trends were recognized for the impact of biofuels (mentions moved from 0.14 times per article in 2000 to 0.38 in 2019) and biogases (from 0.14 to 0.42 mentions). Environmental objectives have guided the interest of researchers, encouraging studies on biomass sources and the optimal use of the energy produced. This analysis aims to describe the research evolution, providing an analysis that can be helpful to predict future scenarios and participation among stakeholders in the sector.

Keywords: renewable energy; bioenergy scenario; biomasses; systematic review

1. Introduction

Bioenergy is renewable energy derived from the treatment of several types of organic sources, which are generically named biomass [1,2]. Biomass is biological material derived, either directly or indirectly, from the transformation of solar energy into chemical energy [3]. It may be constituted of wood, forestry waste, crop residues, manure, urban waste, food industry residues, and the many by-products of agricultural processes [4–7]. International organizations and national governments are increasingly committed to pursuing environmental sustainability policies, setting even more ambitious targets for reducing pollution and the impact of human activities [8,9]. The production of bioenergy obtained from natural and agro-industrial sources represents one of the most critical points of this path [10].

The European Union (EU) has included, in their Sustainable Development Goals (SDGs), " ... 7. Affordable and clean energy ... "; specifying as indicator " ... 7.2.1 Renewable energy share in the total final energy consumption ... " and " ... 7.a.1 International financial flows to developing countries in support of clean energy research and development and renewable energy production, including in hybrid systems ... " [11]. The EU, in the "Renewable Energy Regulation", has established the goal of 32% of energy production from renewable sources by 2030 and reducing greenhouse gas emissions by 40% compared to 1990 [12].

In 2016, bioenergy is the most significant renewable energy source globally, covering 70% of the energy production by renewable sources. In every continent, biomass is the most important renewable energy source; it accounted for 40% of the energy in Oceania and almost 96% in Africa [13]. Biopower (or electricity from biomass) is the third largest renewable electricity generation source, with a share of 571 TWh of electricity produced. Asia is the leader in the sector, with a share of almost 40% of electricity from biomass produced [13]. In the transport sector, the primary renewable sources are liquid biofuels. From 2000 to 2017, biofuel production registered a significant growth: From 16 to 143 billion L. The 86% of the production of biofuel and bioethanol is concentrated in the U.S. and Brazil, with a production share of 87% [13]. Biofuels could help reduce greenhouse gases and many countries have set targets for the production and use of these resources. Ahorsu et al. [14] discussed the relevance of biomass for different generations of biofuels, also showing the main bioethanol producers: USA, Brazil, Europe, China, and Canada.

The widespread use of biomass determines numerous research areas for each phase of the energy supply chain: Biomass production, transport [15,16], treatments and digestion [17], energy production [18] and distribution [19], and plant planning and management [20,21], as well as the social, economic [22], and environmental [23] impacts that the use of biomass implies. Many review articles have been written from 2016 to 2019 to gather the periodical progress in the topic and identify possible future goals in the research. Long et al. [24] reviewed the results of previous studies that had investigated biomass resources and the estimation of their bioenergy potential, finding values of energy potential for 2050 between 96 and 161 Exajoule (EJ). Ferrarini et al. [25] assessed the potential impact of bioenergy buffers, linear areas placed around cultivated fields and watercourses with perennial herbaceous crops or wood biomass, and the biomass supply chain on ecosystem services. Pulighe et al. [26] studied the exploitation of marginal lands in the Mediterranean area as lands to cultivate energy crops. Authors examined the environmental impact of crops in order to assess the ecological costs of cultivations: Mekonnen et al. [27] quantified the consumption of green, blue, and gray water of global crop production for the period 1996–2005.

The research has revealed that the long-term exploitation of bioenergy buffers on previous croplands is more advisable than on grasslands, in order to sustains the long-term provision of multiple ecosystem services: climate, water quality, biodiversity regulation, and soil health. Quadir et al. [28] presented a series of case studies to show the potential economic and environmental benefits of restoration of salt-affected lands. These areas can be dedicated to food production with particular crops, or to bioenergy crops. Kluts et al. [29] reviewed European land studies on bioenergetic potentials and suggested that a more comprehensive approach, combining energy crop production with land demand for food/feed, is necessary for the identification of sustainable courses for European bioenergy production requires a more integrative approach, combining land demand for food, feed, and energy crop production. Kuhmaier and Erber. [30] reviewed the research trends regarding the comminution and transport of forest biomass in Europe. According to their review, future research should be focused on customizing the product quality, taking into consideration the user's requirements and on developing simulation and automatization tool for the co-ordination of chippers and trucks by simulation and automatization tools. Ba et al. [31] focused the attention on the Operations Research perspective studying recent research on models for biomass supply chains models and underlined the importance of multi-disciplinary research teams with the contribution of industrial engineering departments. Pari et al. [32] studied the harvesting technologies available in Europe to manage and take advantage of pruning. These residues could power approximately 200–500 kW electric power plants, with an annual output of 0.8 TWh. Garcia et al. [33] evaluated the biomethane potential and the chemical characteristics of a large number of organic biomasses obtained in the agro-industrial sector. Balussou et al. [34] analyzed possible future developments of the German biogas plant capacity up to 2030, taking into consideration technical, economic, and normative conditions, underlining how this sector is strictly connected to political choices. The model results show rapid growth of small-scale manure plants and large-scale bio-waste plants in the German biogas market. Scarlat et al. [35] studied the biogas market in Europe (in particular, biofuels), analyzing production and consumption trends. Subsequently, they examined a model on a European scale to quantify the biomass potential deriving from livestock activities and the relative optimal location of the exploitation plants [36]. The theoretical biogas potential of manure was estimated, according to the analysis, at 26 billion m^3 biomethane, while the realistic biogas potential, counting on collectable manure, was assessed at 18 billion m^3 biomethane in Europe. These values are compatible with the construction of 13,866–19,482 new biogas plants could be built in Europe, with a total installed capacity between 6144 and 7145 MWe, and with an average capacity between 315 and 515 kWe. Seay et al. [37] reviewed the latest research in the supply-chain, process simulation, discrete event simulation and risk assessment into a sustainable point of view for integrated biorefining. Manfren et al. [38] presented a selection of currently available systems for the planning and design of distributed generation, and analyzed them together their opportunities in an optimization framework; they determined the optimal solutions for providing energy services through distributed generation by adopting a multicriteria perspective. Particular attention should be given to fuel consumption due to biomass transport: Ruiz et al. [39] quantified that the maximum cost of logistics is 11.05 € per ton. An analysis of the Italian situation of biogas plants was presented by Benato and Macor [40]; they investigated the construction and operation management costs of six plants and measured the composition of the emissions produced.

Preliminary models which are able to perform the described procedure have been implemented and are currently being tested. McCormick et al. [41] presented an overview of the bioeconomy and bioenergy, examining it from a political point of view. They focused on two important topics: the involvement of communities and stakeholders in the decisional process and huge attention by the government and industry to innovation, in order to achieve sustainable development of the bioeconomy. Bioenergy research is inter-disciplinary, with connections in many different areas. Indeed, the published articles affect specific sectors in many journals. The various and numerous publications in the sector require a systematic and updated bibliographic review, which is the focus of this study.

Due to the vastness and the importance of the topic, the analysis was carried out using a quantitative method based on text mining techniques, following the guidelines presented by Cogato et al. [42]: (i) Inter-disciplinary, studying the topic from a general point of view; (ii) clearly communicating the state-of-the-art and the research gaps; and (iii) supporting the study and work of the researchers and stakeholders. The use of bibliometric analysis to describe publications trends is widespread also in the bioenergy sector: Weinand [43] described the evolution of the research in local planning of energy system between 1991 and 2019 by analyzing 1235 articles; De La Cruz-Lovera [44] focused attention on the contribution of international institutions in the area of energy saving, analyzing 20,095 articles on the Scopus database from 1939 to 2018. The aim of the present analysis is to provide a comprehensive review of the state-of-the-art of the literature concerning bioenergy in Agriculture and Biological Sciences field. The specific objectives of this work are: (i) Describe the temporal trend of publications over the years; (ii) identify in which field the research has been mainly directed; and (iii) analyze the most important links between topics. A quantitative analysis represents the most effective methodology to perform the above-mentioned objectives.

2. Materials and Methods

A bibliometric analysis was carried out by selecting documents indexed by the SCOPUS database, using the advanced search to define the field of interest. This allows for showing how the research has developed and changed, following changes in society and, in some cases, determining them. Given the large number of publications, it is possible to hypothesize the influences, economic trends, and/or political decisions on the subject [45,46].

A text mining process was used to perform the analysis. The words appearing in the title, keywords, and abstract were analyzed using the textual modification instruments in Block Note, the frequency functions in Microsoft Excel, and the graphic representation in Gephi (Gephi® Consortium, Compiegne, France), an open-source software for network analysis. Text mining is a process which derives significant numeric indices from text by analyzing unstructured (textual) information. The statistical analysis of these indices provides the key to text interpretation, obtaining considerable and high-quality information [42–44].

2.1. Article Selection

The analysis was based on the term "bioenergy". To include also its derived forms, the script "bioenergy *" was used for the research on SCOPUS. With the initial examination, the program selected the articles that contain the string "bioenerg" or its derived terms (here and in the following the asterisk "*" indicate lemma declination as, in this case (e.g., bioenergy, bioenergies, bioenergetic, and so on) in the title, in the keywords, or in the abstract. Some filters were applied for a more pertinent selection of the articles. The review articles were excluded, and the field was limited at "Agricultural and Biological Sciences". As we were expecting, many articles (more than ten thousand) resulted from the search. This reflects the great interest in the topic and the interdisciplinarity of the matter (Table 1).

Table 1. Scripts for extraction of research papers.

Step	Script	Number of Papers
Initial research	TITLE-ABS-KEY (bioenerg *)	40,364
Filter application	TITLE-ABS-KEY (bioenerg *) AND (LIMIT-TO (DOCTYPE,"ar")) AND (LIMIT-TO (SUBJAREA,"AGRI"))	10,274

To better understand the evolution of the research, data was selected year by year, adding a time filter at the query. The script used was *"TITLE-ABS-KEY (bioenerg*) AND (LIMIT-TO (PUBYEAR, 2019)) AND (LIMIT-TO (DOCTYPE, "ar")) AND (LIMIT-TO (SUBJ AREA, "AGRI"))"*, substituting the value for the year of research. The analysis was performed from 2000 to 2019 and included a total of 9504 papers. To perform the download, data relating to the title, keywords, and the full abstract were selected and the .csv extension was chosen.

2.2. Article Elaboration

The text extracted was saved as a .txt file. The first step was tokenization, the procedure in which the sentences are broken into pieces, removing punctuation marks, hyphens, and brackets, reducing the text only to its single words. The result of tokenization was a list of single words. Further elaboration was required to convert all letters to lowercase and to identify and convert all terms that can be written in two ways (e.g., bioenergy/bio-energy or bioenergetic/bio-energetic).

The final list of terms was exported to Microsoft Excel. The software allowed us to order the terms and count how many times each one appeared. This kind of elaboration allowed us to identify the more frequent terms in each year. Using Excel, the 100 most relevant words (occurring in at least 4% of the analyzed papers) were identified and used for the subsequent analysis. Finally, the results were processed with the software Gephi (Gephi® Consortium, Compiegne, France), which is a free tool that allows for the creation of a graphic representation of an association of terms. The representation is a graph in which the nodes are the connected terms (eventually with a specific weight) and the vectors—directed or undirected—are the connections between the terms. The conceptual flux of the analysis is represented in Figure 1.

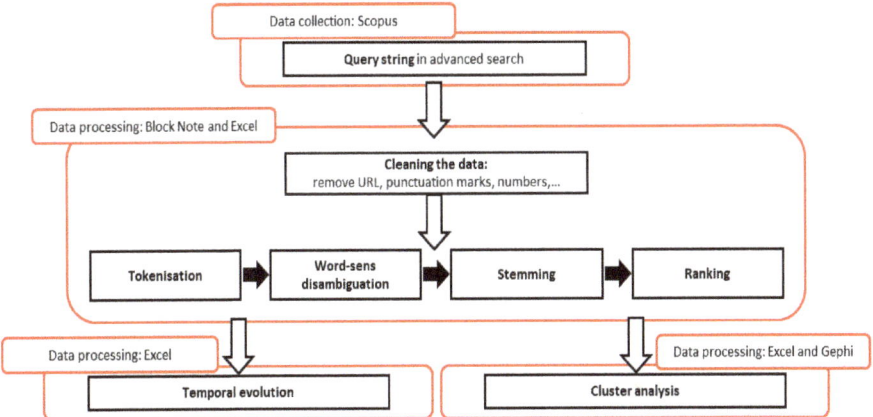

Figure 1. The conceptual flux of the analysis: model and used software.

2.2.1. Combination Matrix

With the 100 most used terms, a word–word connection matrix was built. The matrix had 100^2 cells and, so, 10,000 couples. Starting from this matrix, we built a connection matrix in which, for each couple of words $\{w_1 w_2\}$, the number of articles that contained both the terms was indicated. The connections are not directional, as the value of $\{w_n w_m\}$ was the same as that of $\{w_m w_n\}$. Moreover, the number $\{w_n w_n\}$ was exactly the number of articles the word w_n appeared in. As a result of the matrix, 4950 couples of terms were obtained; the value that corresponds to the k-combinations from a given set of n elements, with k-value of 2 and n-value of 100.

2.2.2. Clusters Definition

Cluster analysis, or clustering, is defined as the task of grouping a set of elements in such a way that the objects in the same group (cluster) share one or more features that make them more similar to each other than to those in other groups. When the object of the analysis is a multidisciplinary topic, cluster analysis makes it possible to investigate the relationship between two or more fields in which the topic is used. By clustering, the most relevant settings and connections are identified. Moreover, it is possible to describe how these rankings and relationships develop and modify over the years.

The bioenergy production phases were chosen as criteria to shape the clusters. Five clusters were identified: Environment, Field, Biomass, Process, and Energy. The number and topic of the Clusters were chosen to adequately cover all aspects of the theme, avoiding an over-fragmentation of the sets. Multiplying the number of clusters could increase time fluctuations and make it challenging to identify trends. The 100 most relevant words previously found were inserted into one of these groups, whichever was more suitable. For more specific analysis, some sub-clusters were created (e.g., crops), as type of produced energy. These groups covered very particular fields, and the included words had a very similar field of application. Some of the included lemmas were not in the top 100 by relevance but, due to their particular significance and pertinence to the sub-cluster, they were included in the analysis: This is the case of some secondary crops (e.g., rice, wheat or barley) or some energy terms (such as heat or methane). It is worth noting that alternative energy sources (e.g., wind or solar power) have not been considered in the analysis. Indeed, the occurrence of related lemmas is almost zero (<1%). Table 2 shows the cluster composition.

Table 2. Cluster composition.

Cluster	Lemmas
Environment	Biodiversity, carbon, ecological, ecosystem, emission, environment, environmental, greenhouse, habitat, impact, land, natural, sustainability, sustainable
Field	Breeding, climate, crop, cultivation, field, harvest, harvesting, population, productivity, rotation, season, soil, species, water, yield
Biomass	Agricultural, animal, biomass, cellulose, corn, feedstock, fish, food, foraging, forest, forestry, grass, lignin, lipid, maize, miscanthus, nitrogen, oil, organic, panicum, perennial, protein, residue, resource, sorghum, sugar, switchgrass, tree, wood
Process	Acid, availability, biological, chemical, composition, cost, cycle, diet, dry, economic, efficiency, feeding, management, metabolic, metabolism, model, nutrient, physiological, physiology, plant, policy, process, respiration, supply, temperature, transport, treatment
Energy	Bioenergy, bioenergetic, biofuel, energ, energetic, ethanol, fossil, fuel, gas, potential, power, production, renewable

3. Results

3.1. Analysis of the Trends

The first consideration concerns the number of articles published per year in the Bioenergy topic and its ratio with the total number of publications in the Agricultural and Biological Science field. As can be seen in Figure 2, the number of publications in the field of Bioenergy registered a slight increase between 2000 and 2006, and then accelerated rapidly until 2017. In the following year, a 12% drop in publications was shown, a stable value in the last year.

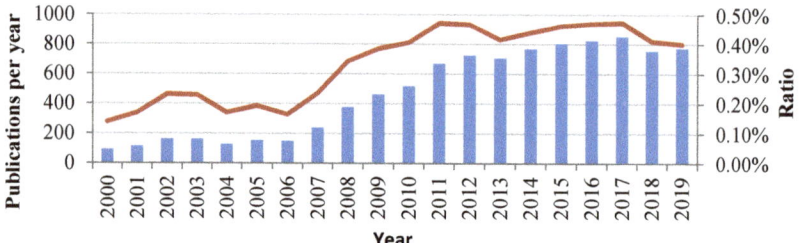

Figure 2. Publications per year (blue histogram) and ratio between publications in the Bioenergy topic in the sector "Agri" and total publications in Agricultural and Biological Science field (red line).

The variations in the number of articles depend both on the increase in the interest of the researchers on the subject and on the overall growth in publications. To clarify this aspect, in Figure 2, the ratio between the Bioenergy articles and total publications in the Agricultural and Biological Science field is represented. It is interesting to note that, from 2006 to 2011, the ratio between the two values tripled; indicating that, in that period, the interest in the topic Bioenergy increased. Since 2011, the ratio has been almost constant, which means that the variations in the articles on the Bioenergy topic are mainly linked to the total number of publications.

To clarify this aspect, a broader analysis was developed. On SCOPUS, articles with the string "bioenerg *" in the Title-Abstract-Keywords and the limitation of "AR" (without the restriction of the sector "Agri") were identified. This series of articles was compared to the total number of publications in the Agricultural and Biological Science field. Results are shown in Figure 3.

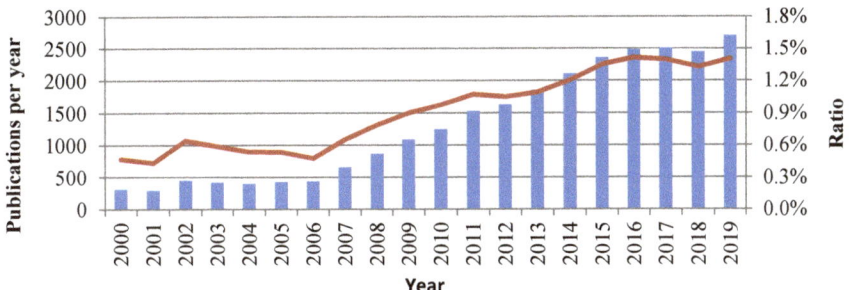

Figure 3. Publications per year (blue histogram) and ratio between publications in the Bioenergy topic and total publications in Agricultural and Biological Science field (red line).

Figure 3 shows a more regular growth of both indicators. The values of articles with the term "Bioenerg*", not limited to the "Agri" sector, have steadily increased from 2006 to 2019, except for a weak decrease in 2018. A comparable trend was shown by the ratio between the value of the same set of articles and the total articles in the "Agricultural and Biological Science" field. The diagrams obtained indicate that, between 2005 and 2006, interest in the bioenergy theme began to increase, occupying even more importance in the efforts of researchers. Interestingly, the Kyoto Protocol entered into force on 16 February 2005, so it is conceivable that it influenced the interests of researchers, encouraging them to find solutions to reduce CO_2 emissions, in order to comply with the agreement.

A further incentive may have been given by the 2009 United Nations Climate Change Conference (commonly known as the Copenhagen Summit) for climate change mitigation. Following this pattern, a slowdown starting from 2016 can be noted. The Paris Agreement in 2015 seems not have made a substantial contribution to research in the renewable energy sector. A confirmation of this trend came by comparing the publications with the term Bioenergy with the total publications on SCOPUS in the same period. Figure 4 shows that the total number of publications has a steady but slower growth than publications with the bioenergy theme.

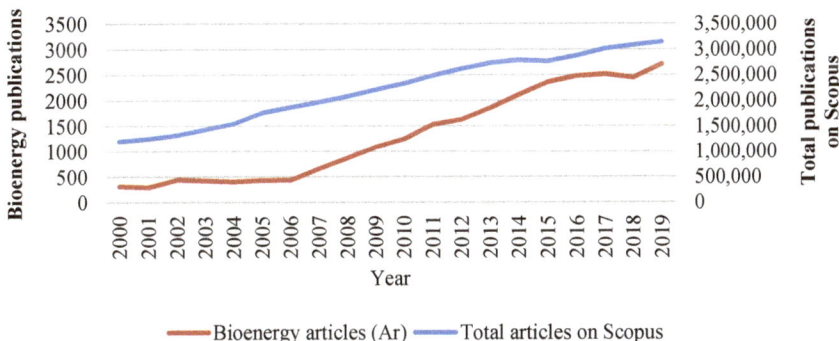

Figure 4. Trends of Bioenergy publications and total publications on SCOPUS.

Another quantitative research performed was the analysis of the affiliations and the international collaborations. Countries of all the continents contributed to the publications on the theme. The United States is the most important contributor, with 39% of the total publications. The top five contributors provide about 49% of the publications (Figure 5). Countries with the highest growth in the last 20 years were Brazil (eight publications from 2000 to 2004 and 301 from 2015 to 2019) and China (12 publications from 2000 to 2004 and 370 from 2015 to 2019).

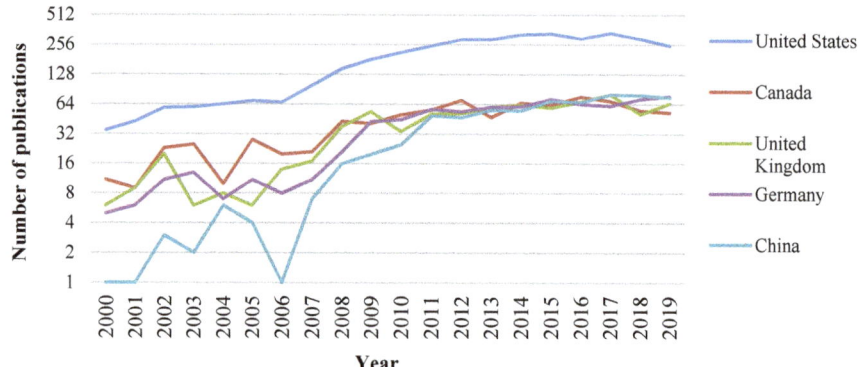

Figure 5. Top five contributors in the last 20 years. The *y*-axis is represented in log $_2$ scale.

The international research collaboration was analyzed. The most relevant collaborations are between the USA and five countries: Canada (201 articles), China (150 articles), the U.K. (106 articles), Germany (101 articles) and Australia (96 articles). Sixth and seventh positions are between the U.K. and Germany (88 articles) and Canada (71 articles) (Table 3).

Table 3. Top 20 international research cooperation.

Countries	Collaborations	Countries	Collaborations
Canada–USA	201	France–U.K.	56
China–USA	150	Germany–Netherlands	54
U.K.–USA	106	Mexico–USA	54
Germany–USA	101	Japan–USA	50
Australia–USA	96	Australia–U.K.	47
Germany–U.K.	88	South Korea–USA	46
Canada–U.K.	71	Canada–France	43
France–USA	70	Italy–U.K.	43
India–USA	64	Netherlands–U.K.	43
Brazil–USA	56	Netherlands–USA	43

3.2. Research on Most Recurrent Terms

Using .txt files and .xlsx files, a ranking of the top words used year-by-year was created. For each term, the number of occurrences in which it was cited in the title, abstract, and keywords was calculated. The ranking is different in different years. The results regarding the words belonging to Cluster "Field" are shown in Table 4.

Table 4. Variation of top used words of cluster "Field".

Year	2000	2001	2002	2003	2004	2005	2006	2007	2008	2009	2010	2011	2012	2013	2014	2015	2016	2017	2018	2019	Temporal Evolution
breeding	4	16	48	30	28	20	15	27	62	112	117	85	105	115	140	122	142	171	149	161	
climate	10	10	5	6	9	20	35	20	70	111	179	201	203	257	337	329	323	357	209	232	
crop	16	17	19	24	17	35	44	44	87	133	177	463	322	358	464	550	566	587	488	427	
cultivation	1	2	0	0	5	8	7	6	3	19	44	120	90	96	100	125	112	124	133	132	
field	30	16	20	40	34	28	27	34	82	94	98	109	109	127	145	246	174	200	188	169	
harvest	4	15	17	24	17	23	25	27	47	24	55	180	138	212	210	242	254	238	154	137	
harvesting	22	10	18	12	19	23	32	45	37	54	146	234	172	271	244	234	308	207	213	123	
population	21	25	45	47	33	49	61	101	205	171	178	184	218	203	183	178	201	227	213	159	
productivity	8	4	11	20	4	25	15	22	42	63	104	111	115	209	157	198	194	183	171	172	
rotation	2	11	5	20	7	6	3	12	17	28	57	123	101	113	140	150	141	152	107	117	
season	4	10	40	34	32	9	19	22	55	71	89	80	109	129	123	116	126	154	91	92	
soil	26	21	23	9	7	37	34	56	128	157	230	447	436	561	537	779	963	989	862	799	
species	35	78	98	109	73	128	112	185	340	382	353	483	553	542	552	573	643	642	515	601	
water	38	70	102	106	70	101	61	81	149	218	257	198	363	309	386	521	415	622	365	528	
yield	5	7	24	35	5	22	26	47	81	113	125	272	289	353	387	470	423	546	432	370	

To classify the terms in the two considered decades, the weighted average of the values over several years was made. For each year, the ratio between the occurrences of a term and the total number of articles in the Bioenergy field was created. The overall score of a term (Equation (1)) was obtained by the weighted mean of the values over the years, giving higher weight to the most recent years to better focus the attention on the current situation:

$$S_T = \frac{\sum_{i=1}^{20} w_i \cdot \frac{o_i}{B_i}}{\sum_{i=1}^{20} w_i}, \tag{1}$$

where w_i is the weight of the ith year, o_i is the number of occurrences of the given term in the ith year, and B_i is the number of articles in the Bioenergy topic in the ith year.

3.3. Cluster Analysis

The first 100 terms among the pre-processed ones were grouped into five conceptual clusters. The weight of a cluster was determined by the sum of the weights of the terms that belong to it. This weight was calculated using the ratio between the occurrences of the terms in a given year and the total articles in the Bioenergy topic in the same year.

The broader cluster was that with the theme "Biomass", which included all words regarding the possible sources of biomass and their characteristics (e.g., "protein", "nitrogen", "organic", "feedstock", and so on). The most important sources of biomass in the cluster were, in descending order: food (8.2%), fish (7.7%), forest (6.0%), wood (4.6%), animal (3.3%), switchgrass (3.1%), agricultural (3.0%), and miscanthus (2.9%). Other significant clusters were Energy (24.6%) and Process (23%), as shown in Table 5. The cluster "Energy" included the terms and the concepts linked to the step of energy production, while the cluster "Process" considered the phase of treatment of the biomass resources, including the economic and management aspects. Features regarding production and resource conditions were included in the cluster "Field", while environmental and sustainability concepts were listed in the "Environment" cluster.

By the results of the analysis, production and treatment were the sectors in which researchers have mainly focused during the last 20 years. Considering the selected words, the sources of biomass (i.e., food, fish, wood, switchgrass, miscanthus, grass, sorghum, oil, corn, residue, panicum and maize) occupied about 37.7% of the occurrences. The terms "emission" and "greenhouse" (mainly related to the greenhouse gases) influenced the cluster for about 12.3%. In the cluster "Energy", specific terms such as "fuel" and "biofuel" presented an impact of 6.3%; meanwhile, other topics such as "electricity" and "biogas" were not even among the most frequent words.

In the "Process" cluster, an important contribution was given by terms relating to chemical and biological aspects: temperature (8.0%), metabolism (7.2%), metabolic (5.0%), feeding (4.5%), diet (4.2%), composition (3.6%), physiological (2.7%), nutrient (2.4%), respiration (2.4%), physiology (2.2%), biological (2.2%), treatment (1.9%), and chemical (1.7%). It is worth noting that some of these terms are important parameters in the production process of biofuels and biogas: the same process is also deeply influenced by the specific implemented crops, which; however, were included in the generic "biomass" cluster for the diverse meaning and use they might have in research papers. The residual contribution consisted of technical and economic terms. The "Field" cluster was made up of terms with fewer occurrences than the others, but it indicated that there was interest in the biomass production aspects. The environmental issue seems to have had minor importance (10.5% of the total), which is likely to tend to increase in the coming years.

The percentage of occurrences of the clusters per number of articles in the bioenergy field were compared. Observing the trends over the last 20 years (Table 7), it is noteworthy to observe that the percentage weight of the "Process" cluster has steadily decreased, from 26.5% to 21.0%. The "Energy" cluster has suffered a comparable, but less accentuated, reduction—from 25.8% to 21.5%. Both the "Field" and "Environment" clusters have been continuously growing; the cluster "Field" from 10.4% to

18.6% (therefore, an increase of about 79%), and "Environment" from 7.3% (the 2001 value was taken, as that in 2000 seemed to be out of scale) to 11.8% (therefore, increasing by 62.3%). It appears that environmental and sustainability issues have been of increasing interest in research, a consequence of the ecological policies promoted by national governments and international institutions.

Table 5. Main clusters: clusters reported by highest frequency terms.

Cluster	Lemmas and Relative Occurrence [%]	Cluster [%]
Biomass	Biomass 21.4%, food 8.2%, fish 7.7%, forest 6.0%, wood 4.6%, protein 4.4%, nitrogen 3.5%, foraging 3.4%, animal 3.3%, switchgrass 3.1%, agricultural 3.0%, miscanthus 2.9%, forestry 2.5%, resource 2.3%, lipid 2.3%, organic 2.1%, feedstock 1.8%, grass 1.8%, tree 1.7%, sorghum 1.7%, oil 1.5%, corn 1.5%, residue 1.4%, perennial 1.4%, lignin 1.3%, panicum 1.1%, source 1.1%, sugar 1.1%, maize 1.0%, cellulose 0.9%	26.6%
Energy	Energy 30.8%, bioenergy 16.7%, production 13.7%, bioenergetic 9.4%, potential 7.5%, fuel 3.7%, energetic 3.3%, gas 2.9%, power 2.9%, biofuel 2.6%, ethanol 2.3%, renewable 2.3%, fossil 1.9%	24.6%
Process	Model 10.3%, temperature 8.0%, metabolism 7.2%, plant 6.5%, metabolic 5.0%, management 4.7%, feeding 4.5%, diet 4.2%, efficiency 4.1%, cost 3.8%, composition 3.6%, supply 3.2%, availability 3.2%, acid 3.2%, physiological 2.7%, cycle 2.6%, dry 2.6%, economic 2.4%, nutrient 2.4%, respiration 2.4%, physiology 2.2%, biological 2.2%, transport 2.0%, treatment 1.9%, chemical 1.7%, process 1.7%, policy 1.7%	23%
Field	Species 16.7%, soil 12.6%, water 12.1%, crop 9.1%, yield 7.4%, population 7.0%, climate 5.5%, harvesting 5.1%, field 4.8%, harvest 4.2%, breeding 4.0%, productivity 3.6%, season 3.4%, rotation 2.5%, cultivation 2.0%,	15.2%
Environment	Carbon 17.9%, environmental 12.5%, land 10.3%, emission 6.8%, habitat 6.3%, impact 5.8%, ecosystem 5.8%, greenhouse 5.6%, CO_2 4.9%, environment 4.7%, ecological 4.6%, sustainable 4.5%, natural 4.4%, biodiversity 3.0%, sustainability 2.9%	10.5%

3.4. Interrelationships Between Terms

The objective was to provide specific information on how main topics belonging to the same or different clusters were addressed together, so interrelations of the terms were studied. Therefore, each of the already mentioned 100 most frequent words in the title, keywords, and abstract section was coupled with each of the remaining 99 words, generating 4950 possible combinations. Such combinations were studied in terms of occurrences on analyzed 20 years bibliography and graphically represented, generating a very complicated net of relationships (Figure 6). The same combinations occurrences were also represented in a table format (Table 6).

Table 6. Relationships between the terms of the clusters.

	Energy	Environment	Field	Process	Biomass
Energy	86.613	84.875	99.820	161.622	149.200
Environment	***	24.656	49.549	68.229	70.549
Field	***	***	29.900	79.833	90.528
Process	***	***	***	65.335	118.926
Biomass	***	***	***	***	65.325

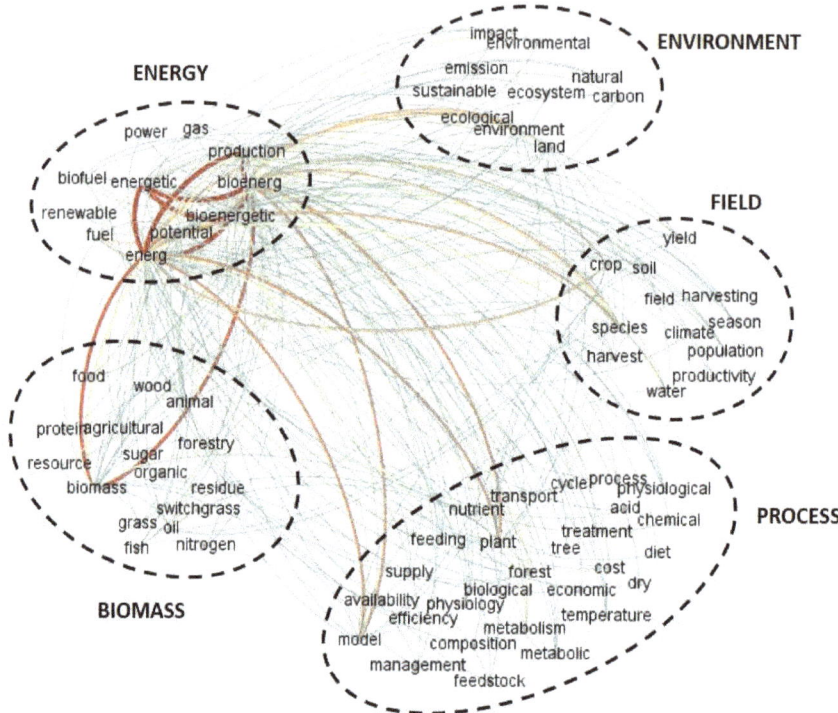

Figure 6. Interrelations between terms in the title, keywords, and abstract sections. Thicker and darker colored lines indicate a more significant number of connections. The circles indicate different clusters. For better visualization, only terms with at least 700 co-occurrences with at least one other term are shown.

Given the research theme, the "Energy" topic was expected to include the terms with the highest number of co-occurrences. It presented the maximum value of co-occurrences both between terms inside the cluster and terms belonging to different clusters. Excluding these groups, the cluster with the maximum number of co-occurrences was "Biomass". This result was also due to the large number of terms belonging to this cluster, including all sources of biomass and energy. The group of words with fewer relationships with other terms was the cluster "Environment", which also presented the minimum value of connections between words inside the same cluster (Table 6).

To better understand the connections between the terms, Figure 6 was exploded, focusing the view on pairs of groups of words. In the first one (Figure 7a), the statistical analysis highlighted that the scientific community has studied every type of energy achieved by biomass.

The analysis of single couples of terms, without considering the cluster they belong to, allowed to show which topics were the most related. The following schemes were elaborated by taking the first 30 couples of terms by relationships. Trivial or non-relevant couples were excluded; for example, "environment–environmental", "fuel–biofuel", and all those that contained the terms "energy" or "bioenergy". The results are summarized in Table 8.

Table 7. Percentage of occurrences in the clusters aggregated by groups of five years.

Year	2000	2001	2002	2003	2004	2005	2006	2007	2008	2009	2010	2011	2012	2013	2014	2015	2016	2017	2018	2019	Temporal Evolution
Energy	25.8	28.3	24.7	25.1	22.9	25.5	24.4	23.8	25.0	24.0	23.7	24.8	23.7	22.4	22.2	21.5	20.9	19.9	20.9	21.5	
Environment	12.3	7.3	7.2	9.1	8.2	9.2	10.4	9.6	9.0	9.2	10.6	10.2	11.6	10.7	10.6	12.1	12.6	11.5	11.0	11.8	
Field	10.4	12.1	14.5	13.2	13.2	14.9	13.8	12.1	15.0	14.2	15.0	15.1	15.2	16.7	16.8	18.8	18.3	19.1	18.2	18.6	
Process	26.5	26.8	29.4	25.1	31.1	25.1	24.9	26.9	27.0	24.4	23.7	20.9	21.5	21.1	21.3	20.3	20.3	20.5	21.5	21.0	
Biomass	25.1	25.4	24.1	27.6	24.6	25.3	26.5	27.6	24.0	28.2	26.9	29.0	28.0	29.1	29.1	27.4	27.8	28.9	28.4	27.1	

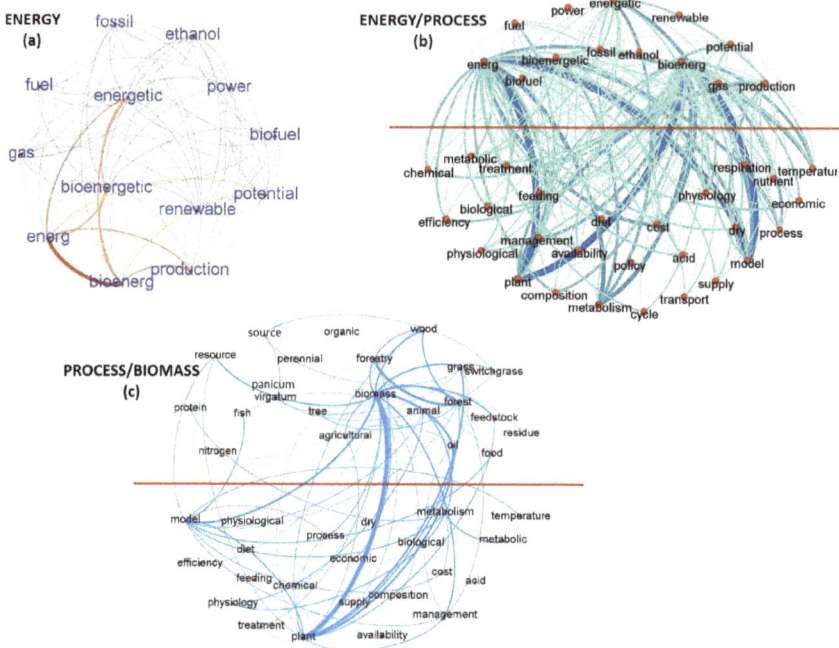

Figure 7. Co-occurrence of topics within the "Energy" cluster (**a**); between the "Energy" and the "Process" clusters (**b**); and between the "Process" and "Biomass" clusters (**c**).

Table 8. Couples of terms with the highest number of relationships during the period 2000–2019, values of the occurrences.

Source	Target	Weight	Source	Target	Weight
biomass	production	2261	crop	potential	1367
biomass	plant	1858	animal	bioenergetic	1358
potential	production	1840	environment	production	1334
plant	production	1794	bioenergetic	food	1296
crop	production	1774	bioenergetic	environment	1271
crop	plant	1700	production	yield	1256
biomass	potential	1686	model	production	1252
biomass	crop	1683	biomass	yield	1250
bioenergetic	species	1682	crop	fuel	1244
bioenergetic	model	1609	bioenergetic	fish	1237
fuel	production	1581	fuel	plant	1220
biomass	fuel	1532	fuel	potential	1219
bioenergetic	metabolism	1479	biofuel	production	1152
plant	potential	1416	bioenergetic	production	1149
land	production	1384	crop	yield	1142

3.5. Temporal Comparison of Related Terms

To describe the evolution of research publications in the bioenergy sector, groups of words with very particular bonds were taken. These groups were constituted by terms that expressed alternative solutions in the study and, by analyzing the variations with which these solutions are cited in the articles, it is possible to understand in which direction the research was addressed.

The first specific cluster considered was related to "Crops" (Figure 8), which included potential biomass sources from agricultural activities. Considering the trend over the last 20 years, a temporal analysis allowed us to identify if there were crops that have gained interest as sources of biomass for energy purposes and if there were others that, on the contrary, are considered less valid at present than in the past.

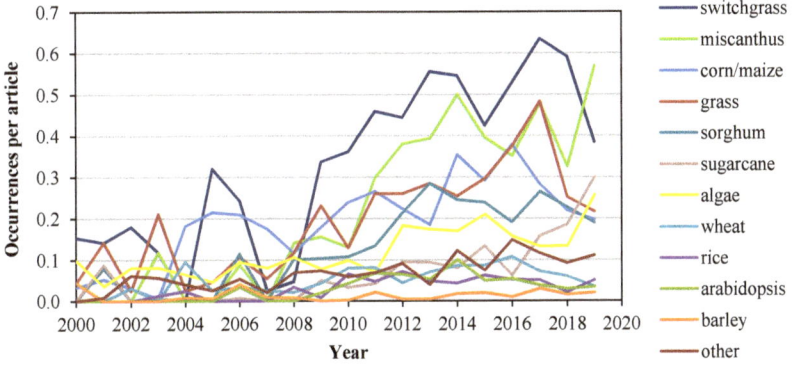

Figure 8. Trend chart of related terms in the "Crops" cluster.

The first general consideration was that citations of crops per article in the bioenergy theme have generally grown over the considered period. In other words, a growing attention has been paid to the selection of specific or alternative crops as potential source for bioenergy production. Above all, Miscanthus has showing the largest evolution, moving from 0 to 0.584 occurrences per article (occ/art), which signifies that there were about 0.58 citations of the term per each article considered to fall under the bioenergy theme. One other significant result is the trend exhibited by the term "sugarcane", the ratio of which increased from 0 to 0.306 occ/art; a result that is particularly important, considering that the production of this product is mainly concentrated in developing countries. The term switchgrass was the most cited for several years (from 0.187 to 0.282 occ/art), although it registered some deep falls. Some types of crops have shown growth over time, albeit with fluctuating trends such as grass (from 0.044 to 0.221 occ/art), corn (from 0.033 to 0.202 occ/art), sorghum (from 0 to 0.195 occ/art), or algae (from 0.099 to 0.256 occ/art); trends and applications of algae were studied by Deviram et al. [47] and by Yang et al. [48], showing growing interest in recent years, particularly in the USA and China. Some other crops have given evidence of an initial interest, but with a loss of relevance in the last years, as in the case of wheat or thale cress (Arabidopsis). Other crops (including Arundo, Beets) have been taken into consideration; however, they still play a weak role in research publications.

The second specific considered cluster was related to "Energy produced" (Figure 9), which included the energy forms that can be obtained using biomass. The relevance of the argument and the benefits and costs associated with each type of utilizations was studied by Guo et al., expecting a growing of the sector in the next future, in particular bioethanol and biogas [49]. This type of analysis makes it possible to understand which kind of produced energy the publications focused on, assessing whether politics or international agreements have had an influence on the research. The occurrences of the terms "heat" and "electricity" were almost constant over the two considered decades. Excluding the first three years, which exhibited an anomalous peak, occurrences of the term "heat" moved from 0.155 to 0.145 occ/art, while the citations of the term "electricity" moved from 0.130 to 0.102 occ/art.

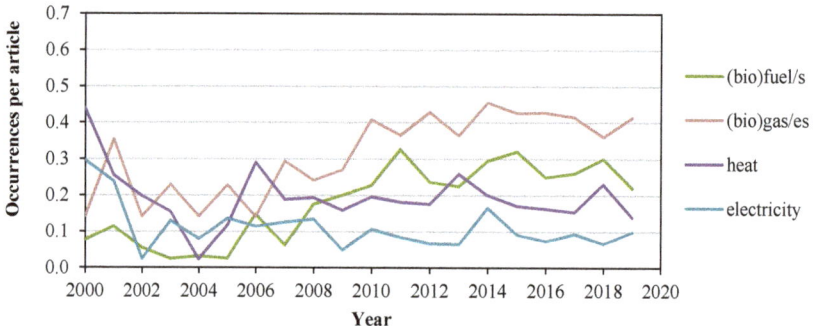

Figure 9. Trend chart of terms related to produced energy.

The slow but steady growth of the other terms related to biogases and biofuels is significant. Indeed, in the first case citations increased from 0.143 to 0.414 occ/art (i.e., with an average increase of 9.5% per year), while in the second case the number of occurrences per article moved from 0.143 to 0.378. For biofuels, a more evident growth can be recognized between 2007 and 2011 (+0.404) occ/art: Such increment might be associated to the increasing impact on economy of crude oil prices (which reached a maximum in July 2008), along with international and in particular European strategies for biofuels, published in 2006 [50].

Biofuels and biogases are detailed also in Figure 10. The most recurrent term in two decades of published research is ethanol, with an average of 0.200 occ/art. On the other hand, a clearly growing interest is being devoted to methane, with a number of occurrences which has moved in the last decade from 0.019 to 0.175 occ/art. Other types of fuels (such as methanol or ethylene) and other types of gases (such as propane, ethane or butane) still exhibit a minor interest for the scientific community, with a total number of citations lower than 0.025 occ/art. Development of the types of renewable sources of energy in recent years has led to specialization in their use. Biomass-derived energy is particularly suitable to be stored and used in case of requirements; more so than the electricity produced by wind farms and solar plants. Furthermore, the objectives of reducing greenhouse gas emissions due to the transport sector can be validly achieved not only by optimizing harvesting process [51] but also by using fuels derived from biomass. These considerations could explain the growing interest in research in the biofuel and methane sector, which are adequate products for storing produced energy and fueling vehicles, and in the comparison of different ways of use of the energy produced [52].

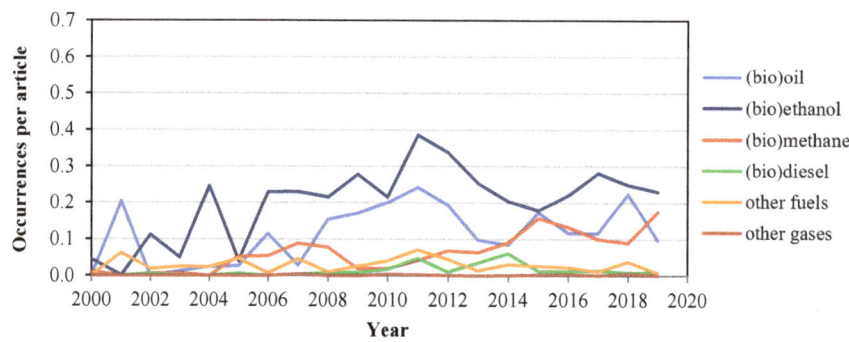

Figure 10. Trend chart of terms related to produced biogases and biofuels.

4. Discussion

The presented research was performed by a text-mining analysis, taking into consideration the title, abstract and keywords of every article. The most critical and frequent terms were identified and analyzed. The most significant relationships were recognized, both between specific terms and aggregated clusters.

The temporal analysis allowed us to describe the evolution of publications; in particular, which topics have gained or lost importance and which relationships have been strengthened.

4.1. Temporal Analysis of Publication Trends

Themes related to bioenergy and its production, management, and use are not recent topics in research. However, interest has risen sharply in recent years, with a growth of about 726% in publications and around 183% by weight of total articles in the Agricultural and Biological Sciences field. Research in the branch has affected every aspect related to the theme in a different way, as was shown in the cluster analysis.

Although the interest of research has been influenced by the economic and environmental policies of countries and international institutions, given the extensive range of topics, it is difficult to establish a link between single events and temporal trends. However, it is legitimate to hypothesize a relationship between the growing number of publications in the bioenergy theme and the even more ambitious targets in renewable energies matters.

4.2. Cluster Analysis and Trends

By the described analysis, it can be seen that the most studied topics were those relating to the phase of the production process. The chemical and biological processes on which the energy production of biomass are based have been the subject of numerous studies. Management and economic aspects seem to have had less quantitative impacts on research works.

The simultaneous growth of the topic "Environment" and reduction of the topic "Process" can be explained by the achievement of a high standard of efficiency in the digestion and transformation processes of biomass into various types of energy. In the meantime, the efforts of researchers have shifted to investigate how these energy sources can be integrated into the overall transformation process of the energy system, from fossil fuel-based to renewable energies-based.

The growth of the "Field" cluster (the highest in the identified clusters) reveals a greater interest in the production phase of the biomass sources. Indeed, the latest goals of international institutions, including the EU directives, have underlined the importance that the collection of biomass does not affect food production. For this reason, crops cultivated for energetic purpose should be avoided, and by-products or wastes of agricultural and livestock activities should be used. Research into the types of plants allows researchers to identify the best way to exploit them for energy purposes.

The most cited crops in the selected articles are miscanthus, switchgrass, and corn, which can all be included in the crop category. It should be investigated whether the use of miscanthus and switchgrass derives from an interest in crops dedicated to energy production or, at least in part, plants that grow spontaneously. Corn is one of the most common crops used for energy purposes. The reduction of the related occurrences in the examined publications can be a positive signal, suggesting that this crop is somehow experiencing a decreasing interest as energy dedicated source, hopefully returning to its food production vocation, at least at a scientific level.

The analysis of the most significant relationships confirmed the decreasing trend of the "Energy" cluster and the growth of the "Biomass", "Field", and "Environment" clusters. This is another sign of the changing interest of research, towards the environmental aspects of bioenergy concerning the technical and processing phases.

5. Conclusions

The last twenty years have seen a growing attention on bioresources for energy applications. In particular, renewed interests have been devoted to specific and different topics in the wide research field of bioenergy science. The present research is aimed at characterizing such evolving trends, highlighting most relevant terms or relations in terms of occurrences in scientific papers.

The most important contributions are concentrated in three macro areas: North America, Western Europe, and China, while the developing countries are actually less represented. Such distribution suggests that political decisions and favorable economic conditions deeply influenced the interest in the topic. As for the contributions of the top five countries, the United States is the most significant contributor for every type of biomass, but it is interesting to note that publications are mainly focused on switchgrass. Considering also data related to rice in China, wheat in Canada, and maize in Germany, it seems that the attention of the research is mainly focused on those crops that are particularly common in the country. Articles with UK affiliation are particularly targeted at miscanthus; the interest in this energy crop indicates the objective of seeking solutions not in competition with food production. Additionally, the results of the review suggest that efforts in the future might be focused both on the biomass production phase and in the analysis of the environmental impacts and benefits, which up to now (compared to process, biomass and energy clusters) have exhibited the lowest percentage of occurrences but on the other hand the highest growth rate.

A systemic approach would be in particular recommendable, where the different elements of the bioenergy process chain from the field to the consumer are studied in a concurrent way, integrating source and process optimization, environmental sustainability, and final users' needs. The use of crops not of interest for food production, as well as the use of wastes from the agricultural and food industries, must be examined in depth. From environmental and economic points of view, studies regarding the integration of bioenergy and other types of renewable energy sources (as e.g., wind, hydro, or solar power) are still lacking and represent another possible goal of research. Combined analyses of integrated energy sources with a systemic approach can potentially further increase environmental benefits, allowing optimization of important factors such as soil or water consumption, use of raw materials, and interaction with anthropogenic activities. To this end, specific and innovative mathematical models would be needed in order to help designing of decision-making tools that allow for more accurate simulation and planning of future scenarios. Political actors and stakeholders will then be able to evaluate the proposed solutions, based on community needs as well as environmental impacts.

Author Contributions: Conceptualization, G.F., A.P., and F.M.; methodology, F.M.; software, G.F.; validation, G.F., A.P., A.-S.N., and F.M.; formal analysis, G.F.; investigation, G.F., A.P., and F.M.; data curation, G.F., A.P., A.-S.N., and F.M.; writing—original draft preparation, G.F.; writing—review and editing, G.F., A.P., A.-S.N., and F.M.; visualization, G.F., A.P., A.-S.N., and F.M.; supervision, A.P. and F.M. All authors have read and agreed to the published version of the manuscript.

Funding: This research received no external funding.

Conflicts of Interest: The authors declare no conflict of interest.

References

1. Appels, L.; Lauwers, L.; Degrève, J.; Helsen, L.; Lievens, B.; Willems, K.; Van Impe, J.; Dewil, R. Anaerobic digestion in global bio-energy production: Potential and research challenges. *Renew. Sustain. Energy Rev.* **2011**, *15*, 4295–4301. [CrossRef]
2. Nizami, A.S.; Ismail, I.M. Life-cycle assessment of biomethane from lignocellulosic biomass. In *Life Cycle Assessment of Renewable Energy Sources*; Springer: New York, NY, USA, 2013; pp. 79–94.
3. Amon, T.; Amon, B.; Kryvoruchko, V.; Machmüller, A.; Hopfner-Sixt, K.; Bodiroza, V.; Hrbek, R.; Friedel, J.; Pötsch, E.; Wagentristl, H.; et al. Methane production through anaerobic digestion of various energy crops grown in sustainable crop rotations. *Bioresour. Technol.* **2007**, *98*, 3204–3212. [CrossRef] [PubMed]
4. Dinuccio, E.; Balsari, P.; Gioelli, F.; Menardo, S. Evaluation of the biogas productivity potential of some Italian agro-industrial biomasses. *Bioresour. Technol.* **2010**, *101*, 3780–3783. [CrossRef] [PubMed]

5. Nizami, A.S.; Orozco, A.; Groom, E.; Dieterich, B.; Murphy, J.D. How much gas can we get from grass? *Appl. Energy* **2012**, *92*, 783–790. [CrossRef]
6. Mattioli, A.; Boscaro, D.; Dalla Venezia, F.; Santacroce, F.C.; Pezzuolo, A.; Sartori, L.; Bolzonella, D. Biogas from residual grass: A territorial approach for sustainable bioenergy production. *Waste Biomass Valoriz.* **2017**, *8*, 2747–2756. [CrossRef]
7. Chiumenti, A.; Boscaro, D.; Da Borso, F.; Sartori, L.; Pezzuolo, A. Biogas from fresh spring and summer grass: Effect of the harvesting period. *Energies* **2018**, *11*, 1466. [CrossRef]
8. Theuerl, S.; Herrmann, C.; Heiermann, M.; Grundmann, P.; Landwehr, N.; Kreidenweis, U.; Prochnow, A. The Future Agricultural Biogas Plant in Germany: A Vision. *Energies* **2019**, *12*, 396. [CrossRef]
9. Visser, L.; Hoefnagels, R.; Junginger, M. The Potential Contribution of Imported Biomass to Renewable Energy Targets in the EU–The Trade-off between Ambitious Greenhouse Gas Emission Reduction Targets and Cost Thresholds. *Energies* **2020**, *13*, 1761. [CrossRef]
10. Chiumenti, A.; Pezzuolo, A.; Boscaro, D.; da Borso, F. Exploitation of Mowed Grass from Green Areas by Means of Anaerobic Digestion: Effects of Grass Conservation Methods (Drying and Ensiling) on Biogas and Biomethane Yield. *Energies* **2019**, *12*, 3244. [CrossRef]
11. United Nations. *Sustainable Development Goals*; UN: New York, NY, USA, 2015.
12. European Union Renewable energy directive. In Proceedings of the European Wind Energy Conference and Exhibition, Brussels, Belgium, 31 March–3 April 2008; pp. 32–38.
13. World Bioenergy Association. *WBA Global Bioenergy Statistics 2017*; World Bioenergy Association: Stockholm, Sweden, 2017.
14. Ahorsu, R.; Medina, F.; Constantí, M. Significance and challenges of biomass as a suitable feedstock for bioenergy and biochemical production: A review. *Energies* **2018**, *11*, 3366. [CrossRef]
15. Delivand, M.K.; Cammerino, A.R.B.; Garofalo, P.; Monteleone, M. Optimal locations of bioenergy facilities, biomass spatial availability, logistics costs and GHG (greenhouse gas) emissions: A case study on electricity productions in South Italy. *J. Clean. Prod.* **2015**, *99*, 129–139. [CrossRef]
16. Shu, K.; Schneider, U.A.; Scheffran, J. Optimizing the bioenergy industry infrastructure: Transportation networks and bioenergy plant locations. *Appl. Energy* **2017**, *192*, 247–261. [CrossRef]
17. Valenti, F.; Zhong, Y.; Sun, M.; Porto, S.M.C.; Toscano, A.; Dale, B.E.; Sibilla, F.; Liao, W. Anaerobic co-digestion of multiple agricultural residues to enhance biogas production in southern Italy. *Waste Manag.* **2018**, *78*, 151–157. [CrossRef] [PubMed]
18. Solarte-Toro, J.C.; Chacón-Pérez, Y.; Cardona-Alzate, C.A. Evaluation of biogas and syngas as energy vectors for heat and power generation using lignocellulosic biomass as raw material. *Electron. J. Biotechnol.* **2018**, *33*, 52–62.
19. Weinand, J.M.; McKenna, R.; Karner, K.; Braun, L.; Herbes, C. Assessing the potential contribution of excess heat from biogas plants towards decarbonising residential heating. *J. Clean. Prod.* **2019**, *238*, 117756. [CrossRef]
20. Resch, B.; Sagl, G.; Trnros, T.; Bachmaier, A.; Eggers, J.B.; Herkel, S.; Narmsara, S.; Gündra, H. GIS-based planning and modeling for renewable energy: Challenges and future research avenues. *ISPRS Int. J. Geo Inf.* **2014**, *3*, 662–692. [CrossRef]
21. Valenti, F.; Porto, S.M.C.; Dale, B.E.; Liao, W. Spatial analysis of feedstock supply and logistics to establish regional biogas power generation: A case study in the region of Sicily. *Renew. Sustain. Energy Rev.* **2018**, *97*, 50–63. [CrossRef]
22. Patrizio, P.; Leduc, S.; Chinese, D.; Dotzauer, E.; Kraxner, F. Biomethane as transport fuel: A comparison with other biogas utilization pathways in northern Italy. *Appl. Energy* **2015**, *157*, 25–34. [CrossRef]
23. Mirkouei, A.; Haapala, K.R.; Sessions, J.; Murthy, G.S. A mixed biomass-based energy supply chain for enhancing economic and environmental sustainability benefits: A multi-criteria decision-making framework. *Appl. Energy* **2017**, *206*, 1088–1101. [CrossRef]
24. Long, H.; Li, X.; Wang, H.; Jia, J. Biomass resources and their bioenergy potential estimation: A review. *Renew. Sustain. Energy Rev.* **2013**, *26*, 344–352. [CrossRef]
25. Ferrarini, A.; Serra, P.; Almagro, M.; Trevisan, M.; Amaducci, S. Multiple ecosystem services provision and biomass logistics management in bioenergy buffers: A state-of-the-art review. *Renew. Sustain. Energy Rev.* **2017**, *73*, 277–290. [CrossRef]

26. Pulighe, G.; Bonati, G.; Colangeli, M.; Morese, M.M.; Traverso, L.; Lupia, F.; Khawaja, C.; Janessen, R.; Fava, F. Ongoing and emerging issues for sustainable bioenergy production on marginal lands in the Mediterranean regions. *Renew. Sustain. Energy Rev.* **2019**, *103*, 58–70. [CrossRef]
27. Mekonnen, M.; Hoekstra, A. The green, blue and grey water footprint of crops and derived crop products. *Hydrol. Earth Syst. Sci.* **2011**, *15*, 1577–1600. [CrossRef]
28. Quadir, M.; Quillérou, E.; Nangia, V.; Murtaza, G.; Singh, M.; Thomas, R.J.; Drechsel, P.; Noble, A.D. Economics of salt-induced land degradation and restorantion. *Nat. Resour. Forum* **2014**, *38*, 282–295. [CrossRef]
29. Kluts, I.; Wicke, B.; Leemans, R.; Faaij, A. Sustainability constraints in determining European bioenergy potential: A review of existing studies and steps forward. *Renew. Sustain. Energy Rev.* **2017**, *69*, 719–734. [CrossRef]
30. Erber, G.; Kühmaier, M. Research trends in European forest fuel supply chains: A review of the last ten years (2007–2017)—Part two: Comminution, Transport & Logistics. *Croat. J. For. Eng.* **2018**, *38*, 269–278.
31. Ba, B.H.; Prins, C.; Prodhon, C. Models for optimization and performance evaluation of biomass supply chains: An Operations Research perspective. *Renew. Energ.* **2016**, *87*, 977–989. [CrossRef]
32. Pari, L.; Suardi, A.; Santangelo, E.; García-Galindo, D.; Scarfone, A.; Alfano, V. Current and innovative technologies for pruning harvesting: A review. *Biomass Bioenergy* **2017**, *107*, 398–410. [CrossRef]
33. Garcia, N.; Mattioli, A.; Gil, A.; Frison, N.; Battista, F.; Bolzonella, D. Evaluation of the methane potential of different agricultural and food processing substrates for improved biogas production in rural areas. *Renew. Sustain. Energy Rev.* **2019**, *112*, 1–10. [CrossRef]
34. Balussou, D.; McKenna, R.; Möst, D.; Fichtner, W. A model-based analysis of the future capacity expansion for German biogas plants under different legal frameworks. *Renew. Sustain. Energy Rev.* **2018**, *96*, 119–131. [CrossRef]
35. Scarlat, N.; Dallemand, J.F.; Fahl, F. Biogas: Developments and perspectives in Europe. *Renew. Energy* **2018**, *129*, 457–472. [CrossRef]
36. Scarlat, N.; Fahl, F.; Dallemand, J.F.; Monforti, F.; Motola, V. A spatial analysis of biogas potential from manure in Europe. *Renew. Sustain. Energy Rev.* **2018**, *94*, 915–930. [CrossRef]
37. Seay, J.; Badurdeen, F. Current trends and directions in achieving sustainability in the biofuel and bioenergy supply chain. *Curr. Opin. Chem. Eng.* **2014**, *6*, 55–60. [CrossRef]
38. Manfren, M.; Caputo, P.; Costa, G. Paradigm shift in urban energy systems through distributed generation: Methods and models. *Appl. Energy* **2011**, *88*, 1032–1048. [CrossRef]
39. Ruiz, J.A.; Juárez, M.C.; Morales, M.P.; Muñoz, P.; Mendívil, M.A. Biomass logistics: Financial & environmental costs. Case study: 2 MW electrical power plants. *Biomass Bioenergy* **2013**, *56*, 260–267.
40. Benato, A.; Macor, A. Italian biogas plants: Trend, subsidies, cost, biogas composition and engine emissions. *Energies* **2019**, *12*, 979. [CrossRef]
41. McCormick, K.; Kautto, N. The Bioeconomy in Europe: An Overview. *Sustainability* **2013**, *5*, 2589–2608. [CrossRef]
42. Cogato, A.; Meggio, F.; De Antoni Migliorati, M.; Marinello, F. Extreme Weather Events in Agriculture: A Systematic Review. *Sustainability* **2019**, *11*, 2547. [CrossRef]
43. Weinand, J. Reviewing municipal energy system planning in a bibliometric analysis: Evolution of the research field between 1991 and 2019. *Energies* **2020**, *13*, 1367. [CrossRef]
44. De La Cruz-Lovera, C.; Perea-Moreno, J.; De La Cruz-Fernandez, L.; Montoya, F.; Alcayde, A.; Manzano-Agugliaro, F. Analysis of research topics and scientific collaborations in energy saving using bibliometric techniques and community detection. *Energies* **2019**, *12*, 2030. [CrossRef]
45. Yu, D.; Meng, S. An overview of biomass energy research with bibliometric indicators. *Energy Environ.* **2018**, *29*, 576–590. [CrossRef]
46. Chen, H.; Ho, Y.S. Highly cited articles in biomass research: A bibliometric analysis. *Renew. Sust. Ener. Rev.* **2015**, *49*, 12–20. [CrossRef]
47. Deviram, G.; Mathimani, T.; Anto, S.; Ahmed, T.; Ananth, D.; Pugazhendhi, A. Applications of microalgal and cyanobacterial biomass on a way to safe, cleaner and a sustainable environment. *J. Clean. Prod.* **2020**, *253*, 119770. [CrossRef]
48. Yang, X.; Wu, Y.; Yan, J.; Song, H.; Fan, J.; Li, Y. Trends of microalgal biotechnology: A view from bibliometrics. *Chin. J. Biotechnol.* **2015**, *31*, 1415–1436.

49. Guo, M.; Song, W.; Buhain, J. Bioenergy and biofuels: History, status and perspective. *Renew. Sustain. Energy Rev.* **2015**, *42*, 715–725. [CrossRef]
50. Commission of the European Communities. *Communication from the Commission. An EU Strategy for Biofuels {SEC(2006) 142}*; Commission of the European Communities: Brussels, Belgium, 2006; p. 29.
51. Boscaro, D.; Pezzuolo, A.; Sartori, L.; Marinello, F.; Mattioli, A.; Bolzonella, D.; Grigolato, S. Evaluation of the energy and greenhouse gases impacts of grass harvested on riverbanks for feeding anaerobic digestion plants. *J. Clean. Prod.* **2017**, *172*, 4099–4109. [CrossRef]
52. Pöschl, M.; Ward, S.; Owende, P. Evaluation of energy efficiency of various biogas production and utilisation pathways. *Appl. Energy* **2010**, *87*, 3305–3321. [CrossRef]

© 2020 by the authors. Licensee MDPI, Basel, Switzerland. This article is an open access article distributed under the terms and conditions of the Creative Commons Attribution (CC BY) license (http://creativecommons.org/licenses/by/4.0/).

Article

Characteristic Properties of Alternative Biomass Fuels

Martin Lisý [1,*], Hana Lisá [1], David Jecha [1], Marek Baláš [1] and Peter Križan [2]

[1] Faculty of Mechanical Engineering, Brno University of Technology, 60190 Brno, Czech Republic; lisa@fme.vutbr.cz (H.L.); jecha@fme.vutbr.cz (D.J.); balas.m@fme.vutbr.cz (M.B.)
[2] Faculty of Mechanical Engineering, Slovak University of Technology in Bratislava, 811 07 Bratislava, Slovakia; peter.krizan@stuba.sk
* Correspondence: lisy@fme.vutbr.cz; Tel.: +420-541-142-582

Received: 19 February 2020; Accepted: 16 March 2020; Published: 19 March 2020

Abstract: Biomass is one of the most promising renewable energy sources because it enables energy accumulation and controlled production. With this, however, the demand for biofuels grows and thus there is an effort to expand their portfolio. Nevertheless, to use a broader range of biofuels, it is necessary to know their fuel properties, such as coarse and elemental analysis, or lower heating value. This paper presents the results of testing the fuel properties of several new, potentially usable biofuels, such as quinoa, camelina, crambe, and safflower, which are compared with some traditional biofuels (wood, straw, sorrel, hay). Moreover, the results of the determination of water content, ash, and volatile combustible content of these fuels are included, along with the results of the elemental analysis and the determination of higher and lower heating values. Based on these properties, it is possible to implement designs of combustion plants of different outputs for these fuels.

Keywords: biomass analysis; alternative biofuels; emissions

1. Introduction

Nowadays, there is increasing pressure on the use of renewable sources of fuel in domestic boilers. The primary renewable energy source is plant biomass [1]. A promising form of biomass is energy crops, which are usually compressed into pellets for combustion [2]. The number of pellets made of alternative non-wood material, so-called agropellets, is continuously increasing. Agropellets are produced by pressing agricultural commodities, such as energy plants, rapeseed and cereal straw, waste, oilcake, and others [3]. The combustion of agrofuels generates minimal greenhouse gases and other potentially hazardous emissions under optimal conditions relative to conventional fuels [4]. Biomass is even considered neutral from the point of view of carbon dioxide production since the amount of carbon dioxide produced by combustion is comparable to the amount consumed by plants as they grow. The amount of these substances released during combustion is influenced by the composition of the fuel, the type of combustion equipment used, the setting of the combustion process, etc. One of the factors that significantly affects the combustion efficiency and potential emissions production is the characteristics of the biomass combusted. In addition to solid biofuels, there are also liquid and gaseous biofuels that are the product of solid biofuel transformation processes; however, this study does not focus on them.

Biomass is composed of organic and inorganic substances containing mainly carbon, hydrogen, and oxygen. In addition to these essential elements, there are also often nitrogen, chlorine, iron, and alkali metals [5].

On the contrary, sulfur and heavy metals are only present in trace amounts compared to fossil fuels. The more of these elements the biomass contains, the higher the number of harmful substances will be released during its combustion. Moreover, the amount of these elements in biomass is greatly influenced by the type of biomass and the place of cultivation.

Emissions from biomass combustion can then be divided into three main groups:

- Pollutants from incomplete combustion: CO, C_xH_y, tar, soot, unburnt hydrocarbon particles, hydrogen, and incompletely oxygenated nitrogen compounds (HCN, NH_3, N_2O).
- Pollutants from complete combustion: nitrogen oxides (NO, NO_2) and CO_2.
- Pollutants from trace elements of impurities: incombustible dust particles, sulfur, chlorine compounds, and trace metals (Cu, Pb, Zn, Cd) [5].

The most monitored pollutants are carcinogenic, poisonous, and greenhouse gases. The most important pollutants are characterized in the following passage.

The quality of the combustion process determines the formation of carbon dioxide. The combustion of biomass is characterized by long-flame CO burning. Undesirable cooling results in the release of pure carbon (soot), resulting in significant heat losses. For this reason, the combustion and post-combustion chambers for biomass in the boiler bodies are much larger than for fossil fuels, and secondary or tertiary air is supplied to the flames. This results in improved combustion in terms of the chemistry of the reaction, which leads to a significant reduction in CO and unburned chemicals. In terms of sulfur oxides, biomass is considered ecological fuel compared to fossil fuels because the sulfur content from which sulfur oxides are produced during combustion processes is present only in low concentrations in biomass. Furthermore, the fuel releases large amounts of water vapor and hydrogen, with which sulfur reacts to form hydrogen sulfide (H_2S) [5,6].

Usually, about 0.5–5% of nitrogen is present in biomass [1–4]. All nitrogen content is converted into NO_x compounds during combustion. At temperatures of 700–800 °C, mainly N_2O is produced, which contributes to the greenhouse effect. At temperatures above 1000 °C the formation of NO prevails, which is unstable and oxidizes to NO_2, which is involved in the creation of photochemical smog, possibly due to a reaction with water to form acid rain (HNO_3) [7]. Domestic boilers, however, usually do not reach temperatures that lead to the formation of NO to such an extent [5]. Nevertheless, the values of NO_x emissions produced by the combustion of different biomass types with varying contents of nitrogen show an apparent effect of the increased nitrogen content in non-woody biomass on total NO_x emissions [8].

Chlorine is present in biomass in the form of inorganic and organic compounds. The fundamental problem caused by these substances in the flue gas is their reactivity and the high ability to corrode the materials they come into contact with. It is released into the environment during the combustion of fuels containing chloride (e.g., coal and some plant materials and wastes). Chlorine reacts with airborne water vapor to form hydrogen chloride. Hydrogen chloride gas is rapidly converted to hydrochloric acid, which contributes to the formation of acid rain [9].

One of the critical factors in terms of the optimization of the combustion process, construction of the fireplace, and distribution of combustion air distribution into primary, secondary, and possibly tertiary air is the proportion of volatile combustible material [10]. Increased portions of volatile combustible materials and a lack of secondary or tertiary air will lead to an increase in unburned chemicals and products of incomplete combustion (CO, C_xH_y) [3,5,10,11].

Emissions of particulate matter (PM) are also a significant problem in combustion. The formation of PM during biomass combustion is closely related to the release of inorganic substances and alkali metals from the fuel. These substances are fuel ash, and therefore the formation of PM is closely associated with the composition of fuel ash, specifically and predominantly with the number of alkali metals in the ash [12]. The polluting particles themselves are usually composed of the K, Cl, and S elements, which form aerosols and alkali metal sulfates, chlorides, and carbonates. The critical element in the composition of the dust particles is potassium, which is usually found in the form of K_2SO_4, KCl, and K_2CO_3 [13,14]. PM emissions may also be related to the phosphorus content of the fuel. Combustion of agropellets with a high phosphorus content produces PM consisting of the chlorides mentioned above, carbonates, and sulfates, plus an increased amount of phosphates [15].

Since all the emissions above and fuel behavior in combustion processes are related to the biomass composition, it is always necessary to know its properties, such as moisture, ash content, elemental composition, or lower heating value to optimize it.

This study aimed to investigate the fuel properties (such as coarse and elemental analysis, or lower heating value) of several new, potentially usable biofuels, such as quinoa, camelina, crambe, safflower, and compare them with some traditional biofuels (wood, straw, sorrel, hay). The obtained data can contribute to the expansion of the biofuel portfolio in energy production.

2. Materials and Methods

The section summarizes the subsections containing the description of tests, procedures of determination, processing of measured data, and formulas used for calculation of the monitored values. Determination of dry matter, water content, ash amount, and loss during annealing, determination of volatile combustible content, elemental analysis (C, H, N, S), determination of calorific value using the calorimetric method, and calculation of the lower heating value were performed.

For determination of the dry matter and water content of solid biofuels, three different gravimetric procedures were used based on standards ČSN EN ISO 18 134-1–3 [16–18], which were used depending on available sample amount. ČSN EN ISO 18 134-1 is a reference method that was used when a large amount of sample was available. The method in the calculation also included the so-called buoyancy effect on the hot sheet on which the analyzed sample was dried. The sample was weighed with an accuracy of 0.1 g. The result was calculated using the formula (1):

$$W_1 = \frac{(m_{1,2} - m_{1,3}) - (m_{1,4} - m_{1,5})}{(m_{1,2} - m_{1,1})} \times 100 \quad (\%), \tag{1}$$

where:

$m_{1,1}$—mass of empty sheet for sample (g),
$m_{1,2}$—mass of sample sheet before drying (g),
$m_{1,3}$—mass of sample sheet after drying (g),
$m_{1,4}$—reference sheet mass before drying (g), and
$m_{1,5}$—reference sheet mass after drying (g).

ČSN EN ISO 18 134-3 is a method that was used when only a limited amount of sample was available. A smaller sample volume was compensated for in this method by higher weighing accuracy requirements. The weighing was carried out only with wholly cooled samples. Both methods mentioned so far utilized oven drying at 105 °C until there was a constant mass. In the second case, the result was calculated according to Equation (2):

$$W_3 = \frac{(m_{3,2} - m_{3,3})}{(m_{3,2} - m_{3,1})} \times 100 \quad (\%), \tag{2}$$

where:

$m_{3,1}$—mass of empty crucible with lid (g),
$m_{3,2}$—mass of crucible with sample and lid before drying (g), and
$m_{3,3}$—mass of crucible with sample and lid after drying (g).

To determine the ash content of solid biofuels and the loss on annealing, a procedure based on the standard ČSN EN ISO 18 122 (Solid biofuels – Determination of ash content) [19] was used, where the sample was annealed in the furnace at 550 °C until a constant sample mass was reached. The result was then calculated as a percentage for both the raw and the anhydrous sample according to Equations (3) and (4):

Determination of ash content in the anhydrous sample:

$$A_d = \frac{m_3 - m_1}{m_2 - m_1} \times 100 \times \frac{100}{100 - M_{ad}}, \tag{3}$$

where:

m_1—mass of empty dish (g),
m_2—mass of dish with test portion (g),
m_3—mass of dish with ash (g), and
M_{ad}—the water content of the test portion used for the determination (%).

Determination of ash content in the raw sample:

$$A_r = \frac{m_3 - m_1}{m_2 - m_1} \times 100, \qquad (4)$$

where:

m_1—mass of empty dish (g),
m_2—mass of dish with test portion (g), and
m_3—mass of dish with ash (g).

The determination of the volatile combustible solid biofuels content was performed gravimetrically according to the standard ČSN EN ISO 18 123 (Solid biofuels–Determination of volatile combustible content) when the biofuel sample was annealed at 900 °C for 7 min in a porcelain crucible with a lid inside an oven [20]. The resulting mass percent of volatile combustible in the sample was calculated using the following Equations (5) and (6):

Determination of volatile combustible content in an anhydrous sample:

$$V_d = \left[\frac{100(m_2 - m_3)}{(m_2 - m_1)} - M_{ad}\right] \times \left(\frac{100}{100 - M_{ad}}\right), \qquad (5)$$

where:

m_1—mass of empty crucible with lid (g),
m_2—mass of crucible with sample and lid before heating (g),
m_3—mass of crucible with sample and lid after heating (g), and
M_{ad}—the percentage of the mass of the sample water content (%).

Determination of volatile combustible content in a raw sample:

$$V_d = \left[\frac{100(m_2 - m_3)}{(m_2 - m_1)} - M_{ad}\right], \qquad (6)$$

where:

m_1—mass of empty crucible with lid (g),
m_2—mass of crucible with sample and lid before heating (g),
m_3—mass of crucible with sample and lid after heating (g), and
M_{ad}—the percentage of the mass of the sample water content (%).

Furthermore, the percentage of carbon, hydrogen, nitrogen, and sulfur in the sample was determined using elemental analysis and the oxygen content was calculated. The elementary analyzer Vario Macro cube CHNS (Elementar company) was used for the analysis, working on the principle of sample combustion in a catalytic tube, separation of different gases from monitored components by adsorption-desorption on columns, and subsequent detection using a thermal conductive detector [21]. The measured concentrations of individual elements in the original sample were also recalculated for combustible and dry matter according to the following Equations (7)–(9):

Determination of elemental content in a biofuel sample:

From the measured concentration values in the original sample (wt%) of carbon C^a, hydrogen H^a, nitrogen N^a, and sulfur S^a in the raw sample, the oxygen concentration O^a was calculated assuming that the elements C, H, N, S, and O constituted all the combustible content in the sample:

$$O^a = 100 - (C^a + H^a + N^a + S^a) - A^a, \qquad (7)$$

where A^a is the ash content in the original sample (wt%).

When determining the concentrations of C, H, N, S, and O in a combustible content, it was necessary to assume that these elements together made up all the combustible content and water in the original sample. The water in the original sample consisted of only the elements H and O. From the molar masses of H and O, it was possible to determine the mass fraction of the given elements in water (H_2O):

$$H_{H_2O} = \frac{2 \times M(H)}{2 \times M(H) + M(O)} \times 100 = \frac{2 \times 1.0079}{2 \times 1.0079 + 15.999} \times 100 = 11.19, \tag{8}$$

$$O_{H_2O} = 100 - H_{H_2O} = 100 - 11.19 = 88.81, \tag{9}$$

where:

H_{H_2O}—hydrogen mass content in water (-),
O_{H_2O}—oxygen mass content in water (-),
$M(H)$—hydrogen molar mass (kg·mol^{-1}), and
$M(O)$—oxygen molar mass (kg·mol^{-1}).

By subtracting water from the original sample, the concentrations of H and O were reduced, while the concentrations of C, N, and S were maintained, as seen in Equations (10) and (11):

$$H^{a,red} = H^a - (H_{H_2O} - w_w) \tag{10}$$

$$O^{a,red} = O^a - (O_{H_2O} - w_w) \tag{11}$$

where:

$H^{a,red}$—reduced hydrogen concentration (wt%),
$O^{a,red}$—reduced oxygen concentration (wt%),
O^a—oxygen concentration in the original sample (wt%),
H^a—hydrogen concentration in the original sample (wt%),
H_{H_2O}—percentage by mass of hydrogen content in water (-), and
O_{H_2O}—percentage by mass of oxygen content in water (-).

Concentrations C^a, N^a, and S^a in the original sample, along with the reduced concentrations $H^{a,red}$ and $O^{a,red}$, together form real ratios related to the combustible content. These have to be recalculated to make up 100% of the combustible content; for a calculation example, see Equations (12) and (13):

$$C^{daf} = \frac{C^a}{C^a + H^{a,red} + N^a + S^a + O^{a,red}} \times 100, \tag{12}$$

$$H^{daf} = \frac{H^{a,red}}{C^a + H^{a,red} + N^a + S^a + O^{a,red}} \times 100, \tag{13}$$

where:

C^{daf}—the concentration of carbon in the combustible content (wt%),
H^{daf}—hydrogen concentration in the combustible material (wt%),
C^a—carbon concentration in the original sample (wt%),
$H^{a,red}$—reduced hydrogen concentration (wt%),
N^a—nitrogen concentration in the original sample (wt%),
S^a—sulfur concentration in the original sample (wt%), and
$O^{a,red}$—reduced oxygen concentration (wt%).

Subsequently, the remaining concentrations were calculated for N^{daf}, S^{daf}, and O^{daf} in the combustible content similarly.

A simple relation was used to convert the concentrations of the elements C, H, N, S, and O in the combustible content to the concentrations of individual elements in the dry matter (only the sample relation for C is described here):

$$C^{dr} = \frac{w_{daf}^{dr} \times C^{daf}}{100}, \tag{14}$$

where:
C^{dr}—the concentration of carbon in dry matter (wt%),
C^{daf}—carbon concentration in the combustible content (wt%), and
w_{daf}^{dr}—combustible content in dry matter (wt%).

The concentrations H^{dr}, N^{dr}, S^{dr}, and O^{dr} in the dry matter were subsequently calculated.

Subsequently, the calorific value of the selected materials was determined using an IKA C 200 calorimeter (IKA company) or a 6100 Compansated Calorimeter (Parr Instrument Company) following the standard ČSN EN ISO 18125 (Solid biofuels – Determination of higher and lower heating values). The principle was to burn the weighed analytical sample in an oxygen atmosphere at high pressure in a calorimeter vessel. The measured higher heating values determined by both calorimetric methods indicate the higher heating value of the original sample HHV^a. The following Equations (15) and (16) were used to convert the higher heating value of the original sample HHV^a to the higher heating value of dry matter HHV^{dr} and the higher heating value of the combustible content HHV^{daf} [22]:

$$HHV^{dr} = \frac{HHV^a}{W_{dr}} \times 100, \tag{15}$$

$$HHV^{dr} = \frac{HHV^{daf}}{w_{daf}^{dr}} \times 100, \tag{16}$$

where:
HHV^{daf}—the higher heating value of the combustible content (kJ·kg^{-1}),
HHV^{dr}—higher heating value of dry matter (kJ·kg $^{-1}$),
HHV^a—higher heating value of the original sample (kJ·kg $^{-1}$),
w_{daf}^{dr}—combustible content in dry matter (= loss by annealing in dry matter) (wt%), and
W_{dr}—dry matter content in the sample (wt%).

The lower heating value could then be calculated from the higher heating value using Equation (17). The lower heating value is defined as the higher heating value released by burning 1 kg of fuel minus the condensation heat of the water produced by combustion. In accordance with ČSN EN ISO 18 125 [23], Equation (17) was chosen to determine the lower heating value of the original sample LHV^a:

$$LHV^a = HHV^a - r_{H_2O}^{20°C} \times \left(W + 8.94 \times x_H^a\right), \tag{17}$$

where the concentration of combustible hydrogen in the original sample x_H^a was calculated using Equation (18):

$$x_H^a = \frac{H^{daf} \times w_{daf}^a}{100}, \tag{18}$$

where:
LHV^a—lower heating value of the original sample (kJ·kg^{-1}),
HHV^a—higher heating value of the original sample (kJ·kg^{-1}),
$r_{H2O}^{20°C}$—the evaporation heat of water at 20 °C has a value of 2454 (kJ·kg $^{-1}$),
W—concentration of water in the sample (wt%),
x_H^a—concentration of combustible hydrogen in the original sample (wt%),
H^{daf}—concentration of hydrogen in the original sample (wt%), and
w_{daf}^a—combustible content in the original sample (wt%).

For the calculation of the lower heating value of the dry matter LHV^{dr}, Equation (19) was used:

$$LHV^{dr} = HHV^{dr} - r_{H_2O}^{20°C} \times 8.94 \times x_H^{dr}, \tag{19}$$

where the concentration of hydrogen in dry matter x_H^{dr} was calculated using Equation (20):

$$x_H^{dr} = \frac{H^{dr}}{100}, \tag{20}$$

where:
LHV^{dr}—lower heating value of dry matter (kJ·kg^{-1}),
HHV^{dr}—higher heating value of dry matter (kJ·kg^{-1}),
$r_{H_2O}^{20°C}$—the evaporation heat of water at 20 °C has a value of 2,454 (kJ·kg^{-1}), and
x_H^{dr}—concentration of hydrogen in dry matter (wt%).

The following Equation (21) was used for the conversion from the lower heating value of dry matter to the lower heating value of the combustible content:

$$LHV^{daf} = \frac{LHV^{dr}}{w_{daf}^{dr}} \times 100, \tag{21}$$

where:
LHV^{daf}—the lower heating value of the combustible content (kJ·kg^{-1}),
LHV^{dr}—the lower heating value of the dry matter (kJ·kg^{-1}), and
w_{daf}^{dr}—combustible content in the dry matter (wt%).

3. Results and Discussion

Table 1 summarizes the measured water content values W, which was determined using the gravimetric method described in the previous section. The water content is an important parameter that affect fuel quality. Above all, it directly affects its lower heating value by reducing the dry matter content and by consuming heat to evaporate water during combustion [24]. During combustion, the combustion temperature may fall below the optimum value due to evaporative heat consumption. Consequently, there is a risk of incomplete combustion of fuel and the generation of above-the-limit emissions [25]. If the flue gas temperature drops below the dew point, water condensation will occur, leading to an acceleration of the flue gas corrosion of the combustion device [26]. Ideally, the moisture of the material to be combusted is less than 15% in the case of pellets or less than 20% in the case of loose material. As can be seen from Table 1, the water content ranged from 3.82% to 11.92%, which meant the materials were suitable for combustion. The pellets had very low moisture contents, which partially caused the pellets to crumble and break. Low moisture in a very narrow range of values is influenced by storage in a dry and warm fuel storage environment. The standard deviation and the confidence interval were calculated for the average water content. From the moisture content, the dry matter content in the sample was found range between 88.08% and 96.18%.

After determining the moisture content and dry matter content, the ash contents of the raw and anhydrous samples were determined, and the loss during annealing and the ballast fraction were calculated from these data. After finding the water content, the ash content is another important parameter that characterizes the examined fuel sample. Table 2 shows that the lowest ash content of 0.3% was found in a wood pellet sample, which corresponded to the fact that only wood mass was present almost entirely in this sample. By contrast, in the case of agro-materials, the ash content is higher: hay 4.83%, sunflower 3.92%, and safflower 6.6%. An increased content of ballast substances was evident, which also corresponded to the increased value of the calculated ballast portion. The highest ash content was determined in samples with a high percentage of waste sludge present due to the increased occurrence of heavy metals and other hazardous elements contained in the combusted material. This phenomenon is disadvantageous for the material to be burned because the increased ash content during the combustion makes the boiler operation more challenging in terms of removing the ash from the boiler body and faster filling of the ashbin.

Table 1. Results of the determination of water and dry matter contents in samples of biofuels.

Sample	Water Content W (wt%)	Dry Matter Content W_{dr} (wt%)	Standard Deviation	Confidence Interval
Digestate	8.43	91.57	0.07	0.08
Softwood pellets (spruce)	7.67	92.33	0.04	0.04
Hardwood pellets (beech)	7.84	92.16	0.07	0.08
Composite wood	9.49	90.51	0.05	0.06
Energo compost	9.35	90.65	0.24	0.27
Rape straw	10.62	89.38	0.08	0.09
Wheat straw pellet	7.16	92.84	0.02	0.02
Hay	7.96	92.04	0.08	0.09
Straw 60% + sludge 40%	3.86	96.11	0.03	0.03
Straw 70% + sludge 30%	3.82	96.18	0.02	0.02
Straw 80% + sludge 20%	4.39	95.61	0.04	0.04
Straw 90% + sludge 10%	4.79	95.21	0.09	0.1
Sunflower—peel	7.71	92.29	0.05	0.06
Sunflower—after the oil press	6.09	93.91	0.06	0.07
Sunflower—whole plant	10.58	89.42	0.32	0.36
Mix—seeds rape, sunflower, mustard, husks	11.92	88.08	0.05	0.06
Mustard—seed	5.88	94.12	0.17	0.19
Spruce sawdust + digestate	6.44	93.56	0.09	0.10
Safflower—seed	5.41	94.59	0.09	0.10
Safflower—peel	5.49	94.51	0.09	0.10
Safflower—after the oil press	7.08	92.92	0.12	0.13
Amaranth	6.69	93.31	0.02	0.02
Flax—waste	5.42	94.58	0.08	0.09
Crambe abyssinica	5.49	94.51	0.04	0.04
Camelina—seed	5.95	94.05	0.01	0.01
Camelina—after the oil press	7.10	92.90	0.08	0.09
Spelt—waste	8.06	91.94	0.04	0.04
Cocoa—peel	7.34	92.66	0.03	0.03
Sorrel pellet (whole plant)	8.64	91.36	0.05	0.05
Rye straw	7.69	92.31	0.12	0.13
Quinoa—waste	8.43	91.57	0.03	0.03

Notes: The accuracy of the determination methods were below 0.5%$_{abs}$

The ash content for the selected commodity may also vary depending on the different regions from which it is extracted. In plant and woody materials, the ash content is greatly influenced by the content and composition of substances derived from the soil, whose composition varies in different locations. For this reason, the ash content can only be compared approximately. For example, in Barbanera and Cotana [27], the ash content in the dry matter of the digestate was 12.38%, whereas in the sample digestate we analyzed, 11.31% ash was found. Similar values were found in safflower seed (3.0%) [28], sunflower peel (2.7%) [29], and wheat straw (6.72%) [30].

Another variable characterizing the fuel is the ballast portion B. As mentioned, it is the proportion of substances reducing the lower heating value of the fuel. The ballast ratio values largely correspond to the ash value. As can be seen in Table 2, low amounts of ballast were observed in the case of wood material, with increased values found in the analyzed agro-materials and the highest values were reached for the material containing waste sludge. It was precisely in the waste sludge that the non-combustible components were concentrated, which in turn significantly reduced the lower heating value of the material. For this reason, waste sludge is often used in mixed pellets with varying proportions of woody or plant biomass.

Table 2. Results of the determination of ash content, loss on annealing, and ballast ratio in samples.

Sample	Ash Content in the Raw Sample A_r (wt%)	Ash Content in the Anhydrous Sample A_d (wt%)	Loss on Annealing the Raw Sample W_{LOR} (wt%)	Loss on Annealing the Anhydrous Sample W_{LOD} (wt%)	Ballast Ratio in the Raw Sample B_r (wt%)
Digestate	10.36	11.31	89.64	87.65	18.79
Softwood pellets (spruce)	2.77	3.01	97.23	96.99	10.44
Hardwood pellets (beech)	2.25	2.44	97.75	97.56	10.09
Composite wood	0.30	0.33	99.70	99.67	9.79
Energo compost	19.92	21.97	80.08	75.76	29.27
Rape straw	3.82	4.27	96.18	95.23	14.44
Wheat straw pellet	6.16	6.64	93.84	93.36	13.32
Hay	4.83	5.25	95.17	94.30	12.79
Straw 60% + sludge 40%	27.93	29.05	72.07	69.77	31.79
Straw 70% + sludge 30%	25.07	26.07	74.93	72.90	28.89
Straw 80% + sludge 20%	17.24	18.03	82.76	81.14	21.63
Straw 90% + sludge 10%	10.24	10.75	89.76	88.71	15.03
Sunflower — peel	3.92	4.20	96.08	95.50	10.60
Sunflower— fter the oil press	5.67	6.04	94.33	93.96	11.76
Sunflower—whole plant	3.93	4.39	96.07	95.10	14.51
Mix—seeds rape, sunflower, mustard, husks	6.98	7.93	93.02	91.00	18.90
Mustard—seed	14.68	15.58	85.32	84.42	20.56
Spruce sawdust + digestate	2.53	2.70	97.47	97.11	8.97
Safflower—seed	6.60	7.00	93.40	93.00	12.01
Safflower—peel	2.69	2.97	97.31	97.03	8.18
Safflower-after the oil press	3.68	3.96	96.32	96.04	10.76
Amaranth—whole plant	7.13	7.70	92.87	92.30	13.82
Flax—waste	17.05	17.90	82.95	82.10	22.47
Crambe abyssinica	5.82	6.16	94.18	93.84	11.31
Camelina—seed	12.57	13.36	87.43	86.64	18.52
Camelina—after the oil press	10.09	10.87	89.92	89.14	17.18
Spelt—waste	4.43	4.82	95.57	95.18	12.49
Cocoa—peel	5.92	6.39	94.08	93.10	13.26
Sorrel pellet (whole plant)	4.73	5.18	95.27	94.33	13.37
Rye straw	12.51	13.67	87.49	86.33	20.19
Quinoa—waste	4.76	5.20	95.24	94.80	13.19

Notes: The accuracy of the determination method was below 2.5%$_{abs}$.

The evaluation of the rough analysis of the selected samples is subsequently shown in Figure 1 and Table 3. The content of water, ash, and combustible content in the chosen materials varied greatly, as can be seen from the table below.

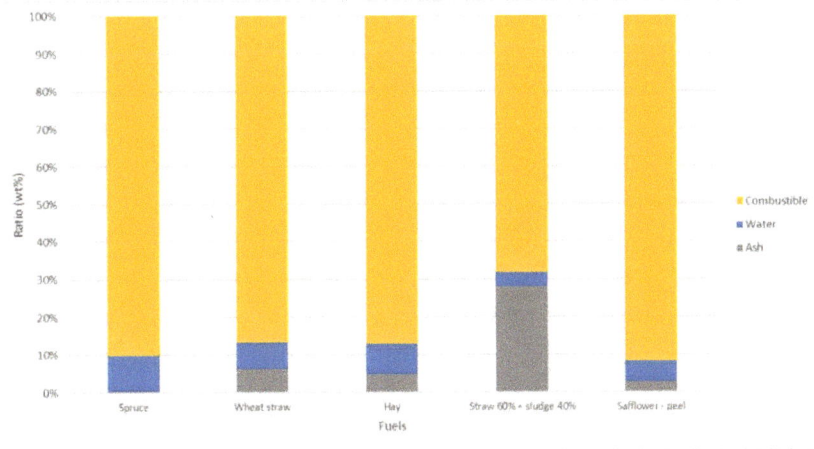

Figure 1. Total ratio of combustible content, water, and ash in the selected materials.

Table 3. Evaluation of the rough analysis of the selected samples.

Proximate Analysis		Spruce	Wheat Straw Pellets	Hay	Straw:Sludge 60:40	Safflower—Peel
Combustible	(wt%)	90.21	86.68	87.21	68.21	91.82
Water	(wt%)	9.49	7.16	7.96	3.86	5.49
Ash	(wt%)	0.3	6.16	4.83	27.93	2.69

After carrying out and evaluating the rough analysis of the materials intended for combustion, the determination of the volatile contents in the raw and anhydrous samples was carried out. The volatile content, together with the solids, make up the total combustible content in the samples. An example is given in the following Table 4.

Table 4. Ratio of volatile content to solids in the combustible.

Ratio of Combustible	Wood	Straw Pellets	Safflower—Peel
Combustible (wt%)	90.21	86.68	91.82
Volatile content (wt%)	75.45	71.37	79.15
Solids (wt%)	14.76	15.31	12.67

As can be seen in Table 5, the volatile content value ranged from 47.49wt% to 81.30wt% for the raw sample and 51.45wt% to 88.42wt% for the anhydrous sample, with average values of 74.4wt% and 79.4wt%, respectively. The volatile content value may vary within one material, as was noted for the safflower. For whole seeds, the value was 73.4wt%. On the other hand, for peels, the volatile content was higher (79.15wt%). In the safflower pellets after the oil press, the volatile content was 74.3wt%. This pellet contained both the seed and the peels. The values of the determined safflower volatile content approximately corresponded to the 83wt% found in another study [28]. A similar trend was observed in the case of camelina.

Table 5. Results of the determination of volatile content.

Sample	Volatile Content in Raw Sample V_r (wt%)	Volatile Content in Anhydrous Sample V_d (wt%)
Softwood pellets (spruce)	76.44	82.79
Hardwood pellets (beech)	76.34	82.84
Composite wood	75.45	83.36
Wheat straw pellet	71.37	76.88
Hay	78.39	89.99
Sunflower—peel	47.49	51.45
Sunflower—after the oil press	75.73	80.64
Mustard—seed	78.16	83.05
Safflower—seed	73.38	71.86
Safflower—peel	79.15	78.85
Safflower—after the oil press	74.34	80.00
Amaranth—whole plant	73.58	78.85
Flax—waste	72.41	76.56
Crambe abyssinica	78.82	83.40
Camelina—seed	80.36	85.45
Camelina—after the oil press	74.15	79.82
Spelt—waste	81.30	88.42
Rye straw	77.02	83.43
Quinoa—waste	74.77	81.65

Notes: The accuracy of the determination method was below 1%$_{abs}$.

After determination of the volatile content, the elemental analysis was carried out to determine the carbon, hydrogen, nitrogen, and volatile sulfur content of the sample, and the calculation of the oxygen content was added. The measured concentrations of individual elements in the original sample were also converted to the content in the combustible and dry matter. The measured and calculated values of the elemental analysis are summarized in the following Tables 6–8.

Table 6. Percentages of carbon, nitrogen, hydrogen, sulfur, and oxygen in the raw (original) samples of combusted material.

Sample	Raw (Original) Sample—Elements (wt%)				
	N^a	C^a	H^a	S^a	O^a
Digestate	1.91	41.68	5.45	0.04	39.60
Softwood pellets (spruce)	0.43	46.65	6.37	0.00	43.79
Hardwood pellets (beech)	0.60	46.33	6.17	0.00	44.65
Composite wood	0.80	48.30	6.08	0.00	44.50
Energo compost	2.13	38.22	4.79	0.33	32.56
Rape straw	0.69	41.72	6.10	0.02	47.02
Wheat straw pellet	0.76	43.81	6.09	0.11	43.07
Hay	0.68	42.20	5.98	0.05	45.84
Straw 60% + sludge 40%	0.40	36.04	4.23	0.01	30.27
Straw 70% + sludge 30%	0.44	37.19	4.36	0.00	31.93
Straw 80% + sludge 20%	0.67	39.10	4.83	0.00	37.37
Straw 90% + sludge 10%	0.39	41.88	5.52	0.00	41.46
Sunflower—peel	0.80	46.48	5.96	0.04	42.51
Sunflower—after the oil press	3.67	50.85	7.97	0.20	31.63
Sunflower—whole plant	0.90	45.24	5.87	0.03	43.57
Mix—seeds rape, sunflower, mustard, husks	2.12	40.32	5.86	0.16	43.58
Mustard—seed	5.02	52.37	8.04	1.08	18.80
Spruce sawdust + digestate	0.61	46.23	6.12	0.00	43.34
Safflower—seed	2.29	52.55	7.52	0.01	31.02
Safflower—peel	1.67	50.82	7.19	0.01	37.61
Safflower—after the oil press	2.70	48.14	6.87	0.04	38.57
Amaranth—whole plant	0.89	40.16	5.71	0.09	46.03
Flax—waste	3.41	46.78	6.75	0.17	25.83
Crambe abyssinica	3.13	54.48	8.28	0.67	27.61
Camelina—seed	4.66	54.81	8.45	0.66	18.06
Camelina—after the oil press	6.17	47.21	7.41	0.90	28.23
Spelt—waste	0.68	44.98	6.05	0.19	43.67
Cocoa—peel	2.39	45.49	6.19	0.11	39.43
Sorrel pellet (whole plant)	1.31	43.28	5.86	0.09	44.29
Rye straw	2.81	41.74	6.79	0.05	36.10
Quinoa—waste	2.87	42.99	6.84	0.18	42.34

Notes: The accuracy of the determination method was below 0.5%$_{abs}$.

Table 7. Percentages of carbon, nitrogen, hydrogen, sulfur, and oxygen in the combustible content.

Sample	Combustible—Elements (wt%)				
	N^{daf}	C^{daf}	H^{daf}	S^{daf}	O^{daf}
Digestate	2.38	51.94	5.62	0.05	40.02
Softwood pellets (spruce)	0.48	52.08	6.15	0.00	41.28
Hardwood pellets (beech)	0.66	51.53	5.88	0.00	41.92
Composite wood	0.85	51.17	5.82	0.00	42.16
Energo compost	3.10	55.65	5.45	0.48	35.32
Rape straw	0.81	49.12	5.78	0.02	44.26
Wheat straw pellet	0.88	50.54	6.10	0.12	42.35
Hay	0.78	48.62	5.86	0.06	44.67
Straw 60% + sludge 40%	0.60	53.72	5.66	0.01	40.01
Straw 70% + sludge 30%	0.63	53.05	5.61	0.00	40.71
Straw 80% + sludge 20%	0.86	50.40	5.59	0.00	43.14
Straw 90% + sludge 10%	0.46	49.59	5.90	0.00	44.05
Sunflower—peel	0.90	52.16	5.85	0.04	41.05
Sunflower—after the oil press	4.16	57.63	8.26	0.23	29.72
Sunflower—whole plant	1.06	53.20	5.51	0.04	40.19
Mix—seeds rape, sunflower, mustard, husks	2.65	50.32	5.65	0.20	41.18
Mustard—seed	6.32	65.93	9.30	1.36	17.09
Spruce sawdust + digestate	0.68	51.45	6.01	0.00	41.87
Safflower—seed	2.60	59.73	7.86	0.02	29.79
Safflower—peel	1.82	55.35	7.16	0.01	35.65
Safflower—after the oil press	3.03	53.94	6.81	0.04	36.17
Amaranth—whole plant	1.03	46.60	5.76	0.10	46.52
Flax—waste	4.40	60.34	7.92	0.22	27.11
Crambe abyssinica	3.53	61.43	8.64	0.76	25.64
Camelina—seed	5.78	67.94	9.64	0.82	15.83
Camelina – after the oil press	7.45	57.00	7.98	1.08	26.49
Spelt—waste	0.78	51.40	5.89	0.21	41.72
Cocoa—peel	2.77	52.73	6.22	0.13	38.15
Sorrel pellet (whole plant)	1.52	50.21	5.68	0.10	42.48
Rye straw	3.52	52.30	7.43	0.07	36.68
Quinoa—waste	3.30	49.47	6.80	0.21	40.22

Notes: The accuracy of the determination method was below 0.5%$_{abs}$.

Table 8. Percentages of carbon, nitrogen, hydrogen, sulfur, and oxygen in the dry matter.

Sample	Dry Matter—Elements (wt%)				
	N^{dr}	C^{dr}	H^{dr}	S^{dr}	O^{dr}
Digestate	2.27	46.06	4.98	0.04	35.49
Softwood pellets (spruce)	0.47	50.64	5.98	0.00	40.14
Hardwood pellets (beech)	0.65	50.37	5.75	0.00	40.98
Composite wood	0.85	51.00	5.80	0.00	42.02
Energo compost	2.42	43.42	4.25	0.37	27.56
Rape straw	0.78	47.03	5.54	0.02	42.37
Wheat straw pellet	0.83	47.42	5.73	0.12	39.75
Hay	0.74	46.07	5.56	0.05	42.33
Straw 60% + sludge 40%	0.42	38.11	4.02	0.01	28.39
Straw 70% + sludge 30%	0.46	39.22	4.15	0.00	30.10
Straw 80% + sludge 20%	0.71	41.31	4.58	0.00	35.37
Straw 90% + sludge 10%	0.41	44.26	5.27	0.00	39.32
Sunflower—peel	0.86	49.97	5.60	0.04	39.32
Sunflower—after the oil press	2.72	37.68	5.40	0.15	19.43
Sunflower—whole plant	1.01	50.87	5.27	0.03	38.43
Mix—seeds rape, sunflower, mustard, husks	2.44	46.33	5.20	0.18	37.91
Mustard—seed	5.39	56.25	7.93	1.16	14.58
Spruce sawdust + digestate	0.66	50.06	5.85	0.00	40.74
Safflower—seed	2.43	55.79	7.34	0.01	27.83
Safflower—peel	1.77	53.86	6.97	0.01	34.69
Safflower—after the oil press	2.92	51.96	6.56	0.04	34.84
Amaranth—whole plant	0.96	43.27	5.35	0.09	43.20
Flax—waste	3.65	50.06	6.57	0.19	22.49
Crambe abyssinica	3.32	57.64	8.11	0.71	24.06
Camelina—seed	5.00	58.86	8.35	0.71	13.71
Camelina—after the oil press	6.40	49.00	6.86	0.93	22.77
Spelt—waste	0.75	49.12	5.62	0.20	39.88
Cocoa—peel	2.59	49.36	5.83	0.12	35.71
Sorrel pellet (whole plant)	1.44	47.61	5.38	0.10	40.28
Rye straw	3.08	45.76	6.50	0.06	32.09
Quinoa—waste	3.14	47.11	6.48	0.20	38.29

Notes: The accuracy of the determination method was below 0.5%$_{abs}$.

It is apparent from Table 6 that wood materials reached very similar values for all monitored elements. The values from the wood samples were close to the measured percentages of elements in the samples of hay and straw, which in terms of elemental analysis, seems to be a suitable fuel that could replace wood pellets. However, a slightly increased sulfur content (up to 0.11wt%) was observed with these samples. The increased sulfur content was also observed for some oilseed samples, such as mustard (1.08wt%), sunflower (up to 0.2wt%), camelina (up to 0.9wt%), and cocoa (0.11wt%). This was similar to information found in other literary sources [29–32]. The combustion of sulfur-containing material releases its volatile content, which subsequently reacts with hydrogen to form hydrogen sulfide, or with oxygen to form sulfur dioxide. The low presence of sulfur in the raw material monitored only meant the formation of a negligible amount of these gaseous emissions in the combustion process.

In Table 7, the contents of the monitored elements in the combustible were recorded. The conversions given in Section 2 were used to obtain these values. Compared to the elements in the raw sample, a slight increase in nitrogen and carbon content, and a decrease in the amount of hydrogen and oxygen, were observed for the combustible content. The change of these values was influenced by the reduction of water and ash in the combustible content.

In the case of Table 7, there was a significant decrease in other elements to the detriment of hydrogen and oxygen. However, as already mentioned, the final concentration of sulfur and nitrogen in the material was mainly influenced by the particular soil composition in which the biomass was grown and the use of fertilizers. Higher sulfur concentrations in pellets increase the SO_2 emissions and may also cause corrosion when sulfur compounds condense on the exchanger surfaces of the boiler [33]. The content of elements in the combustible was converted to the content of elements in the dry matter. The results are summarized in Table 8 below.

After elementary analysis, the higher heating values of individual materials were determined using the calorimetric method. From the higher heating value of the original HHV^a sample, the higher heating value was then calculated in the combustible HHV^{daf} and dry matter HHV^{dr}. The measurement was performed at least three times, and the mean result was calculated from the measured values. The measured and calculated higher heating values fluctuated in a relatively wide range. The amount of woody mass greatly influenced these values in the sample. It is also evident from Table 9 that the presence of oily substances in the material had a significant influence on the value of the higher heating value. This was observed, for example, in the case of camelina and safflower samples. In the case of camelina samples obtained from the same sources, the effect of the oil content was noticeable. The whole seeds reached a significantly higher value of the higher heating value of 25.154 MJ·kg^{-1}, as opposed to the already pressed seeds, which had a value of 20.942 MJ·kg^{-1}. The tables also show that the higher heating value of the combustible was greater than that of the original sample. In fact, in the case of fuel, the carrier of energy was only combustible. The remaining fuel, ash, and water only reduced this energy of combustible content. This was evident, for example, with digestate having a high higher heating value of combustible and a low higher heating value of the original sample. This significant difference was due to the high content of ballast, i.e., ash and water, in the sample.

Table 9. Higher heating value in the original sample of HHV^a, dry matter HHV^{dr}, and combustible HHV^{daf}.

Sample	Higher Heating Value (MJ·kg^{-1})		
	HHV^a	HHV^{dr}	HHV^{daf}
Digestate	15.769	17.222	19.649
Softwood pellets (spruce)	18.207	19.718	20.330
Hardwood pellets (beech)	18.044	19.579	20.069
Composite wood	18.235	20.148	20.215
Energo compost	14.684	16.199	21.382
Rape straw	15.572	17.422	18.292
Wheat straw pellet	17.238	18.567	19.888
Hay	15.790	17.155	18.192
Straw 60% + sludge 40%	13.559	14.150	20.280
Straw 70% + sludge 30%	13.947	14.500	19.891
Straw 80% + sludge 20%	14.467	15.131	18.648
Straw 90% + sludge 10%	15.744	16.535	18.640
Sunflower—peel	17.699	18.966	19.861
Sunflower—after the oil press	22.319	23.766	25.294
Sunflower—whole plant	16.925	18.928	19.904
Mix—seeds rape, sunflower, mustard, husks	15.246	17.308	19.021
Mustard—seed	24.131	25.639	30.369
Spruce sawdust + digestate	17.565	18.773	19.331
Safflower—seed	23.221	24.550	26.397
Safflower—peel	22.677	23.994	24.728
Safflower—after the oil press	20.125	21.658	22.551
Amaranth—whole plant	16.309	17.478	18.937
Flax—waste	20.318	21.483	26.167
Crambe abyssinica	25.351	26.824	28.583
Camelina—seed	25.154	26.746	30.871
Camelina—after the oil press	20.942	22.542	25.289
Spelt—waste	16.947	18.432	19.365
Cocoa—peel	18.078	19.504	20.948
Sorrel pellet (whole plant)	16.211	17.744	18.809
Rye straw	16.835	18.237	21.126
Quinoa—waste	17.700	19.329	20.389

Notes: The accuracy of the determination method was below 0.5%$_{abs}$.

The lower heating values of the original sample LHV^a, dry matter LHV^{dr}, and combustible LHV^{daf} were then calculated from the experimentally determined higher heating values. The lower heating value of the original LHV^a sample is a quantity that indicates the final energy potential of the sample during real combustion. This is the primary parameter for comparing potential fuel, whatever the type of material. The lower heating values showed a similar trend to the higher heating values, i.e., the lower heating values of wood pellets and some oil-containing pellets, such as mustard (22.363 MJ·kg^{-1}), camelina (23.280 MJ·kg^{-1}), and safflower (21.567 MJ·kg^{-1}) were high. On the other hand, digestate samples and the mixture of sawdust with a high waste sludge content showed a low lower heating value. When comparing the higher heating values and the lover heating values of these samples, it was observed that these values were influenced by the high ballast ratio. The results of the lower heating values of all analyzed samples are summarized in the following Table 10.

Table 10. Calculated lower heating values of analyzed samples.

Sample	Lower Heating Value (MJ·kg^{-1})		
	LHV^a	LHV^{dr}	LHV^{daf}
Digestate	14.559	16.129	19.860
Softwood pellets (spruce)	16.807	18.406	20.551
Hardwood pellets (beech)	16.688	18.318	20.374
Composite wood	16.848	18.876	20.925
Energo compost	13.607	15.266	21.582
Rape straw	14.223	16.207	18.942
Wheat straw pellet	15.899	17.311	19.972
Hay	14.470	15.936	18.274
Straw 60% + sludge 40%	12.615	13.269	19.453
Straw 70% + sludge 30%	12.976	13.590	19.113
Straw 80% + sludge 20%	13.396	14.125	18.023
Straw 90% + sludge 10%	14.524	15.380	18.099
Sunflower—peel	16.385	17.737	19.839
Sunflower—after the oil press	20.566	22.581	25.591
Sunflower—whole plant	15.629	17.772	20.787
Mix—seeds rape, sunflower, mustard, husks	13.946	16.167	19.936
Mustard—seed	22.363	23.898	30.083
Spruce sawdust + digestate	16.204	17.490	19.213
Safflower—seed	21.567	22.939	26.071
Safflower—peel	21.096	22.465	24.465
Safflower—after the oil press	18.614	20.219	22.656
Amaranth—whole plant	15.053	16.305	18.920
Flax—waste	18.835	20.042	25.852
Crambe abyssinica	23.530	25.044	28.237
Camelina—seed	23.280	24.913	30.578
Camelina—after the oil press	19.314	21.036	25.400
Spelt—waste	15.617	17.198	19.652
Cocoa—peel	16.711	18.226	21.012
Sorrel pellet (whole plant)	14.918	16.563	19.120
Rye straw	15.342	16.810	21.063
Quinoa—waste	16.195	17.908	20.628

Notes: The accuracy of the determination method was below 0.5%$_{abs}$.

4. Conclusions

With the increasing demand for the use of renewable energy sources, there is scope for using other biofuels as a promising energy source. However, to use a broader range of biofuels, it is necessary to know their fuel properties, such as coarse and elemental analysis or their lower heating value.

This research included 60 tested samples of different biofuels, and this publication presents only selected samples that can be expected to be of potential use, with materials and biofuels remaining as a secondary product of their primary use and processing. The selection of samples also took into account the fact that the possible availability and samples of waste materials from the processing of these crops were preferred. One of the reasons why these biofuels have not been used so far is the fact that there is insufficient knowledge about their properties and possibilities for energy use. These include, for example, crops like quinoa, camelina, cramble, safflower, amaranth, sunflowers, or parts thereof. Their fuel properties are here compared with some traditional biofuels (wood, straw, sorrel, hay). Fuels were also chosen in consultation with agricultural research institutes. These selected fuels are currently being studied intensively in the Czech Republic from an agrotechnical point of view, and they appear to be promising for their expansion in the food industry. The residual parts can then be easily used for energy processing.

Several types of analyses were carried out in the examined samples, which comprehensively characterized the given commodities within the combustion process specifics of particular crops, where the results from individual analyses are discussed directly in the text along with individual results.

The main results of the study can be summarized as follows:

- Some materials examined in this study had not been explored and analyzed yet, where some materials showed great potential for becoming a renewable and sustainable energy source for low-power boilers. A large number of these materials are of waste origin or surpluses from agriculture, and their combustion not only generates energy but also greatly facilitates the solution of disposal or possible waste management problems.
- The moisture content of the analyzed biomass significantly affected the treatment of the material itself and the amount of heat released from a unit amount of the selected material since the lower heating value of the material is reduced by a higher water content of the matrix.
- The use of biomass as fuel also affects the amount of ash formed from combustion. If the material forms a large amount of ash, it is more difficult to remove the ash from the boiler body and to quickly fill the ashpan, which is disadvantageous from a user's point of view. For small boiler bodies, it is, therefore, preferable to use biomass with low ash and low ballast contents as the energy source. This implies that materials with a high ash content (e.g., waste sludge) should be combined with, for example, a readily available woody mass that forms a minimal amount of ash.
- The volatile content and lower heating value also have a significant effect on fuel quality. These are closely related, as a higher volatile content will increase the fuel higher heating value.
- Fuel of a plant origin shows the influence of its growth location. The composition of the soil in which the plant has grown significantly affects the number of elements and their representation in all its parts. For example, increased nitrogen in plants is caused by the use of fertilizers, which directly affects the increased release of nitrogen oxides in the combustion process.

Based on the knowledge of the fuel properties of new biofuels, it is possible to realize the design of combustion devices of different outputs for these fuels, and their use can be expanded in the energy sector.

Author Contributions: Conceptualization, M.L. and M.B.; methodology, H.L.; investigation, H.L., D.J. and P.K.; project administration M.L.; resources, H.L., P.K. and M.B.; supervision M.L., M.B. and D.J.; writing—original draft preparation, H.L. and D.J.; writing—review and editing, M.L. and P.K. All authors have read and agreed to the published version of the manuscript.

Funding: This work was supported by the Ministry of Education, Youth and Sports of the Czech Republic under OP RDE grant number CZ 02 1 01/0 0/0 0/16_019/0000753 "Research centre for low-carbon energy technologies."

Conflicts of Interest: The funders had no role in the design of the study; in the collection, analyses, or interpretation of data; in the writing of the manuscript, or in the decision to publish the results.

References

1. Vorotinskienė, L.; Paulauskas, R. Parameters influencing wet biofuel drying during combustion in grate furnaces. *Fuel* **2020**, *265*, 117013. [CrossRef]

2. Matúš, M.; Križan, P. The effect of papermaking sludge as an additive to biomass pellets on the final quality of the fuel. *Fuel* **2018**, *219*, 196–204. [CrossRef]
3. Hrdlička, J.; Skopec, P. Emission factors of gaseous pollutants from small scale combustion of biofuels. *Fuel* **2016**, *165*, 68–74. [CrossRef]
4. Greinert, A.; Mrówczyńska, M. The use of plant biomass pellets for energy production by combustion in dedicated furnaces. *Energies* **2020**, *13*, 463. [CrossRef]
5. Ochodek, T.; Koloničný, J.; Janásek, P. *Potenciál Biomasy, Druhy, Bilance a Vlastnosti Paliv z Biomasy*; Vysoká škola báňská-Technická univerzita Ostrava: Ostrava, Czech Republic, 2006.
6. Petříková, V.; Punčochář, M. Biomasa—Alternativní Palivo z Hlediska Chemického Složení. Available online: https://biom.cz/cz/odborne-clanky/biomasa-alternativni-palivo-z-hlediska-chemickeho-slozeni (accessed on 19 November 2019).
7. Bešenić, T.; Mikulčić, H. Numerical modellign of emissions of nitrogen oxides in solid fuel combustion. *J. Environ. Manag.* **2018**, *215*, 177–184. [CrossRef] [PubMed]
8. Elbl, P.; Baláš, M.; Vavříková, P.; Lisý, M.; Milčák, P. Gaseous Emissions and Solid Particles from the Combustion of Biomass Pellets in 25kW Automatic Boiler. In Proceedings of the 27th European Biomass Conference and Exhibition, Lisbon, Portugal, 27–30 May 2019; pp. 742–748.
9. Chlor a Anorganické Sloučeniny. Available online: https://www.irz.cz/repository/latky/chlor_a_anorganicke_slouceniny.pdf (accessed on 8 January 2020).
10. Pospíšil, J.; Lisý, M.; Špiláček, M. Optimalization of Afterburner Channel in Biomass Boiler Using CFD Analysis. *Acta Polytech.* **2016**, *56*, 379–387. [CrossRef]
11. Lisý, M.; Pospíšil, J.; Štelcl, O.; Špiláček, M. Optimization of Secondary Air Distribution in Biomass Boiler by CFD Analysis. *Appl. Mechan. Mater.* **2016**, *832*, 231–237. [CrossRef]
12. Carroll, J.; Finnan, J. Emissions and efficiencies from the combustion of agricultural feedstock pellets using a small scale tilting grate boiler. *Biosyst. Eng.* **2013**, *115*, 50–55. [CrossRef]
13. Lamberg, H.; Tissari, J. Fione particle and gaseous emissions from a small-scale boiler fueled by pellets of various raw materials. *Energy Fuels* **2013**, *27*, 7044–7053. [CrossRef]
14. Tissari, J.; Sippula, O.; Kouki, J.; Vuorio, K.; Jokiniemi, J. Fine particle and gas emissions from the combustion oa agricultural fuels fired in a 20 kW burner. 2008, 22, 2033–2042. *Energy Fuels* **2008**, *22*, 2033–2042. [CrossRef]
15. Díaz-Ramírez, M.; Boman, C.; Sebastián, F.; Royo, J.; Xiong, S.; Bostrom, D. Ash characterisation and transformation behavior of the fixed-bed combustion of novel crops: Poplar, brassica, and cassava fuels. *Energy Fuels* **2012**, *26*, 3218–3229. [CrossRef]
16. ČSN EN ISO 18 134-1. *Tuhá Biopaliva—Stanovení Obsahu Vody—Metoda Sušení v Sušárně—Část 1: Celková Aoda—Referenční Metoda*; UNMZ: Praha, Czech Republic, 2016.
17. ČSN EN ISO 18 134-2. *Tuhá Biopaliva—Stanovení Obsahu Vody—Metoda Sušení v Sušárně—Část 2: Celková Voda—Zjednodušená Metoda*; UNMZ: Praha, Czech Republic, 2017.
18. ČSN EN ISO 18 134-3. *Tuhá Biopaliva—Stanovení Obsahu Vody—Metoda Sušení v Sušárně—Část 3: Obsah Vody v Analytickém Vzorku Pro Obecný Rozbor*; UNMZ: Praha, Czech Republic, 2016.
19. ČSN EN ISO 18 122. *Tuhá Biopaliva—Stanovení Obsahu Aopela*; UNMZ: Praha, Czech Republic, 2016.
20. ČSN EN ISO 18 123. *Tuhá Biopaliva—Stanovení Obsahu Prchavé Hořlaviny*; UNMZ: Praha, Czech Republic, 2016.
21. Elementar Analysensysteme GmbH. *[manuál] Návod k obsluze Elementární Analyzátor Vario MACRO Cube*; Operating Instructions; Elementar Analysensysteme GmbH: Langenselbold, Germany, 2009.
22. Rédr, M.; Příhoda, M. *Základy Tepelné Techniky*; SNTL: Praha, Czech Republic, 1991; Volume 677.
23. ČSN EN ISO 18 125. *Tuhá Biopaliva—Stanovení Spalného Tepla a Aýhřevnosti*; UNMZ: Praha, Czech Republic, 2019.
24. Pastorek, Z.; Kára, J.; Jevič, P. *Biomasa: Obnovitelný Adroj Energie*; FCC Public: Praha, Czech Republic, 2004.
25. Effect of Moisture Content, Forest Research. Available online: https://www.forestresearch.gov.uk/tools-and-resources/biomass-energyresources/fuel/woodfuel-production-and-supply/woodfuel-processing/dryingbiomass/effect-of-moisture-content/ (accessed on 25 June 2019).
26. Baláš, M. *Kotle a Aýměníky Tepla*; Akademické Nakladatelství CERM: Brno, Czech Republic, 2013.
27. Barbanera, M.; Cotana, F. Co-combustion performance and kinetic study of solid digestate with gasification biochar. *Renew. Energy* **2018**, *121*, 597–605. [CrossRef]

28. Şensöz, S.; Angin, D. Pyrrolysis of safflower (*Charthamus tinctorius* L) seed press cake in a fixed-beed reactor: Part 2, structural characterisation of pyrolysis bio-oils. *Bioresour. Technol.* **2008**, *99*, 5498–5504. [CrossRef]
29. Arromdeee, P.; Kuprianov, V. A comparative study on combustion of sunflower shells in bubbling and swirling fluidised-bed combustors with a cone-shaped bed. *Chem. Eng. Proc.* **2012**, *62*, 26–38. [CrossRef]
30. Wang, C.H.; Wang, X. the thermal behavior and kinetics of co-combustion between sewage sludge and wheat straw. *Fuel Proc. Tech.* **2019**, *189*, 1–14. [CrossRef]
31. Juszczak, M.; Lossy, K. Pollutant emission from a heat station supplied with agriculture biomass and wood pellet mixture. *Chem. Process Eng.* **2012**, *33*, 233–234. [CrossRef]
32. Harvex, R. Potential Use of Combinable Crop Biomass As Fuel for Small Heating Boilers. 2007. Available online: https://www.researchgate.net/publication/237659583_Potential_use_of_combinable_crop_biomass_as_fuel_for_small_heating_boilers (accessed on 4 May 2019).
33. Vassilev, S.V.; Baxter, D.; Vassileva, C.G. An Overview of the Behaviour of Biomass during Combustion: Part II, Ash Fusion and Ash Formation Mechanisms of Biomass Types. 2019. Available online: https://www.sciencedirect.com/science/article/pii/S0016236113008533 (accessed on 2 July 2019).

© 2020 by the authors. Licensee MDPI, Basel, Switzerland. This article is an open access article distributed under the terms and conditions of the Creative Commons Attribution (CC BY) license (http://creativecommons.org/licenses/by/4.0/).

Article

Briquettes Production from Olive Mill Waste under Optimal Temperature and Pressure Conditions: Physico-Chemical and Mechanical Characterizations

Saaida Khlifi [1], Marzouk Lajili [1], Saoussen Belghith [2], Salah Mezlini [2], Fouzi Tabet [3] and Mejdi Jeguirim [4,5,*]

[1] Ionized and Reactive Media Studies Research Unit (EMIR), Preparatory Institute of Engineering Studies of Monastir (IPEIM), University of Monastir, Monastir 5019, Tunisia; saaida.khlifi@ipeim.rnu.tn (S.K.); marzouk.lajili@ipeim.rnu.tn (M.L.)
[2] Mechanical Engineering Laboratory, National Engineering School of Monastir (ENIM), University of Monastir, Monastir 5019, Tunisia; belghith.saoussen@enim.rnu.tn (S.B.); salah.mezlini@enim.rnu.tn (S.M.)
[3] Institut de Combustion Aérothermique Réactivité et Environnement, UPR3021 CNRS, Université d'Orléans, 45100 Orléans, France; Fouzi.tabet@univ-orleans.fr
[4] Institute des Sciences de Matériaux de Mulhouse, Université de Haute Alsace, 68093 Mulhouse, France
[5] Institut de Science des Matériaux de Mulhouse (IS2M), Université de Strasbourg, 67081 Strasbourg, France
* Correspondence: mejdi.jeguirim@uha.fr; Tel.: +33-389336729

Received: 12 February 2020; Accepted: 2 March 2020; Published: 6 March 2020

Abstract: This paper aims at investigating the production of high quality briquettes from olive mill solid waste (OMSW) mixed with corn starch as a binder for energy production. For this purpose, different mass percentages of OMSW and binder were considered; 100%-0%, 90%-10%, 85%-15%, and 70%-30%, respectively. The briquetting process of the raw mixtures was carried out based on high pressures. Physico-chemical and mechanical characterizations were performed in order to select the best conditions for the briquettes production. It was observed that during the densification process, the optimal applied pressure increases notably the unit density, the bulk density, and the compressive strength. Mechanical characterization shows that the prepared sample with 15% of corn starch shows the best mechanical properties. Moreover, the corn starch binder affects quietly the high heating value (HHV) which increases from 16.36 MJ/Kg for the 100%-0% sample to 16.92 MJ/Kg for the 85%-15% sample. In addition, the kinetic study shows that the binder agent does not affect negatively the thermal degradation of the briquettes. Finally, the briquettes characterization shows that the studied samples with particles size less than 100 µm and blended with 15% of corn starch binder are promising biofuels either for household or industrial plants use.

Keywords: olive mill solid wastes; natural binder; densification; compressive strength; Physico-chemical properties; kinetic parameters

1. Introduction

Biomass feedstocks are recognized as a green energy source since their use for biofuels production could minimize significantly the greenhouse gaseous emissions generated by fossil fuels consumption [1,2]. Indeed, different alternative solid biofuels such as pellets [3,4], briquettes, or logs [5–7] can be produced from different biomass resources. These alternative fuels could be used for various thermochemical conversion processes including pyrolysis, combustion, and gasification [7–10]. In addition, liquid alternative biofuels such as biodiesel and bioethanol can be synthesized using biochemical conversion and extraction techniques [11,12]. Hence, biomass resources coming from agricultural residues, plants, and agri-food by-products are still considered as important bioenergy sources [13,14]. Indeed, after coal and oil the biomass occupies the third place in the energy mapping worldwide [15]. Since the olive mill solid waste (OMSW) is abundant in the Mediterranean basin, leader countries such as Spain, Italy, Greece, and Tunisia have the advantage of valorizing this lignocellulosic biomass type for the energy recovery. It is to be highlighted that the OMSW is produced in an inhomogeneous phase and composed by two or three components (pulp, pits, and olive mill wastewater) depending on a two/three-phase separation process [16,17]. Furthermore, the OMSW composition (organic and inorganic compounds) varies significantly according to the soil cultivation, the rainfall, the degree of ripening, the olive variety, the climatic conditions, the use of pesticides and fertilizers, and also to the trees aging [18,19].

In order to increase the bulk density, the densification process is highly recommended. This technique will not increase only the energy density of pellets/briquettes or logs, but also permit their easy handling, loading, transportation, and storage [20,21]. In addition, for improving the mechanical properties of these solid biofuels it is highly recommended to add carefully other materials as binders such as molasses, starch, and tars [22]. The binder agent is expected to influence certain mechanical properties such as the mechanical strength, the high or low resistance to the compressions, and the low friability index [23]. However, it is necessary to select a binder that could not affect negatively the HHV, the volatile, and the ash contents.

In the literature, few studies have examined the (OMSW) briquetting [24]. Although the availability of these studies, the difference in OMSW characteristics, encourage researchers from each country to conduct their own research in order to optimize the conversion of these wastes into biofuels. Therefore, this investigation is focused on the optimization of the OMSW based briquettes preparation when using a natural binder. The optimized production conditions are selected after the Physico-chemical and mechanical properties determination according to standard methods.

2. Materials and Methods

2.1. Materials Preparation

The OMSW was collected from the Zouila Company (Mahdia, Tunisia) specialized in the second extraction of residual oil in raw OMSW and soap manufacturing situated in the region of Mahdia (Tunisia). Initially, the (OMSW) was dried and the residual oil (3%–5%) was extracted. The raw OMSW can be separated into olive pomace (OP) and olive pits (seeds). It is worth noting that only OP was used in this present study. Before, the densification stage, the raw biomass particles was grinded and sieved into finer particles (<100 μm) in order to ensure more homogeneity. For this purpose, a Hommer Coeffe and spice Grinder mill (Serial N°. 011. 037. 001; ARTNO: 11. 37. 1) was used. Different machines and techniques for the compaction process can be used such as the pelletizing machines, the piston press machines, the screw press, and the roller press machines. All these technologies involve the application of attractive forces between individual particles, formation of solid bridges, capillary pressure, interfacial forces, adhesive and cohesive forces, and mechanical interlocking bonds [23]. As it was reported in the literature, the reduction of the particle sizing increases the material porosity and the number of contact points for inter-particle bonding in the compaction process [25]. Furthermore, the smaller the particle sizes of sample, the lower the relative change in the length of samples as it was mentioned [26]. Hence, the obtained powder of OP was mixed with the corn starch for different mixture compositions; 100%-0%, 90%-10%, 85%-15%, and 70%-30%, respectively. Figure 1 shows the OMSW dried, the corn starch binder and their mixture in the composition 85%-15%. It is to be highlighted that added water to the mixture (up to around 20% moisture dry basis) is preferred in order to produce cohesive forces between the particles which favorite the agglomeration.

Figure 1. The used materials for the briquettes preparation: (**a**) Olive pomace, (**b**) corn starch, and (**c**) the mixture 85%-15%, respectively.

2.2. Densification and Briquettes Production

The thermal-mechanical press process is based on Gottfried Joos Maschinenfabrik GmbH & Co. machine (Stuttgart, Germany). The KG type LAP-100 with limit compression capacity of 1000 kN was chosen for briquettes production. This machine was associated to a specific mold composed of eight cylindrical imprints as it is shown (Figure 2). Every time, the cavity of the imprints should be fulfilled with the same quantities of the mixture and should be leveled off at the top to obtain a smooth surface. The moving down base of the pressing machine rises up the mold. The equipment was placed under a fixed pressure via the control valve for an optimal residence time of 15 min until obtaining the desired briquettes. The machine was wrapped with a heating element for working at a desired temperature. In this case, the samples were prepared at a low temperature of 38 °C in order to forbidden any migration of the extractives such as the residential oil or other low molecular weight molecules to the particle surface and also, to prevent the adhesion mechanisms of the Van der Waals forces causing low briquettes strengths [27]. In order to test the pressure effect on the samples, three pressures values; 100, 125, and 150 MPa were considered. The temperature was maintained at 38 °C for a residence time of 15 min according to the literature data for the commercial pelletization [28]. After the densification process, the obtained wet briquettes were extruded and dried during two weeks at the laboratory conditions (the temperature room was 28 ± 3 °C). Moreover, the densification process depends not only on the particle sizes, the fiber strength of the material, the abrasive components, but also on whether we use or not an additive binder. That is why different briquettes composed of different blends of OP and corn starch were produced. It is to be highlighted that the briquetting process uses different apparatus by comparison to the pelletizing process. Indeed, with briquettes we can go more with the pressure under which samples were prepared. However later, during the energy conversion, pellets give us more freedom for using combustion chambers with different geometries and different feeding systems than briquettes which are more suitable for fixed bed chambers.

Figure 2. The used mould die for the briquetting process.

2.3. Proximate Analysis

The proximate analysis of the prepared samples was conducted following the thermogravimetric analysis (TG) technique by using the thermal analyzer NETZSCH, model STA 449 F3, Jupiter (STA 449 F3, NETZSCH-Gerätebau GmbH, Selb, Germany) Hence, about 50 mg of each sample should be placed in a platinum. First, the initial temperature of 29 °C was maintained for 10 min, and then the samples were heated up to 950 °C at a constant heat rate of 10 °C.min^{-1} under an inert atmosphere of nitrogen with a flow rate of 79 mL/min after reaching the maximum temperature. After that, the temperature should be decreased to 550 °C with a heating rate of 40 °C·min^{-1}. Thereafter, an oxidative atmosphere by supplying 21% oxygen flow rate was supplied. This protocol, allows extracting from the TG curve, moisture, volatiles matters (VM), fixed carbon (FC), and ash contents.

The high heating value (HHV) is the amount of heat released when the sample fuel is completely burnt with oxygen in a calorimeter bomb. For this goal, the HHV was determined based on the ASTM D5865 standards. The parr 1341 oxygen bomb calorimeter was first calibrated using a standard sample of benzoic acid whose known calorific value is 26.4 kJ/kg. A mass of about 1g of the different samples should be used. The bomb should be fulfilled of oxygen under 30 bars. By measuring the variation of temperature, the HHV can be calculated using the following expression:

$$HHV = \frac{W\Delta T - e_1 - e_2 - e_3}{m} \quad (1)$$

where ΔT is the net temperature rise, W is the equivalent energy of the calorimeter determined under standardization, e_1 represents the correction (in calories) for the heat of formation of nitric acid (HNO$_3$), e_2 represents the correction (in calories) for heat of formation of sulfuric acid (H$_2$SO$_4$), e_3 corresponds to the correction (in calories) for the heat of combustion of the fuse wire, and m is the weight of the tested sample.

2.4. Measurement of the Compressive Strength

The compressive strength is a significant parameter in the evaluation of the solid biofuels. This parameter is in relation with the rigidity and durability which make the biofuels storage easier [29]. In order to evaluate the compression resistance as a function of the applied pressure on prepared samples using the Lloyd Instruments (EZ20, AMETEK Company, Berwyn, UK), tests were conducted with the load speed of 1 mm/min [30]. The flat surface of the briquette sample should be placed on the horizontal metal plate of the machine. Then, an increased load is applied at a constant rate until reaching the sample's break. The maximum crushing load force that a briquette can withstand before cracking or breaking corresponds to the so-called compressive strength [27,31].

3. Results and Discussions

3.1. Compressive Strength Measurement

Figure 3 shows that the load required to the briquettes rupture for different binder ratios is significantly different. Table 1 exhibits the compressive strength of biomass briquettes produced at different pressure levels and different binder contents. Indeed, the compressive strength exhibits maximum of 4015 and 4581 kN in the case of the 85%-15% sample when imposing 125 and 150 MPa, respectively. Moreover, it is very remarkable that the compressive strength falls significantly when increasing the binder content (30%) in the samples. This result proves that the starch corn gets its maximum efficiency at 15%. Moreover, for the 100 MPa imposed pressure, the 100%-0% sample shows the highest compression strength (1581 kN) and on the contrary the compressive strength decreases significantly when increasing the binder percentage.

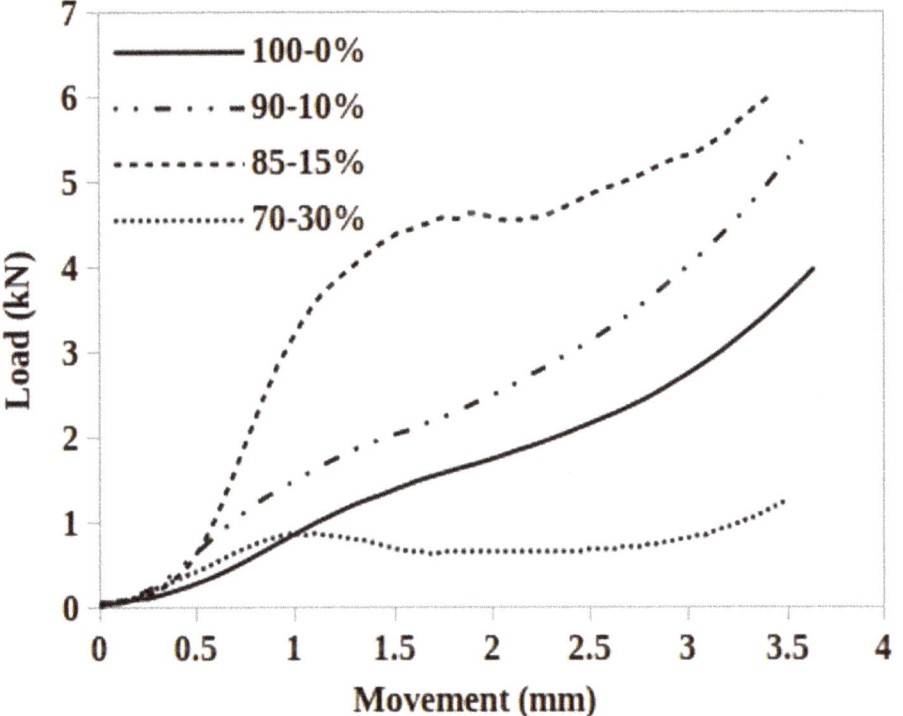

Figure 3. Variation of the load as a function of the displacement.

Table 1. Compressive strength of biomass briquettes produced at different pressure levels and different binder contents.

Pressure (MPa)	100				125				150			
Binder (%)	0	10	15	30	0	10	15	30	0	10	15	30
Compressive Strength (kN)	1581	1039	775	626	1506	1770	4015	130	1557	2090	4581	863

3.2. The Unit Density and the Bulk Density Measurements

Table 2 presents the unit density and the bulk density values of the different briquettes types when prepared at different binder percentages and at different pressure values. The unit density is the averaged ratio of the mass by the volume of each sample. Given that the briquettes are cylindrical, we should determine the mass, the radius, and the height of each sample then, by simple calculation the unit density is calculated as:

$$\rho_u = \frac{m}{V} \qquad (2)$$

Table 2. Unit and bulk density of biomass briquettes produced at different pressure and different binder contents.

Pressure (MPa)	100				125				150			
Binder (%)	0	10	15	30	0	10	15	30	0	10	15	30
Unit Density (kg/dm^3)	2.04	2.83	2.81	2.31	2.41	2.40	2.47	*	2.77	2.84	2.95	3.03
Bulk Density (kg/dm^3)	0.84	0.86	0.88	0.84	0.92	0.93	0.95	*	0.95	0.98	1.20	1.10

However, the bulk density is calculated as the ratio of the briquettes mass when fulfilling a container whose volume is known and proceeding conformingly to the CEN TS15103 standard method [32]. Results indicate that the unit density values of our samples are ranging between 2.04 and 3.7 kg/dm^3. These values seem to be relatively high. This can be explained by the effect of the small particle size, the low moisture content, and also by the high applied pressure [20]. Moreover, the effect of the imposed pressure on the briquettes density is foreseeable when evolving from 100 to 150 MPa. In addition, when increasing the percentage of the binder from 10% to 30% the briquettes unit density grows from 2.8 to 3.0 kg/dm^3 in the case of 150 MPa imposed pressure, whereas, the briquettes bulk density exhibits a small increase from 0.95 to 1.10 kg/dm^3 and 1.20 kg/dm^3 as a maximum value for the 85%-15% sample. This can be explained by the fact that the corn starch particles might have played an important role in fulfilling the void between the particles which increases the inter-particle bonding. Fortunately, the unit density values of the two briquettes types are higher than the minimum value (1.12 Kg/dm^3) claimed by the European standard EN14961-2 during the household use [33].

3.3. High Heating Value Measurement

Table 3 shows the proximate analyses of the prepared briquettes compared with those reported in the literature for other biomass fuels. It can be noticed that the ash content decreases due to the blending operation from 9.49% to 6.72%, but remains relatively higher than the acceptable limit (5%) of European standards. However, the produced ash by combustion for example can be reused for the brick manufacturing. Indeed, the addition of this ash type to bricks preparation helps in terms of reducing thermal conductivity, as it was stated by Eliche-Quesada and Leite-Costa [34]. Moreover, the 85%-15% sample exhibits quite an increase for the HHV which remains within the range of European norms (>16.5 MJ.kg^{-1}) [35]. This can be justified by the quite increase of the VM content during the binder addition [36].

Table 3 also reported the proximate analysis of different biomass from the literature. We remark that the HHV of our prepared samples are lower than the HHV of the pulp (dry basis) or the pomace (except the very wet pomace with 49% moisture content). This may be due to a higher proportion of oxygen and hydrogen, and less carbon. Indeed, Munir et al. [37] found that the amount of energy contained in carbon–oxygen and carbon–hydrogen bonds is lower than in carbon–carbon bonds. The higher oxygen content in the biomass indicates that it will have a higher thermal reactivity than the other biomass [38]. Moreover, Chouchene et al. [39] tested the influence of particle size on combustion properties. They concluded that the more the particles were small, the more they were reactive. In addition, samples having less than 0.5 mm size released a high quantity of volatile matters which is in accordance with the low quantity of produced char. However, the residual ash amounts left during the oxidative pyrolysis of the olive solid wastes increase when decreasing the particle size [40]. Furthermore, using small particle sizing induces no temperature gradient leading to heat transfer limitations [40].

Table 3. Proximate analysis of prepared samples and others from literature.

Biofuels	VM (%wt)	FC (%wt)	Moisture (%wt)	Ash (%wt)	HHV (MJ/kg)
100%-0% (w.b.)	61.86	18.75	9.88	9.49	16.36
85%-15% (w.b.)	64.65	18.39	10.44	6.72	16.92
Pulp [41] (d.b.)	79.10	15.30	6.5	5.60	23.39
OP [42] (d.b.)	65	29.6	10	5.4	*
Dry OP [43] (d.b.)	86.71	7.48	4.52	5.81	19.88
Wet OP [44] (d.b.)	42.35	7.79	49.02	0.84	5.70

w.b.: wet basis; d.b.: dry basis.

3.4. Thermogravimetry Analysis

Figure 4a shows the TG curve as a function of time obtained during slow pyrolysis and followed by the char oxidation for both samples 0%-100% and 85%-15% prepared at 150 MPa, respectively. Figure 4b corresponds to the TG curve as a function of temperature during only the pyrolysis process. Figure 4a can be divided into three steps: The first step corresponds to a mass loss representing mainly the moisture evaporation up to 120 °C. The second step characterizes the devolatilization zone in which the volatiles organic compounds (VOC) are released from the hemicelluloses and a part of the lignin thermal degradation followed by the cellulose degradation. The VOC gas mixture is mainly composed by CH_4, CO_2, CO, H_2 and some traces of other C_nH_m. The second step ended by the rest of lignin degradation yielding to the char formation which is a mixture of fixed carbon and ash. The final step (Step 3 on Figure 4a) corresponds to the char oxidation when injecting oxygen and the temperature is fixed equal to 550 °C. This step ends when obtaining only ash. The obtained curve presents the same behavior as similar works reported in the literature [17,45,46].

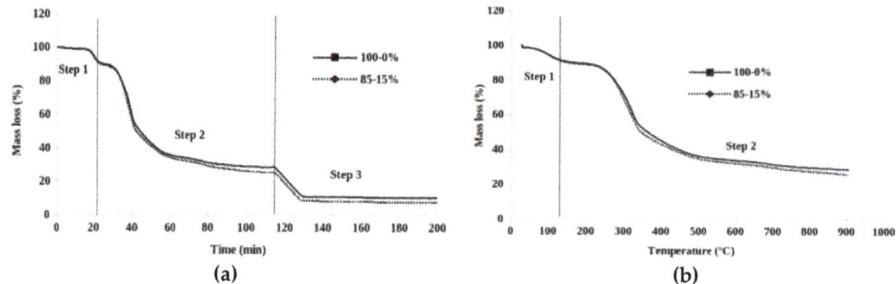

Figure 4. (a) **TG** curve of the 85%-15% and 100%-0% briquettes as a function of the time during pyrolyis step followed by the char oxidation (Step 3); (b) TG curve of the 85%-15% and 100%-0% briquettes as a function of the temperature during only pyrolysis step.

Figure 5 exhibits the derivative versus time of TG curves called (DTG) curves. These curves correspond to the thermal degradation under inert atmosphere of the 100%-0% and the 85%-15% samples respectively. Each one of the superposed curves shows fourth peaks with a different maximum rate of mass loss. The first peak corresponds mainly to the moisture release. Moreover, it can refer to an early stage of residual oil evaporation and light VOC degradation (below 200 °C) [47]. The second one, which is more intense occurring between 200 and 350 °C, is attributed to the hemicelluloses and cellulose degradation. In the present case, the thermal degradation of cellulose dominates the chemistry of pyrolysis (the cellulose content can be three times that of hemicelluloses; 42% and 14%, for example) [48,49]. This causes the replacement of the peak corresponding to the hemicelluloses by a shoulder (observed at the vicinity of 270 °C in our case).

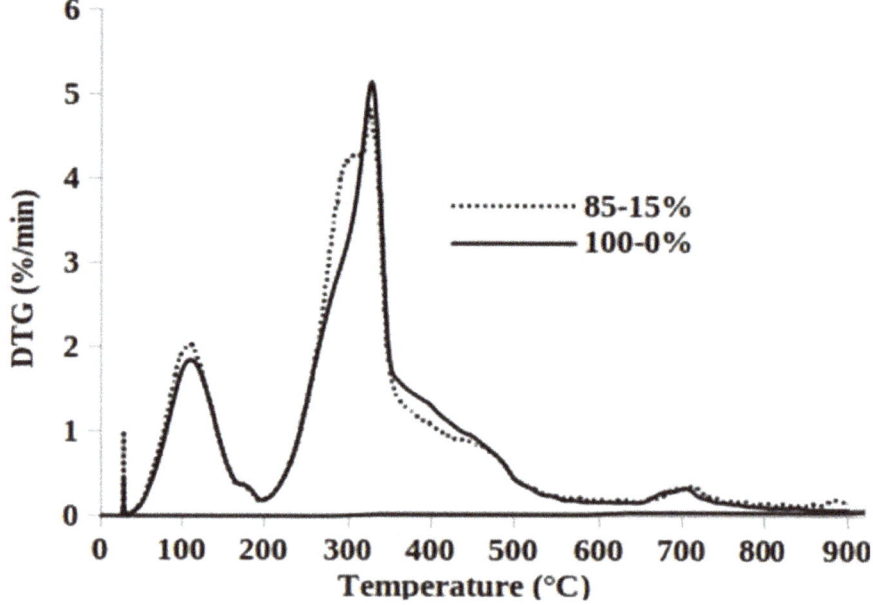

Figure 5. DTG evolution as a function of the temperature.

A similar result was reported by Ghouma et al. [46]. They observed two peaks occurring at 267 and 337 °C corresponding to the thermal decomposition of the hemicelluloses and the cellulose. The third peak appearing at about 425 °C corresponds to the lignin and cellulose thermal degradation [50]. The final peak, occurring at the vicinity of 708 °C, is attributed to the rest of the lignin thermal degradation. Indeed, the lignin thermal degradation is in reality widely spread between 433 and 900 °C [51]. More precisely, serial shoulders have been seen for the Xylan pyrolysis in the range of 350 and 550 °C, and can be attributed to the remaining lignin [52]. It is to be highlighted that the second peak observed for the 85%-15% sample near 295 °C corresponds to the corn starch material. The Pyrolysis process corresponding to the decomposition of the hemicelluloses, the cellulose, and the lignin yields to the char formation. This remaining char will be oxidized in the following step. Indeed, during the char oxidation at 550 °C, the increased porosity of the char particles permits more diffusion of the oxygen. Therefore, the readily combustible part of lignin could react with the diffusing oxygen, even though the reactivity of lignin was low [53].

3.5. Mean Reactivity during Pyrolysis and Char Oxidation

Taking into consideration the fact that the peak intensity is directly proportional to the reactivity R_{DTG} [54], while the corresponding temperature is inversely proportional to the reactivity T_{DTG} [55], the mean reactivity R_M of each sample can be easily calculated, when considering all peaks and shoulders appearing on the DTG curves, using the following expression:

$$R_M = 100 \frac{\sum R_{DTG}}{T_{DTG}} \quad (3)$$

As it is mentioned in Table 4, the 85%–15% sample for which the DTG curve exhibits the highest value displayed the highest reactivity. This may be due to the increase of the surface area, and therefore to a higher concentration of carbon active sites per unit weight. The effect of the heating rate was previously studied [46]. Moreover, as it was stated by Qiang et al. [56] and Tenfei et al. [57], the use of binder in pellets preparation ensures lower ignition temperature, wider temperature interval, and higher oxidation activity. More precisely, when using lignin and $Ca(OH)_2$ the produced pellets show lower compression energy consumption, moisture uptake, enhanced mechanical strength, and promoted combustion performance. It was found that a high heat flux in a high heating regime intensified the Boudouard reaction, while delayed the thermal decomposition of the studies biomass. This is may be due to the combined effects of the heat transfer at the different heating rates and also to the kinetics of gasification. In fact, with the low heating rate (10 °C·min^{-1}) in this study, the initial reaction temperature and the maximum gasification rate were decreased compared to other reported studies [55]. It is necessary to point out that the yield of carbon monoxide produced did not change significantly [55].

Table 4. Determination of the mean reactivity of the biofuels during the pyrolysis and the char combustion.

Samples	Thermal Degradation	Peaks of Temperature (°C)	R_{DTG} (mg·min^{-1})	R_M (%.min^{-1}.°C^{-1})
100%-0%	Pyrolysis	118-180-329-398-448-	0.896-0.176-2.582-0.672	0.897
	Char Combustion	537-544-548	0.636-0.635-0.590	0.228
85%-15%	Pyrolysis	102-180-295-319-458-707-882-904-902	1.039-0.167-2.149-2.460-0.424-0.182-0.108-0.043-0.028	1.199
	Char Combustion	593-568-552	0.660-0.661-0.598	0.227

3.6. Kinetic Study and Pyrolysis Parameters Determination

The kinetic study of the pyrolysis process for both studied samples (100%-0% and 85%-15%) was carried out on the hypothesis where the solid-state material was heated at a constant heating rate;

$\beta = \frac{dT}{dt}$ equal to 10 K min^{-1}. This condition was widely considered in the literature [4,32,46,58,59]. This technique uses the combined kinetics three-parallel-reaction (CK-TPR) model. In the present study, a single step reaction mechanism (Equation (4)) was assumed to describe the pyrolysis kinetics of the lignocellulosic biomass. Isoconversional methods are believed to estimate the apparent activation energy (E) and the pre-exponential factor (A), when the rate of the mass loss is related to the mass and to the temperature according to Equation (5).

$$\text{Biomass} \rightarrow \text{volatile gas} + \text{char} \quad (4)$$

$$d\alpha/dT = K(T) \cdot f(\alpha) = \frac{A}{\beta} \cdot \exp\left(\frac{-E_a}{RT}\right)(1-\alpha)^n \quad (5)$$

where A denotes the pre-exponential factor (s^{-1}), E$_a$ is the activation energy (kJ/mol), R is the ideal gas constant (R = 8.31 J·mol^{-1}·K^{-1}), T is the temperature (K), t is time (s), α is the conversion rate varying between 1 and 0, and n is the reaction order.

The distributed activation energy model (DAEM) assumes infinity irreversible first order reactions (f(α) = 1−α) happening independently during the solid-state pyrolysis [60]. Moreover, based on the Coats-Redfern method, Equation (5) could be written as the following [61]:

$$\ln\left(\frac{-\ln(1-\alpha)}{T^2}\right) = \ln\left(\frac{AR}{\beta E_a}\right) - \frac{E_a}{RT} \quad (6)$$

Figure 6 shows the variation of $Y = \ln\left(\frac{-\ln(1-\alpha)}{T^2}\right)$ versus $X = \frac{1}{T}$. It is to be noticed that an increase of temperature yields to an increase of Y. This result is may be due to the change in pyrolytic mechanisms so that different stages could be observed with the two sample types (100%-0% and 85%-15%) [58]. The apparent activation energy (E$_a$) and the pre-exponential factor (A) were calculated using the slopes of the linear trends of the curve and all results were consigned in Table 5.

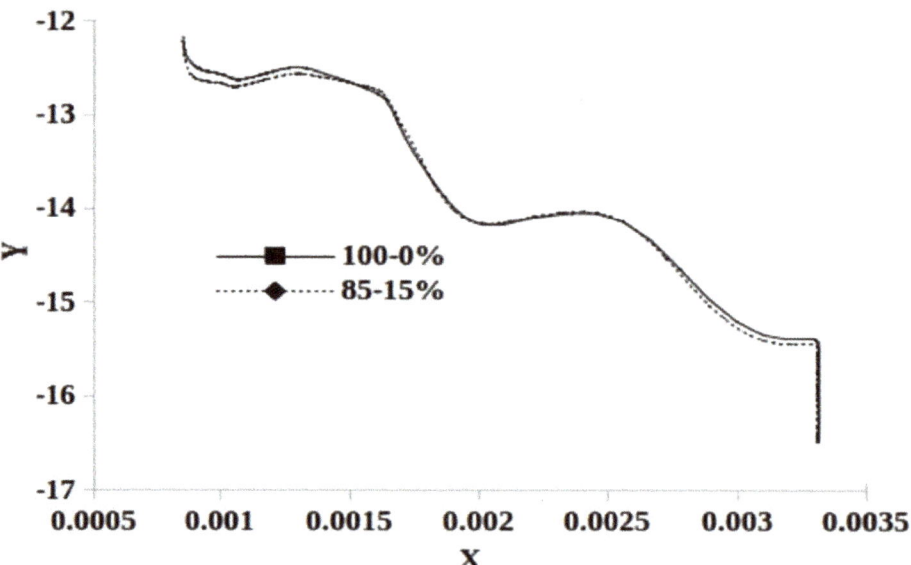

Figure 6. Kinetic analysis during the thermal decomposition of the 85%-15% and 100%-0% samples. The activation energy is calculated from the slope of linear trends.

Table 5. Kinetic parameters determination during the nonisothermal pyrolysis.

Samples	Range of Temperature (°C)	Linear Equation	Ea (kJ·mol^{-1})	A (min^{-1})
85%-15%	54–126	$f(x) = -2518.82x - 7.707$	20.94	11.32
	135–225	$f(x) = 343.72x - 14.85$	n.d.	n.d.
	230–353	$f(x) = -4201.99x - 5.98$	34.93	106.26
	360–438	$f(x) = -595.3x - 11.77$	4.94	0.05
100%-0%	60–134	$f(x) = -2232.08x - 8.47$	18.55	4.38
	151–225	$f(x) = 376.062x - 14.93$	n.d.	n.d.
	235–368	$f(x) = -3895.77x - 6.558$	32.38	55.27
	378–498	$f(x) = -898.09x - 11.3$	7.46	0.11

n.d.: not dtermined

Table 5 shows that the activation energy of the 85%-15% sample for different pyrolysis steps were 20.94, 34.93, and 4.94 kJ·mol^{-1}. The activation energy reaches its maximum during the hemicelluloses and cellulose decomposition. It was concluded that the activation energies present relatively low values, which could be due to the small particle, so that the pyrolysis reactions start at low temperatures. The 85%-15% sample presents a less range of temperature during the decomposition of the three main components than the 100%-0% sample, and this is the reason of the higher reactivity [62]. Moreover, we can notice that, when moving from the first to the second stage, the temperature increases and the activation energy takes a maximum value. At this stage the higher activation energy refers to the diffusing effect of volatiles and gas. Then, when moving from the second to the third stage it is notable that the activation energy decreases. This means that the chemical reaction becomes easier. Similar trends of this activation energy evolution were reported in the literature [63].

4. Conclusions

In this work, we studied the mechanical properties and the thermal-chemical properties of biofuels briquettes. Two sorts blended with a natural binder and non-blended briquettes were densified under a high pressure. The results we obtained show that the briquettes produced from olive pomace blended with corn starch as a binder can be densified into high quality briquettes showing acceptable parameters for a future thermal use. Indeed, the proximate analysis of the prepared briquettes, show less moisture contents. Moreover, the addition of corn starch as a natural binder conducts an improvement on the sample's compressive strength, as well as to a reduction of the ash content and to a quite increase of the high heating value. TG tests were carried out followed by a kinetic study for the two samples 100%-0% and 85%-15%. We concluded that the binder increases the thermal degradation of the briquettes during the nonisothermal pyrolysis, while increasing the activation energy. Next, a numerical simulation and modeling of the combustion of pyrolysis VOC in a cocurrent reactor will be conducted using the OpenFOAM software (18.12+, OpenCFD Limited, Bracknell, UK).

Author Contributions: Conceptualization, M.L. and S.M.; Methodology, M.L. and M.J.; Formal Analysis, M.L. and S.B.; Investigation, M.L., F.T. and M.J.; Resources, M.L. and S.M.; Data Curation, S.K., M.L. and S.M.;

Writing-Original Draft Preparation, S.K., S.B. and M.L.; Writing-Review & Editing, M.L. and M.J.; Supervision, M.L.; Project Administration, M.L. All authors have read and agreed to the published version of the manuscript.

Funding: This research received no external funding.

Acknowledgments: Saaida Khlifi and Marzouk Lajili would like to express their thanks to Brahim Sarh and Toufik Boushaki, as well as the technical team of ICARE Laboratory of Orleans University (France) for the good technical assistance during the experimental tests realization.

Conflicts of Interest: The authors declare no conflict of interest.

References

1. Popp, J.; Lakner, Z.; Harangi-Rákos, M.; Fári, M. The effect of bioenergy expansion: Food, energy, and environment. *Renew. Sust. Energy Rev.* **2014**, *32*, 559–578. [CrossRef]
2. Demirbaş, A. Biomass resource facilities and biomass conversion processing for fuels and chemicals. *Energy Convers. Manag.* **2001**, *42*, 357–1378. [CrossRef]
3. McKendry, P. Energy production from biomass (Part 1): Overview of biomass. *Bioresour. Technol.* **2002**, *83*, 37–46. [CrossRef]
4. Limousy, L.; Jeguirim, M.; Dutournié, P.; Kraeim, N.; Lajili, M.; Said, R. Gaseous products and particulate matter emissions of biomass residential boiler fired with spent coffee ground pellets. *Fuel* **2013**, *107*, 323–329. [CrossRef]
5. Merpati Mitan, N.M.; Azmi, A.H.; Nur Fathiah, M.N.; Se, S.M. Binder application in durian peels briquette as a solid biofuel. *Appl. Mech. Mater.* **2015**, *761*, 494–498. [CrossRef]
6. Limousy, L.; Jeguirim, M.; Labbe, S.; Balay, F.; Fossard, E. Performance and emissions characteristics of compressed spent coffee ground/wood chip logs in a residential stove. *Energy Sustain. Dev.* **2015**, *28*, 52–59. [CrossRef]
7. Khlifi, S.; Lajili, M.; Tabet, F.; Boushaki, T.; Sarh, B. Investigation of the combustion characteristics of briquettes prepared from olive mill solid waste blended with and without a natural binder in a fixed bed reactor. *Biomass Convers. Bior.* **2019**. [CrossRef]
8. Fachinger, F.; Drewnick, F.; Gieré, R.; Borrmann, S. How the user can influence particulate emissions from residential wood and pellet stoves: Emission factors for different fuels and burning conditions. *Atmos. Environ.* **2017**, *158*, 216–226. [CrossRef]
9. Ferreira, S.; Monteiro, E.; Brito, P.; Vilarinho, C. A Holistic Review on Biomass Gasification Modified Equilibrium Models. *Energies* **2019**, *12*, 160. [CrossRef]
10. Zribi, M.; Lajili, M.; Escudero Sanz, R.F. Hydrogen enriched syngas production via gasification of biofuels pellets/powders blended from olive mill solid wastes and pine sawdust under different water steam/nitrogen atmospheres. *Int. J. Hyd. Energy* **2019**, *44*, 11280–11288. [CrossRef]
11. Zhang, F.; Fang, Z.; Wang, Y.T. Biodiesel production directly from oils with high acid value by magnetic $Na_2SiO_3@Fe_3O_4$/C catalyst and ultrasound. *Fuel* **2015**, *150*, 370–377. [CrossRef]
12. Hahn-Hägerdal, B.; Galbe, M.; Gorwa-Grauslund, M.F.; Lidén, G.; Zacchi, G. Bio-ethanol—The fuel of tomorrow from the residues of today. *Trends Biothechnol.* **2006**, *24*, 12. [CrossRef] [PubMed]
13. Hamelinck, C.N.; Faaij, A.P.C. Outlook for advanced biofuels. *Energy Policy* **2006**, *34*, 3268–3283. [CrossRef]
14. Scarlat, N.; Dallemand, J.F.; Monforti-Ferrario, F.; Nita, V. The role of biomass and bioenergy in a future bioeconomy: Policies and facts. *Environ. Dev.* **2015**, *15*, 3–34. [CrossRef]
15. Sahu, S.G.; Sarkar, P.; Mukherjee, A.; Adak, A.K.; Chakraborty, N. Studies on the co-combustion behavior of coal/biomass blends using thermogravimetric analysis. *IJETAE* **2013**, *3*, 131–138.
16. Azbar, N.; Bayram, A.; Filibeli, A.; Muezzinoglu, A.; Sengul, F.; Ozer, A. A review of waste management options in olive oil production. *Crit. Rev. Env. Sci. Technol.* **2004**, *34*, 209–247. [CrossRef]
17. Ouazzane, H.; Laajine, F.; ElYamani, M.; el Hilaly, J.; Rharrabti, Y.; Amarouch, M.Y.; Mazouzi, D. Olive Mill Solid Waste Characterization and Recycling opportunities: A review. *J. Mater. Environ. Sci.* **2017**, *8*, 2632–2650.
18. Niaounakis, M.; Halvadakis, C.P. *Olive Processing Waste Management*; Literature Review and Patent Survey: London, UK, 1 February 2006.
19. Dermeche, S.; Nadour, M.; Larroche, C.; Moulti-Mati, F.; Michaud, P. Olive mill wastes: Biochemical characterizations and valorization strategies. *Process Biochem.* **2013**, *48*, 1532–1552. [CrossRef]

20. Tumuluru, J.S. Effect of process variables on the density and durability of the pellets made from high moisture corn stover. *Biosyst. Eng.* **2014**, *119*, 44–57. [CrossRef]
21. Tumuluru, J.S. Specific energy consumption and quality of wood pellets made from high moisture lodgepole pine biomass. *Chem. Eng. Res. Des.* **2016**, *110*, 82–97. [CrossRef]
22. Kaliyan, N.; Morey, R.V. Natural binders and solid bridge type binding mechanisms in briquettes and pellets made from corn Stover and switch grass. *Bioresour. Technol.* **2010**, *101*, 1082–1090. [CrossRef] [PubMed]
23. Zanella, K.; Concentino, V.O.; Taranto, O.P. Influence of the type of Mixture and Concentration of Different Binders on the Mechanical Properties of "Green" Charcoal Briquettes. *Chem. Eng. Trans.* **2017**, *57*, 199–204.
24. Christoforou, E.; Fokaides, P. A review of olive mill solid wastes to energy utilization techniques. *Waste Manag.* **2016**, *49*, 346–363. [CrossRef] [PubMed]
25. Ganvir, K.D.; Tulankar, P.G.; Bawankar, P.P.; Rahimkar, K.T.; Singh, S.L. Analysis and Comparison of Biomass Pellets with Various Fuels. *IJSRD* **2017**, *5*, 1272–1274.
26. Davies, R.M.; Davies, O.A. Effect of Briquetting Process Variables on Hygroscopic Property of Water Hyacinth Briquettes. *Hindawi Publ. Corp. J. Renew. Energy* **2013**, 1–5. [CrossRef]
27. Stelte, W.; Holm, J.K.; Sanadi, A.R.; Barsberg, S.; Ahrenfeldt, J.; Henriksen, U.B. A study of bonding and failure mechanisms in fuel pellets from different biomass resources. *Biomass Bioenergy* **2011**, *35*, 910–918. [CrossRef]
28. Adapa, P.K.; Tabil, L.G.; Schoenau, G.J.; Sokhansanj, S. Pelleting Characteristics of Fractionated and Sun-Cured Dehydrate Alfalfa Grinds. *Appl. Eng. Agric.* **2004**, *20*, 813–820. [CrossRef]
29. Swietochowski, A.; Lisowski, A.; Dabrowska-Salwin, M. Strength of briquettes and pellets from energy crops. *Eng. Rural Dev.* **2016**, *5*, 25–27.
30. Sprenger, C.J.; Tabil, L.G.; Soleimani, M.; Agnew, J.; Harrison, A. Pelletization of Refuse-Derived Fuel Fluff to Produce High Quality Feedstock. *J. Energy Resour. Technol.* **2018**, *140*. [CrossRef]
31. Richards, S.R. Physical testing of fuel briquettes. *Fuel Process. Technol.* **1990**, *25*, 89–100. [CrossRef]
32. Lajili, M.; Limousy, L.; Jeguirim, M. Physico-chemical properties and thermal degradation characteristics of agropellets from olive mill by-products/sawdust blends. *Fuel Process. Technol.* **2014**, *126*, 215–221. [CrossRef]
33. Mehdipour, I.; Khayat, K.H. Effect of supplementary Cementitious Material Content and Binder Dispersion on Packing Density and Compressive Strength of Sustainable Cement Paste. *ACI Mater. J.* **2016**, *113*, 361–372.
34. Eliche-Quesada, D.; Leite-Costa, J. Use of bottom ash from olive pomace combustion in the production of eco-friendly fired clay bricks. *Waste Manag.* **2016**, *48*, 323–333. [CrossRef] [PubMed]
35. Eija Alakangas, V.T.T. *New European Pellets Standards*; EUBIONET 3: Swedish, Finland, March 2011.
36. Long, Y.; Meng, A.; Chen, S.; Zhou, H.; Zhang, Y.; Li, Q. Pyrolysis and Combustion of Typical Wastes in a Newly Designed Macro Thermogravimetric Analyzer: Characteristics and Simulation by Model Components. *Energy Fuels* **2017**, *31*, 7582–7590. [CrossRef]
37. Munir, S.; Daood, S.S.; Nimmo, W.; Cunliffe, A.M.; Gibbs, B.M. Thermal analysis and devolatilization kinetics of cotton stalk, sugar cane bagasse and shea meal under nitrogen and air atmospheres. *Bioresour. Technol.* **2009**, *100*, 1413–1418. [CrossRef]
38. Haykiri-Acma, H.; Yaman, S. Effect of co-combustion on the burnout of lignite/biomass blends: A Turkish case study. *Waste Manag.* **2008**, *28*, 2077–2084. [CrossRef]
39. Chouchene, A.; Jeguirim, M.; Khiari, B.; Zagrouba, F.; Trouvé, G. Thermal degradation of olive solid waste: Influence of particle size and oxygen concentration. *Resour. Conserv. Recy.* **2010**, *54*, 271–277. [CrossRef]
40. Vamvuka, D.; Kakaras, E.; Kastanaki, E.; Grammelis, P. Pyrolysis characteristics and kinetics of biomass residuals mixtures with lignite. *Fuel* **2003**, *82*, 1949–1960. [CrossRef]
41. Miranda, T.; Esteban, A.; Rojas, S.; Montero, I.; Ruiz, A. Combustion Analysis of Different Olive Residues. *Int. J. Mol. Sci.* **2008**, *9*, 512–525. [CrossRef]
42. Borello, D.; De Caprariis, B.; De Filippis, P.; Di Carlo, A.; Marchegiani, A.; Marco Pantaleo, A.; Shah, N.; Venturini, P. Thermo-Economic Assessment of a olive pomace Gasifier for Cogeneration Applications. *Energy Procedia* **2015**, *75*, 252–258. [CrossRef]
43. Alrawashdeh, K.A.b.; Slopiecka, K.; Alshorman, A.A.; Bartocci, P.; Fantozzi, F. Pyrolytic Degradation of Olive Waste Residue (OWR) by TGA: Thermal Decomposition Behavior and Kinetic Study. *J. Energy Power Eng.* **2017**, *11*, 497–510.
44. Bartocci, P.; D'Amico, M.; Moriconi, N.; Bidini, G.; Fantozzi, F. Pyrolysis of olive stone for energy purposes. *Energy Procedia* **2015**, *82*, 374–380. [CrossRef]

45. Tamošiūnas, A.; Chouchène, A.; Valatkevičius, P.; Gimžauskaitė, D.; Aikas, M.; Uscila, R.; Ghorbel, M.; Jeguirim, M. The Potential of Thermal Plasma Gasification of Olive Pomace Charcoal. *Energies* **2017**, *10*, 710. [CrossRef]
46. Ghouma, I.; Jeguirim, M.; Guizani, C.; Ouederni, A.; Limousy, L. pyrolysis of Olive Pomace: Degradation kinetics. Gaseous analysis and char characterization. *Waste Biomass Valorization* **2017**, *8*, 1689–1697. [CrossRef]
47. Chiti, Y.; Salvador, S.; Commandré, J.M.; Broust, B. Thermal decomposition of bio-oil: Focus on the products yields under different pyrolysis conditions. *Fuel* **2012**, *102*, 274–281. [CrossRef]
48. Abed, I.; Paraschiv, M.; Loubar, K.; Zagrouba, F.; Tazerout, M. Thermogravimetric investigation and thermal conversion kinetics of typical North African and Middle Eastern lignocellulosic wastes. *Bioresources* **2012**, *7*, 1200–1220.
49. Shen, D.K.; Gu, S.; Bridgwater, A.V. The thermal performance of the polysaccharides extracted from hardwood: Cellulose and hemicellulose. *Carbohydr. Polym.* **2010**, *82*, 39–45. [CrossRef]
50. Varma, A.K.; Mondal, P. Physicochemical Characterization and Pyrolysis Kinetic Study of Sugarcane Bagasse Using Thermogravimetric Analysis. *J. Energy Resour. Technol.* **2016**, *138*, 11. [CrossRef]
51. Ounas, A.; Aboulkas, A.; El harfi, K.; Bacaoui, A.; Yaacoubi, A. Pyrolysis of olive residue and sugar cane bagasse: Non-isothermal thermogravimetric kinetic analysis. *Bioresour. Technol.* **2011**, *102*, 11234–11238. [CrossRef]
52. Cheng, K.; Winter, W.T.; Stipanovic, A.J. A modulated-TGA approach to the kinetics of lignocellulosic biomass pyrolysis/combustion. *Polym. Degrad. Stabil.* **2012**, *97*, 1606–1615. [CrossRef]
53. Gani, A.; Naruse, I. Effect of cellulose and lignin content on pyrolysis and combustion characteristics for several types of biomass. *Renew. Energy* **2007**, *32*, 649–661. [CrossRef]
54. Gil, M.V.; Oulego, P.; Casal, M.D.; Pevida, C.; Pis, J.J.; Rubiera, F. Mechanical durability and combustion characteristics of pellets from biomass blends. *Bioresour. Technol.* **2010**, *101*, 8859–8867. [CrossRef]
55. Vamvuka, D.; Karouki, E.; Sfakiotakis, S. Gasification of waste biomass chars by carbon dioxide via thermogravimetry. Part I: Effect of mineral matter. *Fuel* **2011**, *90*, 1120–1127. [CrossRef]
56. Hu, Q.; Shao, J.; Haiping, Y.; Yao, D.; Wang, X.; Chen, H. Effect of binders on the properties of bio-char pellets. *Appl. Energy* **2015**, *157*, 508–516. [CrossRef]
57. Wang, T.; Wang, Z.; Zhai, Y.; Li, S.; Liu, X.; Wang, B.; Li, C.; Zhu, Y. Effect of molasses binder on the pelletization of food waste hydrochar for enhanced biofuels pellets production. *Sustain. Chem. Pharm.* **2019**, *14*, 100183. [CrossRef]
58. Zribi, M.; Lajili, M. Study of the Pyrolysis of Biofuels Pellets Blended from Sawdust and Oleic by-Products: A Kinetic Study. *IJRER* **2019**, *9*, 561–571.
59. Cai, J.; Chen, Y.; Liu, R. Isothermal kinetic predictions from nonisothermal data by using the iterative linear integral isoconversional method. *J. Energy Inst.* **2014**, *87*, 183–187. [CrossRef]
60. Vyazovkin, S.; Burnham, A.K.; Criado, J.M.; Pérez-Maqueda, L.A.; Popescu, C.; Sbirrazzuoli, N. ICTAC Kinetics Committee recommendations for performing kinetic computations on thermal analysis data. *Thermochimca Acta* **2011**, *520*, 1–19. [CrossRef]
61. Wang, Y.; Sun, Y.; Wu, K. Effects of Waste Engine Oil Additive on the Pelletizing and Pyrolysis Properties of Wheat Straw. *Bioresources* **2019**, *14*, 537–553.
62. Guo, J.; Lua, A.C. Kinetic study on pyrolytic process of oil-palm solid waste using two-step consecutive reaction model. *Biomass Bioenergy* **2001**, *20*, 223–233. [CrossRef]
63. Wang, X.; Hu, M.; Hu, W.; Chen, Z.; Liu, S.; Hu, Z.; Xiao, B. Thermogravimetric kinetic study of agricultural residue biomass pyrolysis based on combined kinetics. *Bioresour. Technol.* **2016**, *219*, 510–520. [CrossRef] [PubMed]

© 2020 by the authors. Licensee MDPI, Basel, Switzerland. This article is an open access article distributed under the terms and conditions of the Creative Commons Attribution (CC BY) license (http://creativecommons.org/licenses/by/4.0/).

Article

Effects of Forces, Particle Sizes, and Moisture Contents on Mechanical Behaviour of Densified Briquettes from Ground Sunflower Stalks and Hazelnut Husks

Cimen Demirel [1,2], Gürkan Alp Kağan Gürdil [1], Abraham Kabutey [2] and David Herak [2,*]

1. Department of Agricultural Machinery, Faculty of Agriculture, Ondokuz Mayis University, 55139 Samsun, Turkey; cimendemirel@gmail.com (C.D.); ggurdil@omu.edu.tr (G.A.K.G.)
2. Department of Mechanical Engineering, Faculty of Engineering, Czech University of Life Sciences Prague, Kamycka 129, 16521 Prague, Czech Republic; kabutey@tf.czu.cz
* Correspondence: herak@tf.czu.cz; Tel.: +420-224-383-181

Received: 21 April 2020; Accepted: 14 May 2020; Published: 17 May 2020

Abstract: Using the uniaxial compression process, the mechanical behaviour of densified briquettes from ground sunflower stalks and hazelnut husks was studied under different forces (100, 200, 300, and 400 kN), particle sizes (0, 3, 6, and 10 mm), and moisture contents (sunflower; 11.23%, 14.44%, and 16.89% w.b.) and (hazelnut; 12.64%, 14.83%, and 17.34% w.b.) at a constant speed of 5 mm min^{-1}. For each test, the biomass material was compacted at a constant volume of 28.27×10^{-5} m^3 using a 60 mm-diameter vessel. Determined parameters included densification energy (J), hardness (kN·mm^{-1}), analytical densification energy (J), briquette volume (m^3), bulk density of materials (kg·m^{-3}), briquette bulk density (kg·m^{-3}), and briquette volume energy (J·m^{-3}). The ANOVA multivariate tests of significance results showed that for ground sunflower stalk briquettes, the force and particle size interactions had no significant effect ($p > 0.05$) on the above-mentioned parameters compared to the categorical factors, which had a significant effect ($p < 0.05$) similar to the effects of forces, moisture contents, and their interactions. For ground hazelnut husk briquettes, all the factors and their interactions had a significant effect on the determined parameters. These biomass materials could be attractive for the briquette market.

Keywords: biomass densification; mechanical compaction; processing factors; briquette durability; multivariate tests of significance

---

1. Introduction

Agricultural residues in the form of straws, grasses, stalks, and husks (among others) are excellent sources for biofuel production [1–3]. One of the major limitations of using biomass as a feedstock is its low bulk density, which ranges from 80 to 100 kg/m^3 for agricultural straws and grasses and from 150 to 200 kg/m^3 for woody resources such as wood chips and sawdust [4] Inefficient transportation and large volume requirements for storage are some of the challenges associated with biomass energy usage [5]. Biomass densification for both bioenergy and animal feed utilization has been the approach to mitigate the cost of transportation, handling, and storage [6,7]. Additionally, densified biomass improves fuel feeding in co-firing operations and provides an increased regulation of combustion, thus reducing particulate emissions [8–10]. Densification is widely used in biomass industries, animal feed making, and pharmaceutical industries, and it is classified into pelletization, briquetting, and extrusion [10,11]. Biomass densification is defined as the compression or compaction of biomass to remove inter-and intra-particle voids [5,12]. Generally, the densification of materials requires two stages to take place: particle rearrangement and deformation [1,13–16]. According to [17], as cited in [1], in the first stage, particles rearrange to bring themselves closer together and to reduce voids; little stress is needed to overcome interparticle and particle-to-wall friction. The particles retain their properties, and elastic deformation mainly occurs during this phase [18]. In the second stage, with increasing applied pressure, most of the air is removed from the particulate mass and the elastic–plastic deformation of particles occurs [13–16,18,19].

Recently, studies have been conducted on biomass briquette densification to improve the performance of briquetting technology and to determine the optimum processing factors for producing quality briquettes for energy purposes [2,11,20–26]. The energy requirement for the densification of biomass primarily depends upon the pressure applied and the moisture content of the material to be compressed, as well as the physical properties of the material, including particle size and initial bulk density [8]. The sustainability of biomass densification depends on the energy consumption, emissions, and cost integrated with densification itself and the application of the densified biomass in the combustion or gasification process [6,27]. The machinery for biomass densification is experiencing greatly increasing interest as a result of the concern for its easier mechanical handling of biomass residues, lower storage, and transport space. However, the performance of the briquetting technology is influenced by several operating factors such as pressure, biomass type, particle size, quality, moisture content, feed rate, the forward speed of the machine, field conditions, feeding mechanisms, and power [2]. To optimize the operating factors and the design of new technology for producing biomass briquettes, it is imperative to study and understand the mechanical and rheological behaviours of biomass materials under uniaxial compression [1,28–33].

This information regarding briquette densification from ground sunflower stalks and hazelnut husks with the processing factors is inadequate in the literature. There is also an increasing need to source alternative fuels, especially for cooking to reduce deforestation in the rural areas of developed, developing, and underdeveloped countries. Therefore, these biomass materials could be economically attractive for fuel applications. The objectives of the study were to (i) experimentally and theoretically describe the force and deformation curves of densified briquettes from ground sunflower stalks and hazelnut husks and (ii) to calculate the densification energy (J), hardness (kN/mm), analytical energy (J), briquette volume (m^3), bulk density of materials (kg/m^3), briquette bulk density (kg/m^3), and briquette volume energy (J/m^3).

2. Materials and Methods

2.1. Samples, Milling and Particle Size Distributions

Sunflower stalks and hazelnut husks (Figure 1A,B) were brought from Samsun, Turkey to the Faculty of Engineering, Department of Mechanical Engineering, Czech University of Life Sciences Prague, Prague, Czech Republic. The biomass materials were ground using a hammer mill with a 5.5 kW motor (9FQ-40C, Pest Control Corporation, s.r.o., Vlčnov, Czech Republic) (Figure 1C). The particle size distributions of the ground biomass materials were determined according to the American Society of Agricultural Engineering (ASAE) S319.3 standard [34]. Based on the standard procedure, 100 g of the ground materials were successively placed on top of a sieve shaker (AS 200, Retsch, Haan, Germany) (Figure 1D) of four sieve opening sizes in the order of 0–3, 3–6, 6–10, and/or >6 mm. The sieve shaker was vibrated for 10 min at an amplitude of 3.0 mm/g. Ground material of a particle size of 0–10 and/or >6 mm served as the control.

Figure 1. (**A**) Sunflower stalks and (**B**) hazelnut husks. (**C**) A 5.5 kW motor hammer mill (9FQ-40C, Pest Control Corporation, s.r.o., Vlčnov, Czech Republic) used to grind the biomass materials. (**D**) A sieve shaker (AS 200, Retsch, Haan, Germany) of four sieve opening sizes.

2.2. Determination of Moisture Content and Moisture Conditioning

The initial moisture contents of the ground sunflower stalks and hazelnut husks of 11.23% w.b. and 12.64%, respectively, was determined using the standard oven method [35–37]. The particle size of 0–10 mm of the ground sunflower stalks (Figure 2A) was conditioned to moisture contents of 14.44% and 16.89% (w.b.). The particle size of 0–6 mm of the ground hazelnut husks (Figure 2B) was also conditioned to moisture contents of 14.83% and 17.34% (w.b.). Moisture conditioning equipment (MEMMERT GmbH + Co. KG, Schwabach, Germany) was used. The equipment was equipped with a tube connected to a 2 L gallon on top of it that was filled with distilled water whenever necessary. The samples were loaded into the oven with the parameter settings of a 50 °C temperature with a ±2 °C minimum and maximum to regulate the actual temperature for each relative humidity value between 60% and 90% for 24 h. Afterwards, the samples were put into a conventional oven for 24 h to determine their moisture values.

Figure 2. (**A**) Ground sunflower stalks. (**B**) Ground hazelnut husks of different size distributions.

2.3. Biomass Briquettes Densification

Each of the particle size and moisture content values of the ground biomass materials was densified using a universal compression-testing machine (Tempos, model ZDM 50, Czech Republic) (Figure 3) along with a pressing vessel of diameter 60 mm with a plunger under varying forces (F) between 100 and 400 kN at a speed of 5 mm/min, where the dependencies between the forces and deformation curves were obtained. The initial pressing height (H) of the material was measured at 100 mm using the above mentioned vessel diameter, which remained constant for all tests. Based on this measurement, the volume of the biomass material was calculated to be 28.27×10^{-5} m^3. Two separate experiments were performed for the two types of biomass materials. The first experiment considered the input factors of forces and particle sizes at a constant moisture content of the biomass material ($4 \times 4 = 16 \times 2 = 32$) for two replications. The second experiment considered the input factors of forces and moisture contents at a constant particle size ($4 \times 3 = 12 \times 2 = 24$) for two replications. This makes a total of 56 experiments multiplied by 2 types of materials, thus making 112 experiments. However, the actual number of experiments was 96 as a result of the constant factors. Average values were used in all calculations.

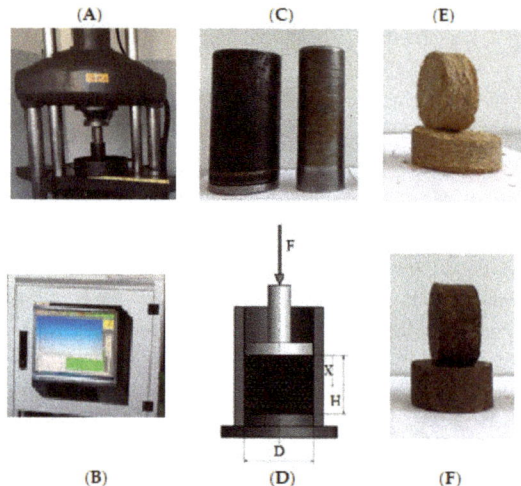

Figure 3. (**A**) Compression test process. (**B**) A computer monitor for data display. (**C**) Pressing vessel with a plunger. (**D**) Schematic of the pressing vessel with a plunger showing the force (F (kN)), deformation (X (mm)), and the initial height of the sample (H (mm)) [28]. (**E**) Densified sunflower briquettes. (**F**) Densified hazelnut briquettes.

2.4. Densification Tests Calculated Parameters

The parameters calculated from the densification tests included densification energy (J), hardness (kN·mm^{-1}), analytical densification energy (J), briquette volume (m^3), bulk density of materials (kg·m^{-3}), briquette bulk density (kg·m^{-3}), and briquette volume energy (J·m^{-3}) with respect to compression forces, particle sizes, and moisture contents using the mathematical equations described in our previous publication on jatropha seedcake briquettes [28]. Other parameters that were directly determined from the densification tests were deformation and briquette thickness. The thickness of the briquettes was measured using a Digital Vernier Caliper. The calculated bulk density at various particle sizes and moisture contents of ground sunflower stalks ranged from 170.99 to 192.54 kg·m^{-3}, whereas that of ground hazelnut husks ranged from 204.43 to 355.09 kg·m^{-3}.

2.5. Statistical Analyses

The calculated parameters were statistically analysed using Statistica 13 software [38] by employing the ANOVA, regression, and correlation methods. The theoretical dependencies between the forces and deformation curves were described using the MathCAD 14 software [39] based on the tangent curve function [28,40–43], where the analytical energies of the densified briquettes were determined.

3. Results

3.1. Deformation, Thickness, Densification Energy and Hardness of Ground Sunflower Stalks Briquettes

The determined amounts of the deformation, thickness, densification energy, and hardness of densified ground sunflower stalk briquettes in relation to the combined effects of forces, particle sizes, and moisture contents are given in Tables 1 and 2. The deformation values ranged from 87.63 to 106.73 mm, and both increased and decreased trends were observed with increased forces, particle sizes, and moisture contents, thus indicating that the deformation values tended to not be affected by the observed factors. Briquettes thickness ranged from 18.45 to 27.34 mm for forces, particle sizes, and moisture contents. Briquette thickness decreased with increased particle sizes but increased with moisture contents for all forces. However, for the particle size of 10 mm at forces 200 and 300 kN, the values increased, which could have been due to the large intercellular air space of the biomass cell walls, which is important for bonding, For all forces, densification energy values increased with increased particle sizes but decreased with moisture contents. The values ranged from 783.88 to 2092 J. Hardness values decreased along with increased forces and particle sizes. However, at force 300 kN, hardness values increased at particle sizes between 0 and 6 mm and then decreased at 10 mm. The values ranged from 1.01 to 4.27 kN·mm^{-1}.

Table 1. Deformation, thickness, densification energy, and hardness of ground sunflower stalks with forces and particle sizes.

Force F (kN)	Particle Size PS (mm)	Deformation X (mm)	Thickness TK (mm)	Densification Energy EN (J)	Hardness HR (kN·mm^{-1})
100	* 0	92.49 ± 6.60	25.72 ± 0.35	924.87 ± 61.72	1.08 ± 0.08
	3	89.19 ± 0.81	27.27 ± 0.06	880.34 ± 44.49	1.12 ± 0.01
	6	90.88 ± 3.45	26.67 ± 3.01	971.22 ± 94.06	1.10 ± 0.04
	10	100.98 ± 19.11	25.67 ± 2.19	1038.46 ± 221.20	1.01 ± 0.19
200	* 0	99.31 ± 1.35	21.99 ± 0.69	1314.96 ± 39.07	2.01 ± 0.03
	3	87.63 ± 4.26	21.70 ± 1.82	1316.38 ± 96.59	2.28 ± 0.11
	6	94.60 ± 2.77	21.45 ± 0.47	1433.54 ± 149.11	2.11 ± 0.06
	10	106.73 ± 15.05	21.57 ± 2.94	1463.52 ± 224.57	1.89 ± 0.27
300	* 0	94.81 ± 2.02	20.15 ± 0.58	1616.09 ± 58.69	3.16 ± 0.07
	3	94.77 ± 5.28	19.92 ± 0.43	1679.75 ± 71.99	3.17 ± 0.18
	6	94.12 ± 4.57	19.56 ± 2.15	1673.74 ± 203.98	3.19 ± 0.16
	10	102.59 ± 11.36	19.72 ± 3.22	1781.96 ± 308.30	2.94 ± 0.33
400	* 0	94.45 ± 8.10	18.45 ± 0.10	1942.61 ± 72.22	4.25 ± 0.36
	3	94.15 ± 9.36	19.95 ± 0.45	1969.24 ± 52.88	4.27 ± 0.42
	6	95.31 ± 2.28	18.78 ± 1.90	1983.79 ± 206.98	4.20 ± 0.10
	10	104.15 ± 15.61	18.68 ± 3.34	2092.96 ± 350.82	3.88 ± 0.58

* 0 (0–10)—control, 3 (3–6), 6 (3–6) and 10 (6–10) mm.

The multivariate results of significance and correlation of the effects of forces/particle sizes and forces/moisture contents on energy and hardness are given in Tables 3 and 4. The effects of the forces, particle sizes, and moisture contents and their interactions on deformation, thickness, densification energy, and hardness were interpreted based on Wilk's lambda value, F-value, and p-value. For all determined parameters, the correlation between thickness, densification energy, hardness, and forces were higher compared to particle sizes and moisture contents.

Table 2. Deformation, thickness, densification energy, and hardness of ground sunflower stalks with forces and moisture contents.

Force F (kN)	Moisture Content MC (% w.b.)	Deformation X (mm)	Thickness TK (mm)	Densification Energy EN (J)	Hardness HR (kN·mm^{-1})
100	11.23	92.49 ± 6.60	25.72 ± 0.35	924.87 ± 61.72	1.08 ± 0.08
	14.44	95.68 ± 2.62	24.97 ± 0.28	786.50 ± 13.98	1.05 ± 0.03
	16.89	102.51 ± 0.59	27.34 ± 0.08	783.88 ± 5.52	1.08 ± 0.08
200	11.23	99.31 ± 1.35	21.99 ± 0.69	1314.96 ± 39.07	2.01 ± 0.03
	14.44	96.12 ± 0.28	21.67 ± 0.32	1104.19 ± 19.45	2.08 ± 0.01
	16.89	96.26 ± 0.88	22.95 ± 0.37	1066.98 ± 16.96	2.01 ± 0.03
300	11.23	94.81 ± 2.02	20.15 ± 0.58	1616.09 ± 58.69	3.17 ± 0.07
	14.44	93.89 ± 6.78	19.48 ± 0.33	1378.49 ± 0.68	3.20 ± 0.23
	16.89	102.56 ± 0.57	21.78 ± 0.42	1326.65 ± 9.39	3.17 ± 0.07
400	11.23	94.45 ± 8.10	18.45 ± 0.10	1942.61 ± 72.22	4.25 ± 0.36
	14.44	100.73 ± 1.88	19.27 ± 0.04	1607.68 ± 15.66	3.97 ± 0.07
	16.89	101.52 ± 0.20	21.10 ± 0.08	1528.55 ± 34.17	4.25 ± 0.36

Table 3. ANOVA multivariate tests of significance of the determined parameters of ground sunflower stalk briquettes.

Effect	Test	Wilks Value	F-Value (-)	Effect df	Error df	p-Value (-)
Effects of F and PS						
Intercept	Wilks lambda	<0.05	6197.566	7	10.000	<0.05
F	Wilks lambda	<0.05	23.061	21	29.265	<0.05
PS	Wilks lambda	<0.05	3.109	21	29.265	<0.05
F × PS	Wilks lambda	>0.05	0.631	63	62.429	>0.05
Effects of F and MC						
Intercept	Wilks lambda	<0.05	3,542,185	7	6.000	<0.05
F	Wilks lambda	<0.05	79	21	17.779	<0.05
MC	Wilks lambda	<0.05	163	14	12.000	<0.05
F × MC	Wilks lambda	<0.05	7	42	31.595	<0.05

F: force (kN); PS: particle size (mm); MC: moisture content (% w.b.); df: degree of freedom.

Table 4. Correlation results of the determined parameters against force, particle size, and moisture content of ground sunflower stalk briquettes.

Determined Parameters	Correlation			
	F	PS	F	MC
Deformation (mm)	0.14	0.40	0.15	0.52
Thickness (mm)	−0.84	−0.04	−0.89	0.29
Densification Energy (J)	0.94	0.14	0.93	−0.31
Analytical Densification Energy (J)	0.74	−0.16	0.66	0.04
Volume (×10^{-5} m^3)	−0.84	−0.04	−0.89	0.29
Bulk Density (kg·m^{-3})	0.89	−0.07	0.93	−0.09
Hardness (kN·mm^{-1})	0.98	0.07	0.99	0.01
Volume Energy (×10^6 J·m^{-3})	0.98	0.12	0.92	−0.33

F: force (kN); PS: particle size (mm); MC: moisture content (% w.b.).

The regression results of the densification energy and hardness of the densified briquettes of ground sunflower stalks with the effects of forces/particle sizes and forces/moisture contents are given in Tables 5 and 6. The results in Tables 5 and 6 represent an example of the regression models of other dependent variables which are highlighted in the discussion section. The models for densification energy and hardness with forces/particle sizes and forces/moisture contents are indicated. The suitability of the models was assessed by the coefficient of determination (R^2), F-values, and p-values.

Table 5. Regression results of energy and hardness of ground sunflower stalk briquettes with the effects of force and particle size.

Effect	Model	R^2 (-)	F-Value (-)	p-Value (-)
\multicolumn{5}{c}{Densification Energy EN (J)}				
Intercept	574.98			
F	3.44	0.90	134.19	<0.05
PS	14.99			
\multicolumn{5}{c}{Hardness HR (kN·mm^{-1})}				
Intercept	0.14			>0.05
F	0.01	0.97	499.04	<0.05
PS	−0.02			<0.05

F: force (kN); PS: particle size (mm).

Table 6. Regression results of densification energy and hardness of ground sunflower stalk briquettes with the effects of force and moisture content.

Effect	Model	R^2 (-)	F-Value (-)	p-Value (-)
\multicolumn{5}{c}{Densification Energy EN (J)}				
Intercept	1303.53			
F	2.862	0.95	199.17	<0.05
MC	−51.44			
\multicolumn{5}{c}{Hardness HR (kN·mm^{-1})}				
Intercept	−0.01			>0.05
F	0.01	0.98	685.66	<0.05
MC	0.001			>0.05

F: force (kN); MC: moisture content (% w.b.).

3.2. Deformation, Thickness, Densification Energy and Hardness of Ground Hazelnut Husks Briquettes

For the ground hazelnut husk briquettes, the values of deformation, thickness, densification energy, and hardness with forces, particle sizes, and moisture contents are given in Tables 7 and 8. Deformation values increased with increased forces and particle sizes, whereas for forces and moisture contents, both increased and decreased amounts were observed. Deformation values ranged from 71.65 to 100.55 mm for all factors. Briquette thickness decreased along with increased forces and particle sizes, but it increased with forces and moisture contents except for force 100 kN, where the values decreased. For the particle size of 3 mm, the thickness values were highest for all forces. Thickness values ranged from 20.18 to 43.81 mm. A similar trend was also observed for the densification energy values, which ranged from 804.11 to 2812.38 J. Hardness values increased along with increased forces but decreased with particle sizes. For forces and moisture contents, the hardness values showed both increasing and decreasing trends. However, for forces 100 and 200 kN, the hardness values slightly increased at the particle size of 3 mm. Hardness values ranged from 1.20 to 4.97 kN·mm^{-1}.

The multivariate results of the significance and correlation of the effects of the forces/particle sizes and forces/moisture contents on the densification energy and hardness of ground hazelnut husks are given in Tables 9 and 10. The effects of the forces, particle sizes and moisture contents and their interactions on deformation, thickness, densification energy and hardness were explained based on Wilk's lambda value, F-value, and p-value. For all determined parameters, the correlation values for thickness, densification energy, and hardness with forces were higher compared to particle sizes and moisture contents, which showed lower values.

Table 7. Deformation, thickness, densification energy, and hardness of ground hazelnut husks with forces and particle size.

Force F (kN)	Particle Size PS (mm)	Deformation X (mm)	Thickness TK (mm)	Densification Energy EN (J)	Hardness HR (kN·mm^{-1})
100	* 0	79.28 ± 0.45	39.75 ± 0.72	1094.46 ± 1.03	1.27 ± 0.01
	3	71.65 ± 2.76	43.81 ± 0.69	1050.07 ± 6.85	1.40 ± 0.05
	6	80.08 ± 1.75	35.49 ± 0.26	1031.56 ± 6.58	1.25 ± 0.03
	10	92.42 ± 0.89	26.69 ± 1.16	804.11 ± 5.44	1.08 ± 0.01
200	* 0	79.63 ± 6.48	33.07 ± 1.03	1689.61 ± 31.52	2.53 ± 0.21
	3	77.12 ±1.78	36.82 ± 0.82	1750.20 ± 3.87	2.60 ± 0.06
	6	90.10 ± 6.52	30.41 ± 0.54	1605.23 ± 19.86	2.23 ± 0.16
	10	93.45 ± 2.16	22.91 ± 0.16	1276.39 ± 8.74	2.15 ± 0.05
300	* 0	79.92 ± 0.30	31.65 ± 0.37	2218.97 ± 7.79	3.75 ± 0.01
	3	79.56 ± 2.38	34.63 ± 0.08	2307.47 ± 6.68	3.77 ± 0.11
	6	89.98 ± 6.27	27.78 ± 0.17	2045.82 ± 3.85	3.35 ± 0.23
	10	100.55 ± 3.19	20.71 ± 0.01	1644.56 ± 14.30	2.99 ± 0.09
400	* 0	80.80 ± 6.07	30.48 ± 0.11	2602.32 ± 50.66	4.97 ± 0.37
	3	85.11 ± 7.42	33.44 ± 0.32	2812.38 ± 7.33	4.72 ± 0.41
	6	87.39 ± 1.36	26.61 ± 0.02	2437.41 ± 53.60	4.58 ± 0.07
	10	97.17 ± 1.34	20.18 ± 0.30	2002.76 ± 76.86	4.12 ± 0.06

* 0 (0–10)—control, 3 (3–6), 6 (3–6) and 10 (6–10) mm.

Table 8. Deformation, thickness, energy, and hardness of ground hazelnut husks with forces and moisture contents.

Force F (kN)	Moisture Content MC (% w.b.)	Deformation X (mm)	Thickness TK (mm)	Energy EN (J)	Hardness HR (kN·mm^{-1})
100	12.64	79.28 ± 0.45	39.75 ± 0.72	1094.46 ± 1.03	1.27 ± 0.01
	14.83	76.21 ± 3.70	38.92 ± 1.37	992.81 ± 16.02	1.32 ± 0.06
	17.34	83.34 ± 0.62	38.76 ± 0.79	926.40 ± 12.13	1.20 ± 0.01
200	12.64	79.63 ± 6.48	33.07 ± 1.03	1689.61 ± 31.52	2.53 ± 0.21
	14.83	85.15 ± 7.30	33.19 ± 1.32	1581.69 ± 8.78	2.36 ± 0.21
	17.34	77.75 ± 6.89	34.73 ± 1.49	1340.72 ± 55.80	2.58 ± 0.23
300	12.64	79.92 ± 0.30	31.65 ± 0.37	2218.97 ± 7.79	3.75 ± 0.01
	14.83	79.62 ± 1.50	32.23 ± 0.49	1986.39 ± 51.85	3.77 ± 0.07
	17.34	87.20 ± 0.98	33.78 ± 1.45	1735.02 ± 17.29	3.44 ± 0.04
400	12.64	80.80 ± 6.07	30.48 ± 0.11	2602.32 ± 50.66	4.97 ± 0.37
	14.83	82.65 ± 0.01	31.74 ± 0.01	2343.60 ± 12.22	4.84 ± 0.00
	17.34	81.65 ± 9.74	35.21 ± 2.79	1875.71 ± 123.63	4.94 ± 0.59

Table 9. ANOVA multivariate tests of significance of parameters of ground hazelnut husk briquettes.

Effect	Test	Wilks Value	F-Value (-)	Effect df	Error df	p-Value (-)
			Effects of F and PS			
Intercept	Wilks lambda	<0.05	1,705,818	7	10.00	<0.05
F	Wilks lambda	<0.05	101	21	29.26	<0.05
PS	Wilks lambda	<0.05	111	21	29.26	<0.05
F × PS	Wilks lambda	<0.05	4	63	62.43	<0.05
			Effects of F and MC			
Intercept	Wilks lambda	<0.05	1,517,95.7	7	6.00	<0.05
F	Wilks lambda	<0.05	58.6	21	17.78	<0.05
MC	Wilks lambda	<0.05	24.7	14	12.00	<0.05
F × MC	Wilks lambda	<0.05	2.9	42	31.59	<0.05

F: force (kN); PS: particle size (mm); MC: moisture content (% w.b.); df: degree of freedom.

Table 10. Correlation results of parameters against force, particle size, and moisture content of ground hazelnut husk briquettes.

Determined Parameters	Correlation			
	F	PS	FR	MC
Deformation (mm)	0.31	0.78	0.19	0.23
Thickness (mm)	−0.49	0.73	−0.76	0.25
Densification Energy (J)	0.92	0.31	0.92	−0.34
Analytical Densification Energy (J)	0.78	0.41	0.92	−0.22
Volume ($\times 10^{-5}$ m^3)	−0.49	0.73	−0.76	0.25
Bulk Density (kg·m^{-3})	0.93	0.21	0.81	−0.09
Hardness (kN·mm^{-1})	0.98	0.17	0.99	−0.03
Volume Energy ($\times 10^6$ J·m^{-3})	0.98	0.14	0.89	−0.36

F: force (kN); PS: particle size (mm); MC: moisture content (% w.b.).

The regression results of the densification energy and hardness of the densified briquettes from ground hazelnut husks with the effects of forces/particle sizes and forces/moisture contents are also given in Tables 11 and 12. The results in Tables 11 and 12 represent an example of the regression models of the other dependent variables that are highlighted in the discussion section. The suitability of the models was evaluated based on the coefficients of determination (R^2), F-values, and p-values.

Table 11. Regression results of densification energy and hardness of the ground hazelnut husk briquettes with the effects of force and particle size.

Effect	Model	R^2 (-)	F-Value (-)	p-Value (-)
Densification Energy EN (J)				
Intercept	792.27			
F	4.88	0.95	273.71	<0.05
PS	−50.29			
Hardness, HR (kN·mm^{-1})				
Intercept	0.41			
F	0.01	0.98	755.87	<0.05
PS	−0.06			

F: force (kN); PS: particle size (mm).

Table 12. Regression results of densification energy and hardness of the ground hazelnut husk briquettes with the effects of force and moisture content.

Effect	Model	R^2 (-)	F-Value (-)	p-Value (-)
Densification Energy EN (J)				
Intercept	2255.82			
F	4.25	0.95	216.22	<0.05
MC	−107.97			
Hardness HR (kN·mm^{-1})				
Intercept	0.37			>0.05
F	0.01	0.98	557.22	<0.05
MC	−0.02			>0.05

F: force (kN); MC: moisture content (% w.b.).

3.3. Effects of Particle Sizes and Moisture Contents on Densification Energy and Hardness of Densified Briquettes

Comparisons of densification energy and hardness of the densified briquettes of the ground sunflower stalks and hazelnut husks with respect to the processing factors are indicated in Figures 4–7.

For all processing factors, briquettes from the ground hazelnut husks required more densification energy than ground sunflower stalk briquettes. Additionally, the hardness of the hazelnut husk briquettes for particle sizes 0–10 and 0–3 mm at forces 300 and 400 kN were higher than that of the sunflower stalk briquettes. However, the hardness of the biomass briquettes for particle sizes 3–6 and 6–10 mm for all forces was similar. This suggests that particle sizes 0–10 and 0–3 mm could be used for producing briquettes for energy purposes. The plots of normality of the dataset of energy and hardness are presented in Figure 8, and they are similar to the other determined parameters. It can be seen that the data showed an approximately normal distribution. The normal distribution of the data was also evaluated by the Shapiro–Wilk test [44]. The statistical results, however, are not presented herein.

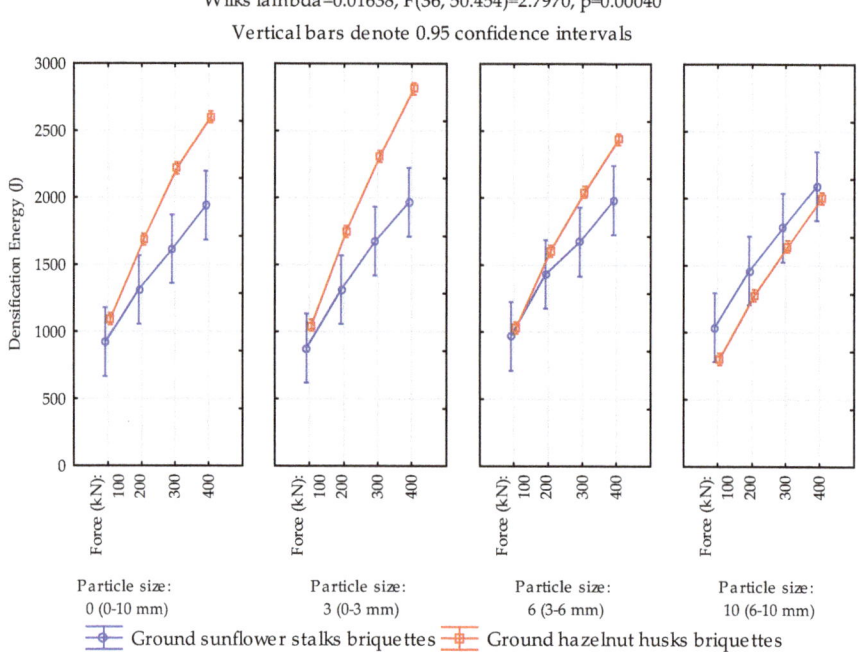

Figure 4. Effects of forces and particle sizes on briquette densification energy.

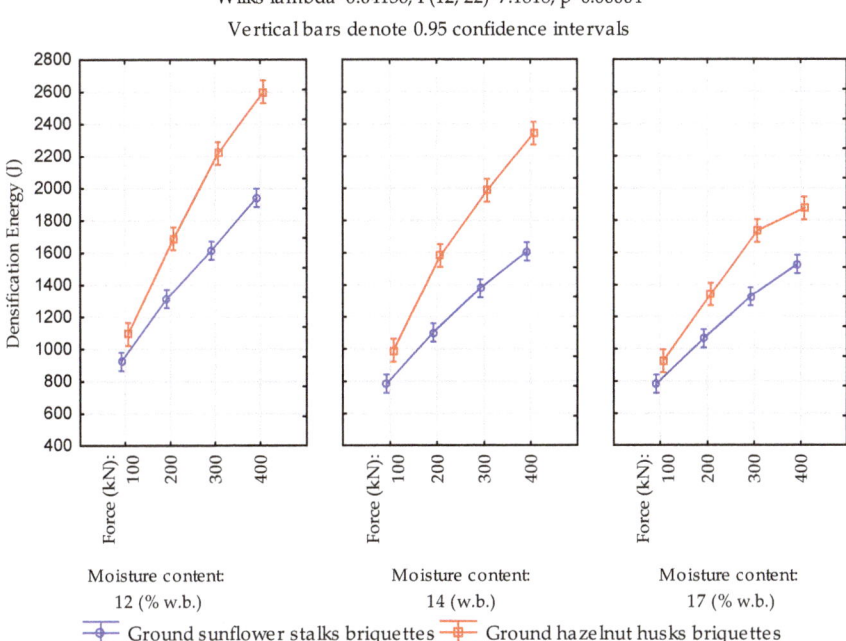

Figure 5. Effects of forces and moisture contents on briquette densification energy.

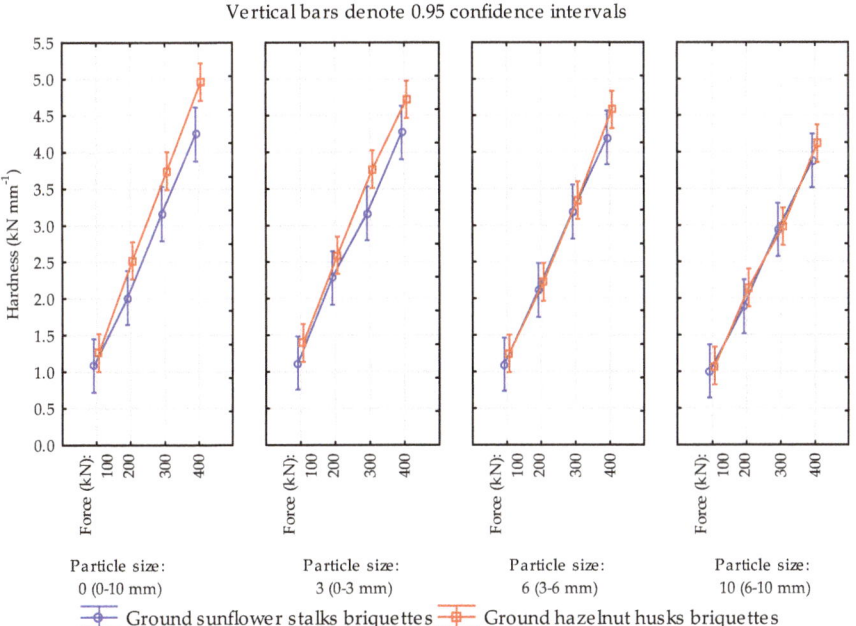

Figure 6. Effects of forces and particle sizes on briquette hardness.

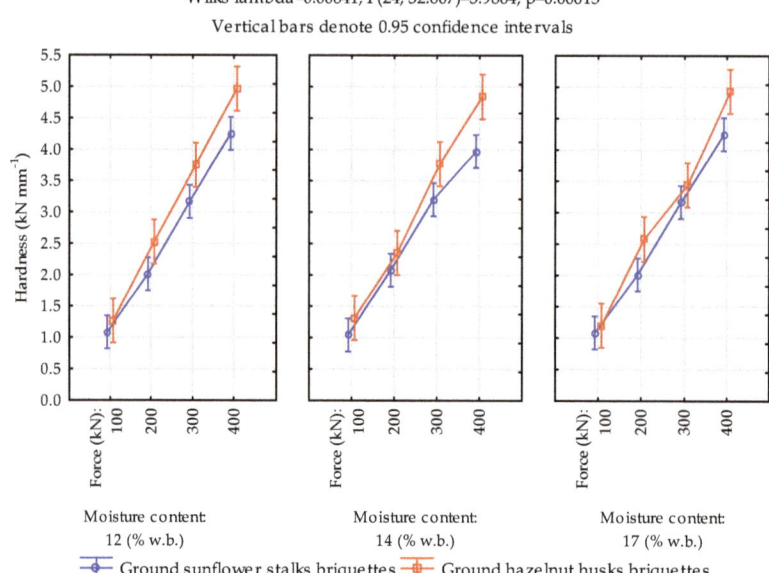

Figure 7. Effects of forces and moisture contents on briquette hardness.

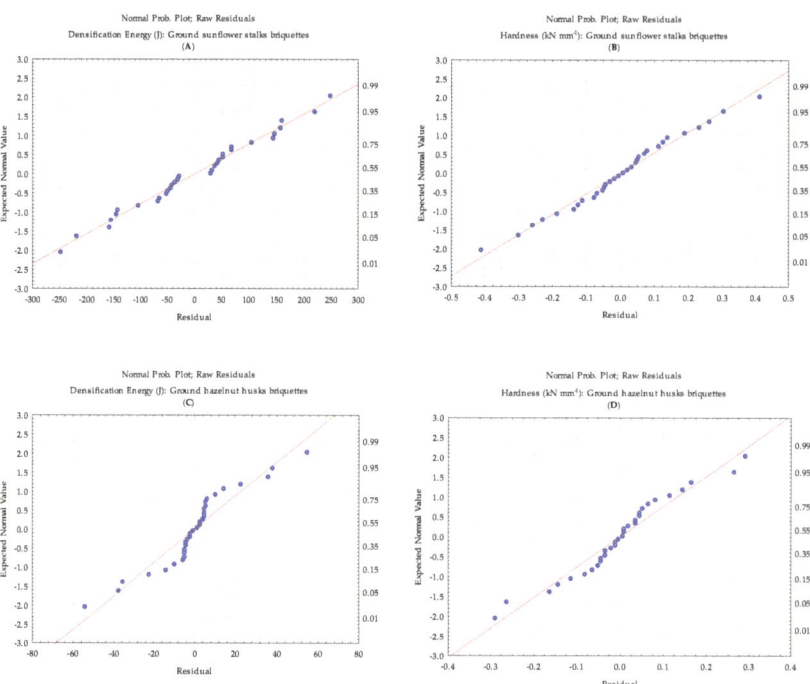

Figure 8. Normal probability plots of densification energy (**A**) and hardness (**B**) of ground sunflower stalk, and densification energy (**C**) and hardness (**D**) of ground hazelnut husk briquettes under the effects of forces and particle sizes similar to other determined parameters.

3.4. Description of Analytical Densification Energy of Biomass Briquettes

The densification energies of the ground sunflower stalk and hazelnut husk briquettes were theoretically described, and the results are given in Tables 13 and 14. The experimental and theoretical curves of particle sizes and moisture contents at a maximum force of 400 kN are illustrated in Figures 9 and 10. The analytical densification energy for the ground sunflower stalk briquettes for forces and particle sizes ranged from 799.33 to 2391.98 J, and for forces and moisture contents, the values ranged from 589.92 to 2536.52 J. At forces 200 and 300 kN for particle sizes 6 and 10 mm, as well as moisture contents 14.44% and 16.89% w.b., the data points were further apart compared to the densification energies. For ground hazelnut husk briquettes, the analytical energies for forces and particle sizes ranged from 657.63 to 3376.84 J, whereas for forces and moisture contents, the values ranged from 748.60 to 3427.44 J. At force 300 kN for particle sizes 0–10 mm (the control) and forces 300 and 400 kN at moisture contents 12.64% and 17.34% w.b., the theoretical data points were very different in comparison with the experimental values. This could have been due to the air spaces and friction between the bulk biomass materials and the walls of the pressing vessel during the densification process. The use of binders such as cassava starch wastewater, rice dust, and okra stem gum [25] could ensure effective compaction for the accurate theoretical description of the experimental data.

Table 13. Analytical densification energy of ground sunflower stalks and hazelnut husks at moisture contents of 11.23% and 12.64% w.b.

Force F (kN)	Particle Size PS (mm)	* Analytical Densification Energy AE (J)	
		Ground Sunflower Stalk Briquettes	Hazelnut Husk Briquettes
100	* 0	830.80 ± 121.6	948.93 ± 9.90
	3	799.63 ± 146.54	963.83 ± 73.87
	6	1072.12 ± 179.14	1009.98 ± 167.98
	10	878.64 ± 118.47	657.63 ± 52.44
200	* 0	1614.12 ± 493.71	1476.58 ± 114.49
	3	1212.15 ± 192.61	1741.12 ± 288.89
	6	1500.36 ± 41.95	1460.79 ± 217.01
	10	892.07 ± 13.43	1301.10 ± 367.25
300	* 0	1360.17 ± 470.35	2892.48 ± 112.92
	3	1416.873 ± 86.28	2332.56 ± 326.83
	6	1201.46 ± 324.27	2045.30 ± 624.73
	10	1560.88 ± 679.50	1386.11 ± 514.52
400	* 0	2318.85 ± 511.33	3097.17 ± 280.00
	3	2391.98 ± 519.70	3376.84 ± 218.21
	6	1541.78 ± 266.91	2318.94 ± 712.73
	10	1903.98 ± 419.61	1629.64 ± 445.34

* 0 (0–10)—control, 3 (3–6), 6 (3–6) and 10 (6–10) mm.

Table 14. Analytical densification energy of ground sunflower stalks and hazelnut husks for particle sizes 0–10 mm.

Force F (kN)	Moisture Content MC (% w.b.)	* Analytical Densification Energy AE (J)	
		Ground Sunflower Stalks Briquettes	Ground Hazelnut Husks Briquettes
100	14.44 [a] (14.83) [b]	695.53 ± 187.54 [a]	1039.33 ± 21.27 [b]
	16.89 [a] (17.34) [b]	589.92 ± 62.52 [a]	748.60 ± 50.37 [b]
200	14.44 [a] (14.83) [b]	606.22 ± 34.69 [a]	1629.72 ± 479.32 [b]
	16.89 [a] (17.34) [b]	1340.93 ± 1004.64 [a]	1115.33 ± 3.49 [b]

Table 14. Cont.

Force F (kN)	Moisture Content MC (% w.b.)	* Analytical Densification Energy AE (J)	
		Ground Sunflower Stalks Briquettes	Ground Hazelnut Husks Briquettes
300	14.44 [a] (14.83) [b]	1361.89 ± 562.43 [a]	2046.46 ± 318.29 [b]
	16.89 [a] (17.34) [b]	2536.52 ± 670.48 [a]	2197.08 ± 370.86 [b]
400	14.44 [a] (14.83) [b]	1750.59 ± 784.10 [a]	3427.44 ± 113.30 [b]
	16.89 [a] (17.34) [b]	1698.35 ± 16.87 [a]	2444.56 ± 110.08 [b]

* 0 (0–10)—control, [a]: Ground Sunflower Stalks Briquettes, [b]: Ground Hazelnut Husks Briquettes.

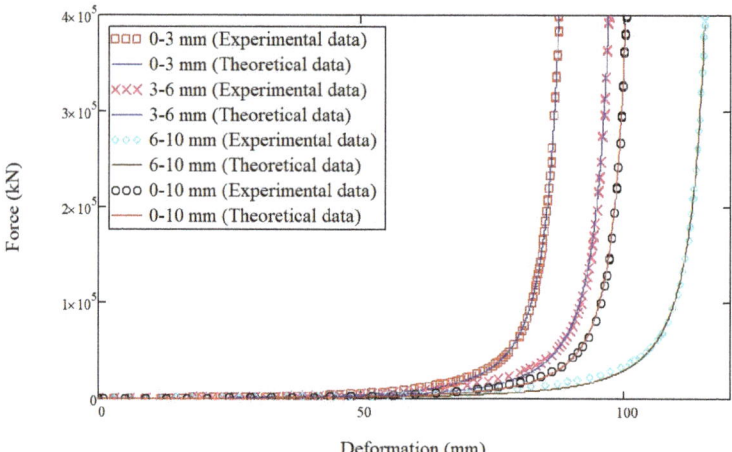

Figure 9. Force–deformation curves of ground sunflower stalk briquettes for particle sizes at a moisture content of 11.23% w.b.

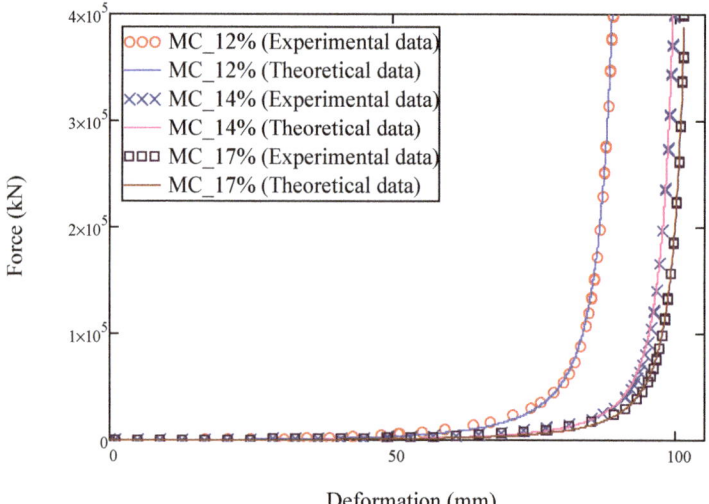

Figure 10. Force–deformation curves of ground sunflower stalk briquettes at different moisture contents for the particle size of 0 (0–10) mm.

3.5. Other Calculated Parameters of Densified Biomass Briquettes (Supplementary Material)

The coefficients of the tangent model (the force coefficient of mechanical behaviour (kN), the deformation coefficient of mechanical behaviour (mm), and the fitting value (-)) and their statistical results for describing the theoretical/analytical force–deformation curves and densification energies (as well as the experimental calculation of the briquette volume, bulk density, and volume energy) are presented in the Supplementary Materials Tables S1–S8 and Figures S1–S3, respectively. For the forces, particle sizes, and moisture contents of the ground sunflower stalk briquettes, the force coefficients of the mechanical behaviour ranged from 1.256 to 3.575 kN, while the deformation coefficients of mechanical behaviour ranged from 0.014 to 0.017 mm^{-1}. On the other hand, for the forces, particle sizes, and moisture contents of the ground hazelnut husk briquettes, the force coefficients of the mechanical behaviour ranged from 2.477 to 24.765 kN, while the deformation coefficients ranged from 0.015 to 0.019 mm^{-1}. The fitting curve exponent of the model was found to be 2 (-), with a high coefficient of determination (R^2) of 99%. The briquette volume decreased along with the increased forces for each particle size and moisture content. However, both increasing and decreasing trends of the volume with particle sizes and moisture contents were noticed for all forces. The polynomial function of (R^2) values between 0.87 and 0.97 suitably described the relationships between briquette volumes and forces, particle sizes, and moisture contents. For all the predictors, the briquette volume from the sunflower stalks ranged from 5.22 to 7.73×10^{-5} m^3 and that of the hazelnut husk briquettes ranged from 5.70 to 11.24×10^{-5} m^3. The bulk density of the briquettes increased along with forces for each particle size and moisture content, but it generally decreased for varying particle sizes and moisture contents at a specific force. The polynomial function of (R^2) values between 0.66 and 0.93 adequately described the relationships between briquette density and forces, particle sizes, and moisture contents. The bulk density of ground sunflower stalk briquettes ranged from 653.29 to 964.41 $kg \cdot m^{-3}$, and that of the ground hazelnut husk briquettes ranged from 765.60 to 1056.69 $kg \cdot m^{-3}$ with respect to the predictors. The briquette volume energy increased along with forces and particle sizes, but it decreased with moisture contents. The polynomial function of (R^2) values between 0.82 and 1 satisfactorily described the relationships between briquette volume energy and forces, particle sizes, and moisture contents. Ground sunflower stalk briquette volume energy ranged from 10.14 to 39.68×10^{-6} $J \cdot m^{-3}$, and that of ground hazelnut husk briquettes ranged from 8.48 to 35.12×10^{-6} $J \cdot m^{-3}$ in relation to the predictors.

4. Discussion

The determined parameters (responses) from the densification tests of the ground sunflower stalk and hazelnut husk briquettes under different processing factors (forces, particle sizes, and moisture contents) were densification energy (J), hardness ($kN \cdot mm^{-1}$), analytical densification energy (J), briquette volume (m^3), bulk density of materials ($kg \cdot m^{-3}$), briquette bulk density ($kg \cdot m^{-3}$), and briquette volume energy ($J \cdot m^{-3}$). The densification curves and energies were theoretically described using the tangent curve model.

For the ground sunflower stalk briquettes, the ANOVA multivariate tests of significance of the effect of the forces and particle sizes on the responses were significant ($p < 0.05$). The interaction effect of the force and particle size on the above-mentioned parameters was not significant ($p > 0.05$). However, based on the univariate results, force did not have significant effect on deformation. The particle size effect was only significant on bulk density and volume energy. In addition, the multivariate tests of significance of the effects of the forces and moisture contents and their interactions on the responses proved significant. Nevertheless, the univariate results showed that only moisture content had a significant effect on deformation. Moisture content and the interactions of force and moisture content also had no significant effects on analytical densification energy and hardness. The correlation between force and the dependent variables was significant, except for deformation, which was not significant. On the other hand, deformation only correlated significantly with particle size and moisture content compared to the other responses, which showed non-significant correlations. The regression results showed that the coefficients of the force and particle size on the models for densification energy, hardness, and volume energy were significant ($p < 0.05$); only the coefficients of the particle size were significant for the thickness, analytical energy, briquette volume, and bulk density models; and for deformation, only the force coefficient was significant. In addition, for the regression results of the interactions of force and moisture content, the coefficients of the force and moisture content were significant for thickness, densification energy, briquette volume, and volume energy. The models for deformation, analytical densification energy, bulk density, and briquette hardness showed only the force coefficients as being significant.

For the ground hazelnut husk briquettes, the ANOVA multivariate tests of significance of the effects of forces, particle sizes, moisture contents, and their interactions with the above-mentioned responses were significant ($p < 0.05$). However, the univariate results showed that the interaction effect of force and particle size on deformation, analytical densification energy, bulk density, and hardness was not significant ($p > 0.05$). The effects of force, moisture content, and interactions on deformation were not significant, but those of densification energy, analytical energy and volume energy were significant. The interaction effects of force and moisture content on thickness and briquette volume were not significant. Briquette bulk density and hardness showed that moisture content, as well as force and moisture content interactions, were non-significant. The correlation between deformation, force, and moisture content were non-significant, similar to the results of ground sunflower stalk briquettes. Densification energy, bulk density, hardness, and volume energy did not significantly correlate with particle size compared to thickness, analytical densification energy, and briquette volume which significantly correlated with particle size. There was no significant correlation between the dependent variables and moisture content. The coefficients of the factors (force and particle size) in the regression models describing all the responses of the ground hazelnut husk briquettes were significant compared to the processing factors (force and moisture content), where only the densification energy, analytical energy, and volume energy were significant. For deformation, all the predictors were not significant, whereas only the moisture content predictor was not significant for thickness, hardness, briquette volume, and bulk density.

Generally, based on the test of the sum of squares whole model against the sum of squares residual model, the factors/predictors had a significant effect on all the responses except for deformation, where the combined effect of force/moisture content and force/particle size had no significant effect. The coefficients of determination (R^2) of the regression models ranged between 30% and 98%.

Furthermore, the densification energy of the briquettes was determined from the area under the force and deformation (densification) curves. Using the tangent curve model [28,40–43], the analytical energy was determined. It is important to state that the application of the tangent model took the physical principles of the uniaxial compression process into account; these principle are that zero force means zero deformation, increasing force causes deformation to reach a maximum limit, and the integral of the force as a function of deformation from the zero to the maximum limit is the energy (that is, the densification energy for biomass materials and deformation energy in the case of bulk oilseeds). The ANOVA results of the tangent model coefficients were significant where the F-critical values were higher than the F-ratio values and/or p-values greater than the alpha level of 0.05, thus confirming the suitability of the tangent curve model for describing the uniaxial compression data.

In the literature, the authors of [29] explained that at low forces, straw bales had a small stiffness that changed with the applied force and the behaviour was almost linear; as the load increased further, a stiffening behaviour was realized. Additionally, the authors of [1] reported that the density of the compacted biomass briquettes from barley, oat, canola, and wheat straw increased with increasing pressure and moisture content. The authors of [8] highlighted that the briquette density of corn stover increased with pressure, whereas low moisture content between 5% and 10% (w.b.) resulted in denser, more stable, and more durable briquettes than the high moisture corn stover content of 15% (w.b.). In a separate study by the authors of [32], the pellet density of wheat straw, barley straw, corn stover, and switchgrass increased as compressive pressure increased at a sample particle size of 3.2 mm and a moisture content of 12% (w.b.). The authors of [44] also mentioned that increased particle size and moisture content decreased the durability of cassava stalk pellets. Additionally, our previous study [28] showed that densified briquettes from jatropha seedcake with a particle size of 10 mm recorded the minimum energy followed by the particle size of 6.7 mm. However, the hardness of the briquettes at a maximum force of 400 kN (pressure of 141.47 MPa) was achieved at a particle size of 6.7 mm followed by the particle size of 5.6 mm. Finally, the authors of [45] stated that corn stover feedstock moisture <34% (w.b.) and preheating >70 °C increased the density and durability of the pellets. The results of the present study are in agreement with published studies on different biomass materials and thus prove the scientific relevance of the work and provide an important contribution to the literature.

5. Conclusions

The effects of processing factors (forces, particle sizes, and moisture contents) on the mechanical behaviour of ground sunflower stalk and hazelnut husk briquettes were studied under uniaxial compression loading. ANOVA multivariate tests of significance, univariate tests, correlation and regression analyses, and normality tests were used to evaluate the statistical significance of the responses. The experimental data (densification curves and energies) were theoretically described using the tangent model by determining the force coefficient of the mechanical behaviour, the deformation coefficient of mechanical behaviour, and the fitting curve value. The coefficients of the model were statistically significant with a high coefficient of determination of 99%. The test of the sum of squares whole model against the sum of squares residual model of the regression analysis showed that the processing factors had a significant effect on all the responses except for deformation, where the combined effect of the force and moisture content and the force and particle size had no significant effect. The coefficients of determination (R^2) of the established regression models ranged between 30% and 98%. The hardness of ground sunflower stalk and hazelnut husk briquettes was achieved at a higher force of 400 kN and particle sizes of 0–10 mm, altogether, and/or 0–3 mm at the moisture contents of 11.23% and 12.64% w.b., respectively. The optimum densification energy and hardness values of the ground sunflower stalk briquettes was between 1942.61 ± 72.22 and 1969.24 ± 52.88 J and between 4.25 ± 0.36 and 4.27 ± 0.42 kN/mm. For the ground hazelnut husk briquettes, the optimum densification energy and hardness values were between 2602.32 ± 50.66 and 2812.38 ± 7.33 J and between 4.72 ± 0.41 kN/mm and 4.97 ± 0.37 kN/mm. The briquette volume decreased along with increased forces for each particle size and moisture content. The bulk density of the briquettes increased along with forces for each particle size and moisture content, but it generally decreased for varying particle sizes and moisture contents at a specific force. The briquette volume energy increased along with forces and particle sizes, but it decreased with moisture contents.

Briquette production from ground sunflower stalks and hazelnut husks could be also attractive for the briquette market. However, binding additives such as cassava starch wastewater, rice dust, and okra stem gum, as well as pre-treatment methods and response surface designs of experiments should be considered in future research to fully understand the mechanical behaviour of the studied biomass materials, among others, to determine the optimum processing conditions for briquette production.

Supplementary Materials: The following are available online at http://www.mdpi.com/1996-1073/13/10/2542/s1, Table S1: Determined tangent curve model coefficients and statistical analysis of ground sunflower stalks briquettes of 11.23 % (w.b.), Table S2: Determined tangent curve model coefficients and statistical analysis of densified briquettes of ground sunflower stalks for particle size 0–10 mm, Table S3: Determined tangent curve model coefficients and statistical analysis of ground hazelnut husks briquettes of 12.64 (% w.b.), Table S4: Determined tangent curve model coefficients and statistical analysis of densified briquettes of ground hazelnut husks for particle size of 0 (0–10) mm (control), Table S5: Ground sunflower stalks briquette volume, bulk density and volume energy at moisture content of 11.23% (w.b.), Table S6: Ground sunflower stalks briquette volume, bulk density and volume energy at particle size of 0 (0–10) mm (control), Table S7: Ground hazelnut husks briquette volume, bulk density and volume energy at moisture content of 12.64% (w.b.), Table S8: Ground hazelnut husks briquette volume, bulk density and volume energy at particle size of 0 (0–10) mm (control), Figure S1: Normal probability plots of densification energy and hardness of ground sunflower stalks and hazelnut husks briquettes under the effects of force and moisture content similar to other determined parameters, Figure S2: Force-deformation curves of ground hazelnut husks particle sizes at moisture content of 12.64 % (w.b.) and Figure S3: Force-deformation curves of ground hazelnut husks at different moisture content for particle size of 0 (0–10) mm.

Author Contributions: Conceptualization, G.A.K.G., D.H., and C.D.; funding acquisition, G.A.K.G.; methodology, C.D., G.A.K.G., and A.K.; data curation, C.D. and A.K.; writing—original draft, C.D., G.A.K.G., A.K., and D.H.; writing—review and editing, C.D., G.A.K.G., A.K., and D.H. All authors have read and agreed to the published version of the manuscript.

Funding: This study was funded by the Scientific and Technological Research Council of Turkey (TÜBİTAK), TÜBİTAK BİDEB 2214-A Doctoral Research Fellowship Program at Abroad with the number 1059B141700726.

Conflicts of Interest: The authors declare no conflict of interest.

References

1. Guo, L.; Wang, D.; Tabil, L.G.; Wang, G. Compression and relaxation properties of selected biomass for briquetting. *Biosyst. Eng.* **2016**, *148*, 101–110. [CrossRef]
2. Mikulandric, R.; Vermeulen, B.; Nicolai, B.; Saeys, W. Modelling of thermal processes during extrusion based densification of agricultural biomass residues. *Appl. Energy* **2016**, *184*, 1316–1331. [CrossRef]
3. Adapa, P.; Tabil, L.; Schoenau, G. Compression characteristics of selected ground agricultural biomass. *Agric. Eng. Int. CIGR J.* **2009**, *9*, 1–19.
4. Rajaseenivasan, T.; Srinivasan, V.; Qadir, G.S.M.; Srithar, K. An investigation on the performance of sawdust briquette with neem powder. *Alex. Eng. J.* **2016**, *55*, 2833–2838. [CrossRef]
5. Van Pelt, T.J. Maize, soybean, and alfalfa biomass densification. *Agric. Eng. Int. CIGR J.* **2003**, *5*, 1–17.
6. Muazu, R.I.; Borrion, A.L.; Stegemann, J.A. Life cycle assessment of biomass densification systems. *Biomass Bioenergy* **2017**, *107*, 384–397. [CrossRef]
7. Shaw, M.D.; Tabil, L.G. Compression, relaxation and adhesion properties of selected biomass grinds. *Agric. Eng. Int. CIGR J.* **2007**, *9*, 1–15.
8. Mani, S.; Tabil, L.G.; Sokhansanj, S. Specific energy requirement for compacting corn stover. *Bioresour. Technol.* **2006**, *97*, 1420–1426. [CrossRef]
9. Sokhansanj, S.; Mani, S.; Bi, X.T.; Zaini, P.; Tabil, L. *Binderless Pelletization of Biomass*; ASAE Paper No. 056061; ASABE: St. Joseph, MI, USA, 2005; p. 1.
10. Li, Y.; Liu, H. High-pressure densification of wood residues to form an upgraded fuel. *Biomass Bioenergy* **2000**, *19*, 177–186. [CrossRef]
11. Gilvari, H.; de Jong, W.; Schott, D.L. Quality parameters relevant for densification of bio-materials: Measuring methods and affecting factors—A review. *Biomass Bioenergy* **2019**, *120*, 117–134. [CrossRef]
12. Balatinecz, J.J. The potential role of densification in biomass utilization. *Biomass Utilization* **1983**, *67*, 181–190.
13. Kaliyan, N.; Morey, R.V. Constitutive model for densification of corn stover and switchgrass. *Biosyst. Eng.* **2009**, *104*, 47–63. [CrossRef]
14. Mani, S.; Tabil, L.G.; Sokhansanj, S. Compaction behavior of some biomass grinds. In *AIC Meeting*; Saskatchewan AIC Paper No. 02–305; CSAE/SCGR Program: Saskatoon, SK, Canada, 2002.
15. Faborode, M.O.; O'Callaghan, J.R. A rheological model for the compaction of fibrous agricultural materials. *J. Agric. Eng. Res.* **1989**, *42*, 165–178. [CrossRef]
16. Faborode, M.O.; O'Callaghan, J.R. Optimizing the compression/briquetting of fibrous agricultural materials. *J. Agric. Eng. Res.* **1987**, *38*, 245–262. [CrossRef]
17. Mani, S.; Tabil, L.G.; Sokhansanj, S. An overview of compaction of biomass grinds. *Powder Handl. Process* **2003**, *15*, 160–168.
18. Cooper, A.R.; Eaton, L.E. Compaction behavior of several ceramic powders. *J. Am. Ceram. Soc.* **1962**, *45*, 97–101. [CrossRef]
19. Nona, K.D.; Lenaerts, B.; Kayacan, E.; Saeys, W. Bulk compression characteristics of straw and hay. *Biosyst. Eng.* **2014**, *118*, 194–202. [CrossRef]
20. Adeleke, A.A.; Odusote, J.K.; Lasode, O.A.; Ikubanni, P.P.; Malathi, M.; Paswan, D. Densification of coal fines and mildly torrefied biomass into composite fuel using different organic binders. *Heliyon* **2019**, *5*, e02160. [CrossRef]
21. Frodeson, S.; Henriksson, G.; Berghel, J. Effects of moisture content during densification of biomass pellets, focusing on polysaccharide substances. *Biomass Bioenergy* **2019**, *122*, 322–330. [CrossRef]

22. Oliveira, H.R.; Bassin, I.D.; Cammarota, M.C. Bioflocculation of cynobacteria with pellets of *Aspergillus niger*: Effects of carbon supplementation, pellet diameter, and other factors in biomass densification. *Bioresour. Technol.* **2019**, *294*, 122167. [CrossRef]
23. Zvicevičius, E.; Raila, A.; Čipliene, A.; Černiauskiene, Z.; Kadžiuliene, Z.; Tilvikiene, V. Effects of moisture and pressure on densification process of raw material from Artemisia dubia Wall. *Renew. Energy* **2018**, *119*, 185–192.
24. Whittaker, C.; Shield, I. Factors affecting wood, energy grass and straw pellet durability—A review. *Renew. Sust. Energ. Rev.* **2017**, *71*, 1–11. [CrossRef]
25. Yank, A.; Ngadi, M.; Kok, R. Physical properties of rice husk and bran briquettes under low pressure densification for rural applications. *Biomass Bioenergy* **2016**, *84*, 22–30. [CrossRef]
26. Chou, C.-H.; Lin, S.-H.; Peng, C.-C.; Lu, W.-C. The optimum conditions for preparing for solid fuel briquette of rice straw by a piston-mold process using the Taguchi method. *Fuel Process. Technol.* **2009**, *90*, 1041–1046. [CrossRef]
27. Caputo, A.C.; Palumbo, M.; Pelagagge, P.M.; Scacchia, F. Economics of biomass energy utilization in combustion and gasification plants: Effects of logistic variables. *J. Biomass Energy* **2005**, *28*, 35–51. [CrossRef]
28. Ivanova, T.; Kabutey, A.; Herak, D.; Demirel, C. Estimation of energy requirement of Jatropha curcas L. seedcake briquettes under compression loading. *Energies* **2018**, *11*, 1980. [CrossRef]
29. Molari, L.; Maraldi, M.; Molari, G. Non-linear rheological model of straw bales behaviour under compressive loads. *Mech. Res. Commun.* **2017**, *81*, 32–37. [CrossRef]
30. Divišová, M.; Herák, D.; Kabutey, A.; Sigalingging, R.; Svatoňová, T. Deformation curve characteristics of rapeseeds and sunflower seeds under compression loading. *Sci. Agric. Bohem.* **2014**, *45*, 180–186. [CrossRef]
31. Lysiak, G. Fracture toughness of pea: Weibull analysis. *Food Eng.* **2007**, *83*, 436–443. [CrossRef]
32. Mani, S.; Tabil, L.G.; Sokhansanj, S. Effects of compressive force, particle size and moisture content on mechanical properties of biomass pellets from grasses. *Biomass Bioenergy* **2006**, *30*, 648–654. [CrossRef]
33. Gupta, R.K.; Das, S.K. Fracture resistance of sunflower seed and kernel to compressive loading. *J. Food Eng.* **2000**, *46*, 1–8. [CrossRef]
34. ASAE. *Method of Determining and Expressing Fineness of Feed Materials by Sieving*; Method S319; American Society of Agricultural and Biological Engineers (ASABE): St. Joseph, MI, USA, 2003.
35. Blahovec, J. *Agromatereials Study Guide*; Czech University of Life Sciences Prague: Prague, Czech Republic, 2008.
36. BS EN ISO 18134-3. *Solid Biofuels—Determination in Moisture Content—Oven Dry Method—Part 3: Moisture in General Analysis Sample*; BSI Standards Publication: Bonn, Germany, 2015; pp. 1–14.
37. ISI. Indian Standard Methods for Analysis of Oilseeds. In *Indian Standard Institute*; IS:3579; ISI: New Delhi, India, 1966.
38. StatSoft Inc. *STATISTICA for Windows*; StatSoft Inc.: Tulsa, OK, USA, 2013.
39. MathSoft Inc. *Parametric Technology Corporation*; MathSoft Inc.: Needham, MA, USA, 2014.
40. Herak, D.; Kabutey, A.; Divisova, M.; Simanjuntak, S. Mathematical model of mechanical behaviour of Jatropha curcas L. seeds under compression loading. *Biosyst. Eng.* **2013**, *114*, 279–288. [CrossRef]
41. Kabutey, A.; Herak, D.; Ambarita, H.; Sigalingging, R. Modeling of linear and non-linear compression processes of sunflower bulk oilseeds. *Energies* **2019**, *12*, 2999. [CrossRef]
42. Sigalingging, R.; Herak, D.; Kabutey, A.; Dajbych, O.; Hrabe, P.; Mizera, C. Application of a tangent curve mathematical model for analysis of the mechanical behaviour of sunflower bulk seeds. *Int. Agrophys.* **2015**, *29*, 517–524. [CrossRef]
43. Marquardt, D.W. An algorithm for the least-squares estimation of nonlinear parameters, SIAM. *J. Appl. Math.* **1963**, *11*, 431–441.
44. Kaewwinud, N.; Khokhajaikiat, P.; Boonma, A. Effect of biomass characteristics on durability of cassava stalk residues pellets. *Res. Agric. Eng.* **2018**, *64*, 15–19.
45. Tumuluru, J.S. Effect of process variables on the density and durability of the pellets made from high moisture corn stover. *Biosysts. Eng.* **2014**, *119*, 44–57. [CrossRef]

© 2020 by the authors. Licensee MDPI, Basel, Switzerland. This article is an open access article distributed under the terms and conditions of the Creative Commons Attribution (CC BY) license (http://creativecommons.org/licenses/by/4.0/).

Article

Valorization of Bio-Briquette Fuel by Using Spent Coffee Ground as an External Additive

Anna Brunerová [1,*], Hynek Roubík [2], Milan Brožek [1], Agus Haryanto [3], Udin Hasanudin [4], Dewi Agustina Iryani [5] and David Herák [6]

[1] Department of Material Science and Manufacturing Technology, Faculty of Engineering, Czech University of Life Sciences Prague, Kamýcká 129, 16500 Prague, Czech Republic; brozek@tf.czu.cz
[2] Department of Sustainable Technologies, Faculty of Tropical AgriSciences, Czech University of Life Sciences Prague, Kamýcká 129, 16500 Prague, Czech Republic; roubik@ftz.czu.cz
[3] Department of Agriculture Engineering, Faculty of Agriculture, University of Lampung, Jl. Sumantri Brojonegoro 1, Bandar Lampung 35145, Republic of Indonesia; agus.haryanto@fp.unila.ac.id
[4] Department of Agro-industrial Technology, Faculty of Agriculture, University of Lampung, Jl. Sumantri Brojonegoro 1, Bandar Lampung 35145, Republic of Indonesia; udinha@fp.unila.ac.id
[5] Department of Chemical Engineering, Engineering Faculty, University of Lampung, Jl. Sumantri Brojonegoro 1, Bandar Lampung 35145, Republic of Indonesia; dewi.agustina@eng.unila.ac.id
[6] Department of Mechanical Engineering, Faculty of Engineering, Czech University of Life Sciences Prague, Kamýcká 129, 16500 Prague, Czech Republic; herak@tf.czu.cz
* Correspondence: brunerova@tf.czu.cz; Tel.: +420-737-077-949

Received: 8 November 2019; Accepted: 15 December 2019; Published: 20 December 2019

Abstract: The present study investigates the quality changes of wood bio-briquette fuel after the addition of spent coffee ground (SCG) into the initial feedstock materials (sawdust, shavings) in different mass ratios (1:1, 1:3). Analysis of SCGs fuel parameter proved great potential for energy generation by a process of direct combustion. Namely, level of calorific value ($GCV = 21.58$ MJ·kg^{-1}), of ash content ($Ac = 1.49\%$) and elementary composition ($C = 55.49\%$, $H = 7.07\%$, $N = 2.38\%$, $O = 33.41\%$) supports such statement. A comparison with results of initial feedstock materials exhibited better results of SCG in case of its calorific value and elementary composition. Bulk density ρ (kg·m^{-3}) and mechanical durability DU (%) of bio-briquette samples from initial feedstock materials were following for sawdust: $\rho = 1026.39$ kg·m^{-3}, $DU = 98.44\%$ and shavings: $\rho = 1036.53$ kg·m^{-3}, $DU = 96.70\%$. The level of such mechanical quality indicators changed after the addition of SCG. Specifically, SCG+sawdust mixtures achieved $\rho = 1077.49$ kg·m^{-3} and $DU = 90.09\%$, while SCG + shavings mixtures achieved $\rho = 899.44$ kg·m^{-3} and $DU = 46.50\%$. The addition of SCG increased wood bio-briquettes energy potential but decreased its mechanical quality. Consequently, the addition of SCG in wood bio-briquette has advantages, but its mass ratio plays an important key role.

Keywords: solid biofuel; waste management; *Coffea* spp.; waste biomass; calorific value; mechanical durability

1. Introduction

A group of plants called *Coffea* L. (*Rubiaceae* family) bears the coffee cherries and contains more than 70 specific species. Nevertheless, only two of them are purposely cultivated as agriculture crops, namely, *Coffea arabica* (75% of the world's production) and *Coffea canephora* (syn. *Coffea robusta*) (25% of the world's production) [1,2]. Coffee beverages have been produced and consumed for more than 1000 years. Currently, 400 billion cups of coffee are consumed every year [3]. Consequently, coffee represents one of the most valuable commodities in the world and the second-largest traded commodity after petroleum [4]. Its production has a significant influence on international relationships, economics, politics, and the trade of many developing countries. The coffee production industry, i.e., plant cultivation, cherries harvest, bean processing, product packaging, sale marketing, and final product transportation, offers job opportunities for millions of people [5].

Brazil belongs to the top countries in coffee production, as well as Vietnam, Indonesia, and Colombia. Together those countries generate more than 50% of the world's coffee production. Specific statistical data provided by the Food and Agriculture Organization of the United Nations (FAO) and by the International Coffee Organization (ICO) related to the coffee industry in the last years are expressed in Table 1.

Table 1. Worldwide coffee production in years 2014–2017 [6,7].

Year	Harvested Area (Ha)	Yield (Hg·Ha^{-1})	Green Bean Amount (Tons)	Coffee Production (In Thousands 60 kg Bags)
2014	10,517,049	8367	8,800,137	154,066
2015	10,951,718	8102	8,872,748	148,559
2016	10,951,718	8594	9,319,855	153,561
2017	10,840,130	8498	9,212,169	159,047

As shown above, the coffee industry contributes to the global market a great deal. Unfortunately, the inevitable result of such large-scale coffee production generates large quantities of agriculture residuals (in liquid and solid form), which results in serious environmental pollution. Those are produced mainly during the treatment of the coffee beans (coffee cherries skin, pulp, husk), as well as the coffee beverage preparation itself, specifically, a spent coffee ground (SCG) [4,5].

SCG can be generated in small-scale within the individuals or small gastronomy units, but also on a large-scale within the manufactories of the coffee industry. Reports displayed that 1 kg of produced coffee beans in a large-scale industry offers approximately 400 g of instant coffee and the rest of the material (600 g) represents the SCG [8]. Fortunately, large-scale manufactories have developed awareness about waste management of their own residues, invest into residues subsequent reusing, thus, adapt the functioning of the manufactory processes within such an idea. Knowledge about proper waste management leads to the awareness, that the residues are not waste materials but raw commodity which can be valorized and can cause manufactory's economical increase within the fuel and energy production issue [5].

Focused on the SCG, its subsequent reusing within the environmental life cycle (small-scale production) or within the economic savings as a replacement of purchased fuel (large-scale production) represent its specific treatment. Prior investigations have proven the SCGs contain specific degradable organic materials that are hardly efficient. Such degradation results in the consumption of a great amount of oxygen. Thus, it is highly inadequate to discharge it into landfills due to its putrefaction [9,10]. Such knowledge also supports the idea of SCG sustainable treatment necessity within the proper waste management issue.

Previous studies have reported several suitable treatment methods of SCG within its subsequent purpose utilization. SCG can be converted into biofuels of different forms (liquid, solid, gaseous) as biodiesel, bioether, biochar, bio-oil, or biogas by using of advanced biotechnological and chemical treatment processes [11,12]. Its utilization for production of other value-added products, such as H_2 or ethanol, was also reported with satisfactory results due to its high content of residual oil (approximately 15%) [13,14]. Moreover, SCG is a valuable resource of fatty acids, polyphenols, amino acids, polysaccharides and minerals suitable for further utilization [15]. Several studies also dealt with the suitability of SCG as an animal feed. Nevertheless, regarding its high content of residual oil, caffeine, and lignin (approximately 25%) such investigations did not prove suitable results [5,8,14,16–19].

SCG can compete with other agro-industrial residues while used as a heating fuel in the industrial boilers due to its high heating potential related to its high content of residual oil [20,21]. On the contrary, the combustion suitability of SCG was discussed by other authors because of its negative effect on the air quality [22].

Within the solid biofuel production, the high content of lignin in SCG indicates its advantage within the densification process because lignin is a natural binder. Such an advantage was investigated in research focused on the production of pellets from SCG mixed with wood sawdust [14,23,24].

Regarding the available literature review, the main aim of performed investigations was to state the suitability of SCG to produce bio-briquette fuel in large-scale within its sustainable and environmentally friendly valorization. The investigation using high-pressure briquetting press simulates the large-scale production of bio-briquette fuel in the commercial sector and reflects the practice. Investigation using SCG and commercial conditions of bio-briquette fuel production is nowadays missing. To achieve such a major aim, several minor aims were developed. Primarily, a chemical analysis of SCG was investigated to determine its suitability for the process of direct combustion and energy generation. Secondary, the SCG was used as a feedstock material to produce solid biofuels, specifically, of bio-briquettes. Within the SCG solid biofuel production issue, two main topics were investigated:

1. How does SCG influence the energy potential of produced bio-briquette samples?
2. How does SCG influence the mechanical quality of produced bio-briquette samples?

2. Materials and Methods

The present chapter describes all characteristics of performed research. It starts with the investigated waste materials, their origin and parameters, followed by the description of used chemical experimental measurements, up to the bio-briquette samples production and testing procedures.

2.1. Investigated Materials

Even though present research was focused mainly on the utilization of SCG (sample A), two other waste materials were also investigated, i.e., larch sawdust (sample B) and spruce shavings (sample C). Extension of research by those two materials was performed as a response to the inconveniences caused by SCG behavior during the densification process (bio-briquette production), which are explained in the further sub-chapter "Bio-briquette samples".

The initial form of chosen materials before experimental measurements in their initial unprocessed form is expressed in Figure 1.

Figure 1. Feedstock samples: (**a**) spent coffee ground (sample A); (**b**) larch sawdust (sample B); (**c**) spruce shavings (sample C).

A microscopic analysis of particle size and shape of chosen materials was investigated within their visible disparity, which in practice results in the heterogeneity of the further created mixtures. Measurements were performed by using the stereoscopic microscope Arsenal, Type 347 SZP 11-T Zoom (Prague, Czech Republic) with a measurement scale of 1 mm and 5 mm; the image analysis is visible in Figure 2.

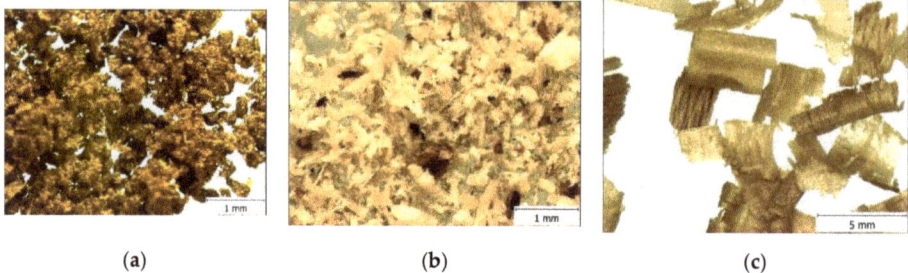

Figure 2. Microscopic analysis of feedstock samples: (**a**) coffee ground; (**b**) larch sawdust; (**c**) spruce shavings.

Worth to be mentioned, that all materials occurred in a form suitable for the densification process regarding their particle size, thus, the milling or crushing processes were not required. It represents a great advantage due to the reduction of electricity demands of such bio-briquette fuel production. On the contrary, the drying process was necessary.

2.2. Fuel Parameters

Chemical composition and energy potential of investigated materials represent important information within the statement of their suitability for energy generation by the process of direct combustion or possibly, for a different sustainable purpose (as was mentioned in the "Introduction" chapter). In total, three different waste materials were subjected to a set of tests, namely, two kinds of wood waste biomass-larch sawdust (sample B) and spruce shavings (sample C) and spent coffee ground (sample A), which represents fruit waste biomass.

The set of performed experimental measurements contained a determination of moisture content Mc (%) and ash content Ac (%) by using thermogravimetric analyzer LECO, type TGA 701 (Saint Joseph, United States). Further, the determination of gross calorific value GCV (MJ·kg^{-1}) by using of isoperibol calorimeter LECO, type AC 600 (Saint Joseph, United States) was performed, while result values of net calorific value NCV (MJ·kg^{-1}) were calculated. Finally, the results of elementary composition as Carbon C (%), Hydrogen H (%), Nitrogen N (%) were carried out by laboratory instrument LECO, type CHN628 + S (Saint Joseph, United States), which uses helium as a carrier gas. The content of Oxygen O (%) was expressed as a difference from the total sum of previously measured elements and ash (in a dry state). All measurements were repeated until the difference between observed results values correspond to the requirements of the standard. The methodology of performed experimental measurements fully followed the instructions of applied mandatory technical standards, see Table 2.

Table 2. List of used standards within material samples chemical composition.

Number	Name	Year
EN ISO 18125	Solid Biofuels-Determination of Calorific Value	2017
EN ISO 16948	Solid Biofuels—Determination of Total Content of Carbon, Hydrogen and Nitrogen	2016
ISO 18122	Solid Biofuels—Determination of Ash Content	2015
EN 18134-2	Solid Biofuels—Determination of Moisture Content—Oven Dry Method—Part 2: Total Moisture—Simplified Method	2015
ISO 1928	Solid Mineral Fuels—Determination of Gross Calorific Value by the Bomb Calorimetric Method, and Calculation of Net Calorific Value	2010

2.3. Bio-Briquette Samples Production

After performed chemical analysis the investigated materials were used as a feedstock for bio-briquette fuel production. Primarily, the materials were compared with the requirements on the feedstock materials for solid biofuel production, specific standards are noted in Table 3.

Table 3. List of used standards within feedstock materials requirements.

Number	Name	Year
EN ISO 17225-1	Solid Biofuels—Fuel Specifications and Classes—Part 1: General Requirements	2015
EN ISO 16559	Solid Biofuels—Terminology, Definitions and Descriptions	2014

Secondary, when the materials were evaluated as a suitable feedstock (waste biomass) for solid biofuel production, they were used for the actual densification process. For bio-briquette samples production a high-pressure hydraulic briquetting press Briklis, type BrikStar 30-12 (shown in Figure 3) (Malšice city, Czech Republic) was used, which works with a piston as a pressing unit. Used briquetting press operates automatically, thus, ensures similar bulk density ϱ (kg·m^{-3}) of produced bio-briquette samples. Produced bio-briquette samples were cylindrically shaped with diameters of 50 mm due to the shape and size of briquetting press die matrix.

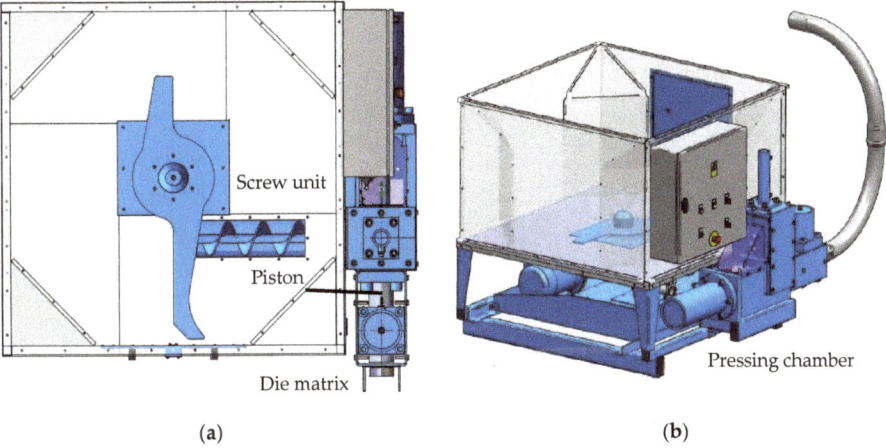

Figure 3. Scheme of used high-pressure hydraulic briquetting press: (**a**) top view; (**b**) side view.

A technical specification of used briquetting press is noted in Table 4 within the better understanding of the impact of such equipment use for bio-briquette in large-scale commercial production.

In the first step, 100% A sample (see Table 4) was used as a feedstock. Further, larch sawdust and spruce shavings have been involved in the investigation and been mixed with the SCG in different mass ratios (1:1, 3:1) to avoid another unsuccessful experiment. Created mixtures represented mixed biomass when one of the materials should be considered as an additive. Using additives within the mixed biomass bio-briquette fuel is common practice to increase specific parameters or properties. In the present case, the wood biomass was used to improve the unsuitable behavior of SCG during the densification process. Specifically created mixtures of feedstock materials are described in Table 5.

Table 4. Technical and basic specifications of used briquetting press.

Parameter	Specification
Operation pressure	80–100 MPa
Pressing chamber	Cylindrical
Pressing unit	Piston
Power	Electricity
Power consumption	4.4 kW
Size	2.91 m^3
Weight	780 kg
Productivity	30 kg·h^{-1}
Bio-briquette shape	Cylindrical
Bio-briquette diameter	50 mm

Table 5. Description and identification of feedstock materials mixture types.

Mixture Types	Mass Ratio	Identification
Spent coffee ground	Pure	100% A
Larch sawdust	Pure	100% B
Spent coffee ground + larch sawdust	1:1	50% A + 50% B
Spent coffee ground + larch sawdust	1:3	25% A + 75% B
Spruce shavings	Pure	100% C
Spent coffee ground + spruce shavings	1:1	50% A + 50% C
Spent coffee ground + spruce shavings	1:3	25% A + 75% C

All created mixtures were successfully used for bio-briquette samples production. Six different types which are expressed in Figure 4 were produced.

Figure 4. Produced bio-briquette samples: (a) 100% B, (b) 25% A + 75% B, (c) 50% A + 50% B, (d) 100% C, (e) 25% A + 75% C, (f) 50% A + 50% C.

All bio-briquette samples were produced with the same diameter (approximately 50 mm), but their length and weight differed. After measurements of all sample dimensions, the average values were calculated and are noted in Table 6.

Table 6. Basic technical parameters of produced bio-briquette samples (in wet basis).

m (g)	h (mm)	Ø (mm)
113.76 ± 14.39	53.94 ± 4.19	50.51 ± 0.20

Notes: m-samples weight, h-samples height, Ø-samples diameter, ±-standard deviation.

2.4. Mechanical Quality Indicators

After measurements of basic parameters were bio-briquette samples subjected to the determination of their mechanical quality. Specific indicators were experimentally tested within the statement of the type of bio-briquette samples with the highest mechanical quality. Experimental measurements were performed within the evaluation of the final mechanical quality of investigated bio–briquette samples and the procedures corresponded to the related standards (see Table 7) or were based on knowledge from practice.

Table 7. List of used standards within bio-briquette samples quality testing.

Number	Name	Year
EN 15234-1	Solid Biofuels—Fuel Quality Assurance—Part 1: General Requirements	2011
EN ISO 17831-2	Solid Biofuels—Determination of Mechanical Durability of Pellets and Briquettes—Part 2: Briquettes	2015

Basic dimension parameters of produced bio-briquette samples were used for the calculation of the first important mechanical quality indicator, a Bulk density ϱ (kg·m^{-3}). Such an indicator describes the ability and suitability of the material for the densification process and resulting in the final quality of products. Following formula was used within performed calculations:

$$\rho = \frac{m}{V} \tag{1}$$

ρ-volume density (kg·m^{-3}), m-bio-briquette samples mass (kg), V-bio-briquette samples volume (m^3).

As an important indicator of bio-briquette fuel mechanical quality within the commercial biofuel sale is considered a Mechanical durability DU (%); such an indicator describes the mechanical strength and ability of the bio-briquette fuel to resists the impacts during the handling, transportation or storage. Within the experimental testing were bio-briquette samples subjected to controlled impacts inside of the special electric rotating dust-proof drum equipped with a rectangular steel partition, see Figure 5.

Before and after experimental testing where all samples weighted and final loss of material (abrasion) was calculated by using of following formula:

$$DU = \frac{m_a}{m_e} \cdot 100 \tag{2}$$

DU-mechanical durability (%), m_a-samples weight after testing (g), m_e-samples weight before testing (g).

To simulate the stress of the bio-briquette fuel in practice, the mechanical indicator of Compressive strength σ (N·mm^{-1}) was applied. Such an indicator did not correspond to any technical standards but is based on previously published papers about strength of products under the pressure [25–28]. Such an indicator plays an important role in the logistics of solid biofuel transportation and storage when bio-briquette fuel is stored above each other. Basically, a maximum increasing load, which can tested bio-briquette sample absorbs before it disintegrates, was measured [29]. Experimental measurements were performed by using a universal testing machine Labortech, type MP Test 5.050 (Opava, Czech Republic) with force meter KAF-S (range of 0–5000 N, used accuracy 0.1 N). After experimental deformation measurement of maximal load F_{max} (N) were used in the following formula to calculate the result values of Compressive strength σ (N·mm^{-1}):

$$\sigma = \frac{F_{max}}{L} \qquad (3)$$

σ-compressive strength in cleft (N·mm^{-1}), F_{max}-maximal load (N), L-bio-briquette sample length (m).

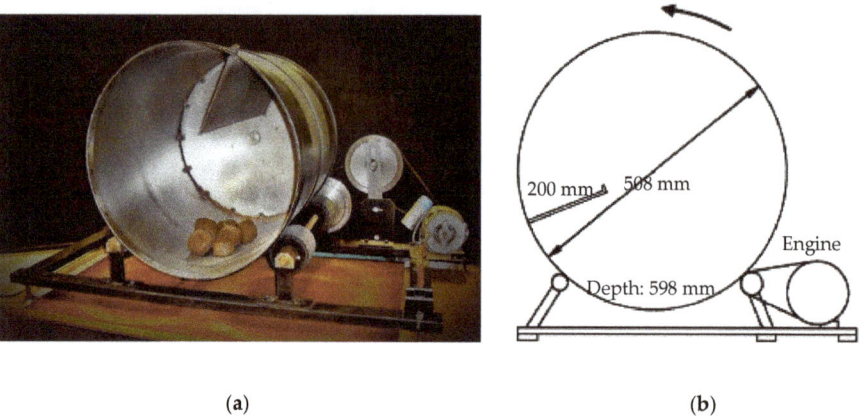

(a)   (b)

Figure 5. Equipment for mechanical durability DU (%) testing: (**a**) in use; (**b**) scheme.

3. Results and Discussion

Present result data corresponds to the investigated issues of feedstock materials parameters and bio-briquette samples quality; consequently, the current chapter is divided in such order as well.

3.1. Fuel Parameters

The first evaluated indicators should primarily describe the suitability of SCG for direct combustion processes within the bio-briquette fuel burning. As Table 8 below describes, the results of the basic chemical parameters of investigated samples were obtained, in both, wet basis and dry basis.

Table 8. Analysis of samples fuel parameters and energy potential (in w.b.).

Biomass Sample	Mc (%)	Ac (%)	GCV (MJ·kg^{-1})	NCV (MJ·kg^{-1})
100% A	9.56 ± 0.15	1.49 ± 0.06	21.58 ± 0:04	19.96
100% B	14.36 ± 0.15	0.43 ± 0.73	17.86 ± 0.03	16.42
100% C	8.25 ± 0.01	0.31 ± 0.16	18.68 ± 0:01	17.27

Notes: w.b.-wet basis, Mc-moisture content, Ac-ash content, GCV-gross calorific value, NCV-net calorific value.

Samples moisture content Mc (%) was stated as a first; results of all samples occurred at a higher level than it is recommended for the bio-briquette fuel production, but the level was still acceptable. Moisture content Mc (%) of properly prepared feedstock material should not exceed 15%. A higher level of moisture content Mc (%) should be considered as a limitation because it results in a lower amount of produced energy (more energy is consumed for vaporizing of moisture during the fuel burning) [30,31]. Nevertheless, such an indicator can be easily improved by the feedstock drying process. Further, an energy potential of samples expressed as a calorific values CV (MJ·kg^{-1}) was investigated. As is visible, calorific value CV (MJ·kg^{-1}) of wood biomass samples occurred at a satisfactory level, typical for wood biomass according to the technical standard EN ISO 17225-1 (2015): Solid Biofuels - Fuel Specifications and Classes-Part 1: General Requirements. On the contrary, desired SCG proved an extremely high level of calorific values CV (MJ·kg^{-1}), which indicated a high potential in energy generation by using combustion processes. Moreover, if the level of moisture content Mc (%) occurred at lower level, the results of calorific values CV (MJ·kg^{-1}) would be even better. Such a positive result was amplified by the observed low level of ash content Ac (%), which is highly appreciated because it indicates the positive behavior of fuel during burning. Moreover, observed ash content Ac (%) was comparable with the ash contents Ac (%) of wood biomass samples, which commonly occurred at low level, but other biomass kinds (herbaceous, fruit, aquatic, mixed) commonly express worst results. Thus, the combination of a high level of calorific values CV (MJ·kg^{-1}) and low level of ash content Ac (%) was evaluated as a significant advantage of SCG samples.

In consequence, SCG represents high-quality feedstock material for bio-briquette production, as well as to produce other types of biofuel intended for energy generation by burning. It indicates that SCG can be used as an additive or as one of feedstock in specific feedstock mixtures within increasing of final mixture calorific value. Such an idea was already investigated in a case of a mixture of SCG with herbaceous biomass (wheat straw) with satisfactory results [32]. As was reported in a different study, addition of SCG in amount of 10% and 25% to beech wood biomass feedstock increased the calorific value CV of final products from initial 18.77 MJ·kg^{-1} to 19.12 MJ·kg^{-1} (10% of SCG) and 20.32 MJ·kg^{-1} (10% of SCG) [33].

For comparison of observed result values with other author's results (sorted from the best result to the worst) were inserted in Table 9 below.

Table 9. Comparison of SCG basic chemical parameters.

Indicator	Result	Reference
Ash content Ac (%)	2.43	[24]
	2.06	[20]
	1.60	[34]
	1.43	[35]
	1.07	[33]
Calorific value CV (MJ·kg^{-1})	19.30	[35]
	21.60	[23]
	22.89	[32]
	23.72–24.07	[36]
	26.00	[37]

The observed high level of calorific value CV (MJ·kg^{-1}) could be caused by the presence of residual oil in the SCG. As the literature reports, the content of residual oil in SCG occurs at the following levels: 13.0% [38], 14.7% [34], 28.3%. Moreover, the calorific value CV (MJ·kg^{-1}) of SCG residual oil occurred at an extremely high level, specifically 36.4 MJ·kg^{-1} [35]. The elementary composition analyses (expressed in Table 10) proved low levels of Oxygen O (%) in the case of SCG, which is required. Result values of both wood biomass materials occurred at a satisfactory level as well; Oxygen O (%) level should occurred around 40%.

Table 10. Analysis of elementary composition in dry basis (in d.b.).

Biomass Sample	C (%)	H (%)	N (%)	O (%)
100% A	55.49	7.07	2.38	33.41
100% B	49.76	6.12	0.10	42.38
100% C	51.08	6.06	0.04	42.48

Notes: C-Carbon, H-Hydrogen, N-Nitrogen, O-Oxygen.

To compare observed data of elementary composition, Table 11 was prepared; noted data originates from other authors' studies and except one value of Nitrogen N (%) (highlighted by bold letters) all values occurred at a similar level as the result value of the present research.

Table 11. Literature review of elementary composition of SCG.

C (%)	H (%)	N (%)	O (%)	Reference
48.67	6.54	2.27	40.03	[24]
52.20	-	2.10	-	[35]
52.50	7.00	3.46	34.80	[39]
53.00	6.80	2.10	38.10	[34]
46.42	6.04	**15.50**	-	[38]
58.50	7.40	1.30	-	[37]
53.05	7.19	1.45	36.20	[20]

Notes: C-Carbon, H-Hydrogen, N-Nitrogen, O-Oxygen.

For a more detailed evaluation of investigated materials chemical analyses where the results were also determined in a dry ash free state, see Table 12, which expresses them in the most exact way. Within this state, the results are expressed without the presence of ash, which can occasionally influence the final result values. Such an influence can be caused by the contamination of samples by dust or external impurities.

Table 12. Analysis of elementary composition and energy potential in a dry ash free state (d.a.f.).

Biomass Sample	C (%)	H (%)	N (%)	O (%)	GCV (MJ·kg^{-1})	NCV (MJ·kg^{-1})
100% A	56.42	7.19	2.42	33.97	24.27	22.71
100% B	50.59	6.22	0.10	43.09	20.08	18.72
100% C	51.26	6.08	0.04	42.62	20.43	19.11

Notes: C-Carbon, H-Hydrogen, N-Nitrogen, O-Oxygen, GCV-gross calorific value, NCV-net calorific value.

3.2. Mechanical Quality

The first practical result, which was observed, was the inability of production of bio-briquette samples from pure SCG, thus, it was concluded that such production is not feasible. Moreover, related to the complications monitored during the briquetting press work, it was not recommended to continue in the procedure due to the high possibility of briquetting press damages. Before such a statement, it must be highlighted that the SCGs were properly prepared for the densification process and fulfilled all requirements (suitable moisture content and particle size), thus, the difficulties were related directly to the characteristics of SCG itself as a reported high content of residual oil.

The second observed result was related to the visual conditions of produced bio-briquette samples. As visible from Figure 6 the homogeneity of samples (as well as feedstock mixture) achieved better results in the case of the A + B mixtures. Such a result was related to the similarity of their particle size. In the case of the A + C mixtures was concluded that the difference in the particle sizes was too significant, thus, the particles could not establish permanent and strong bonds between each other. Such a statement was supported by the observation of particle bonds breaking directly after the bio-briquette samples production, which was reflected as a material loss during sample handling.

(a)   (b)

Figure 6. Visual comparison of produced bio-briquette samples: (a) A + B mixtures; (b) A + C mixtures.

If compare results of tested mechanical quality indicators, all of them proved the higher mechanical quality of bio-briquette samples produced from A + B mixtures. First monitored (calculated) indicator, the bulk density ρ (kg·m^{-3}), proved a satisfactory level of all produced bio-briquette samples if compared with the requirements for commercial sale; ρ should range between 900–1200 kg·m^{-3} [37–39]. Observed data noted in Table 13 provide a clear comparison between all tested bio-briquette samples, while Table 13 provides a comparison between bulk densities ρ (kg·m^{-3}) of bio-briquette fuel produced from different feedstock materials.

Table 13. Mechanical quality indicators of investigated bio-briquette samples.

Biomass Sample	Mc (%)	ρ (kg·m^{-3})	DU (%)	σ (N·mm^{-1})
100% A	-	-	-	-
100% B	13.14 ± 0.68	1026.39 ± 27.08	98.44 ± 0.08	102.78 ± 29.78
50% A + 50% B	11.47 ± 0.42	1112.58 ± 34.83	90.05 ± 1.04	46.07 ± 8.98
25% A + 75% B	13.47 ± 0.17	1042.39 ± 57.86	90.12 ± 0.03	50.85 ± 11.64
100% C	9.2 ± 0.1	1036.53 ± 24.44	96.70 ± 1.00	179.48 ± 24.43
50% A + 50% C	9.9 ± 0.4	956.45 ± 68.40	49.00 ± 0.38	37.09 ± 11.25
25% A + 75% C	10.3 ± 0.1	842.42 ± 69.99	44.00 ± 0.11	31.06 ± 8.87

Notes: Mc-moisture content, ρ-bulk density, DU-mechanical durability, σ-compressive strength, ± -standard deviation.

Further, mechanical durability DU (%) represents the most important quality indicator of bio-briquette fuel, which indicates if the fuel is suitable for commercial production (achievement is mandatory). The lowest acceptable level of mechanical durability is DU > 90%; the next level defining solid biofuel of the highest mechanical durability is DU > 95% [40]. As is visible from Table 12, the bio-briquette samples produced from A + B mixtures achieved the acceptable level and fulfilled mandatory requirements for commercial sale. Bio-briquette samples produced from A + C mixtures exhibited results deeply below the acceptable level of DU (%). Satisfactory results were observed only in the case of 100% C bio-briquette samples. Observed result values can be easily compared with the results of pure wood biomass feedstock samples to evaluate the influence of SCG on the final mechanical quality of samples. A comparison of investigated bio-briquette samples DU (%) with the results of other author's studies is expressed in Table 14.

Moreover, bio-briquette samples produced from A + B mixtures proved a high level of mechanical durability DU (%), despite their higher level of moisture content Mc (%), which is commonly evaluated as a limitation. In general, a high level of moisture content Mc (> 15%) can cause problems during the densification process or within the product's final quality.

The last investigated indicator, the compressive strength σ (N·mm^{-1}), monitored the ability of bio-briquette samples to resist the pressure. In practice, such ability is important during bio-briquette fuel handling and storage when the fuel is packed above each other in multiple layers. Best results were achieved by 100% B and 100% C bio-briquette samples, which was expected. All bio-briquette samples created from feedstock mixtures exhibited worse result, but comparable within the mixture samples. The investigated bio-briquette samples after the deformation testing are expressed in Figure 7.

Compressive strength σ (N·mm^{-1}) of bio-briquette fuel is not stated by any mandatory standards, thus, there are no required levels to achieve. Its evaluation can be performed only by the form of comparison with similar performed measurements, see Table 14.

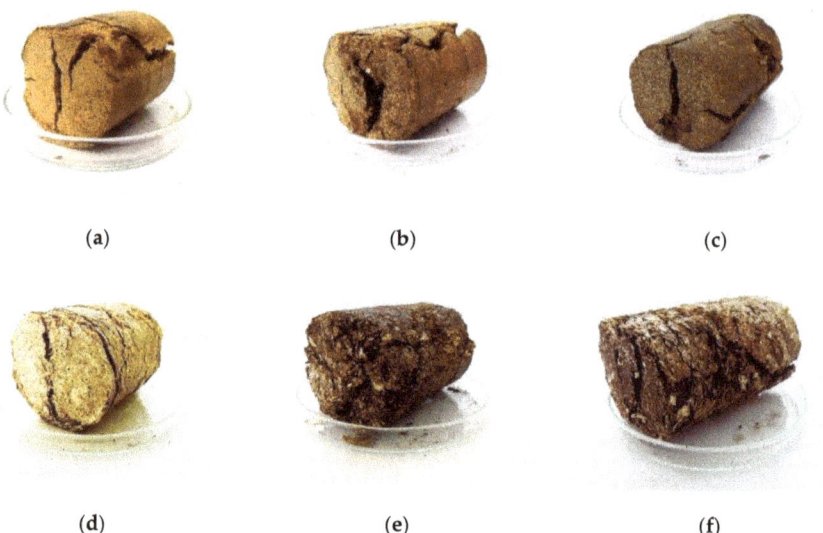

Figure 7. Bio-briquette samples after compressive strength σ (N·mm^{-1}) testing: (**a**) 100% B, (**b**) 25% A + 75% B, (**c**) 50% A + 50% B, (**d**) 100% C, (**e**) 25% A + 75% C, (**f**) 50% A + 50% C.

Table 14. Comparison of reported bio-briquette fuel mechanical quality indicators.

Indicator	Result	Feedstock	Reference
Bulk density ρ (kg·m^{-3})	1,110.00	Coffee pulp	[41]
	986.37	Bamboo fiber	[42]
	964.00	Cotton stalk	[43]
	930.00	Oat straw	[44]
	896.34	Jatoba sawdust	[45]
Mechanical durability DU (%)	98.90	Digestate	[46]
	97.06	Date palm stalks	[47]
	91.40	Energy crop	[48]
	83.46	Wheat straw	[49]
	77.60	Jatoba sawdust	[45]
Compressive strength σ (N·mm^{-1})	176.10	Plane tree chips	[50]
	112.10	Japanese knotweed	[51]
	58.73	Waste paper	[52]
	47.05	Jatoba sawdust	[45]
	32.00	Poppy husk	[53]

As data noted in Table 14 indicates, bio-briquette samples investigated in present research proved a satisfactory level of bulk density ρ (kg·m^{-3}), if compare with other types of bio-briquette fuel. In the case of mechanical durability DU (%) can be indicated that bio-briquette samples from A + B mixtures and 100% C material correspond to the highest level of DU (%) across different biomass bio-briquette fuel. On the contrary, it can be concluded that observed results of bio-briquette samples produced from 100% A and A + C mixtures occurred at a very low level, even if compare with various types of other bio-briquette fuel. Finally, the compressive strength σ (N·mm^{-1}) comparison expressed that 100% B and 100% C occurred at a satisfactory level of such indicator, while all mixed bio-briquette samples occurred at a lower level.

4. Conclusions

In conclusion, performed investigations proved the suitability of SCG for processes of direct combustion within the energy production, however, proved that SCG in pure unmixed form is not usable for the briquetting process. The creation of feedstock separate mixtures of SCG with two different wood biomasses (sawdust, shaving) improved such inappropriate properties of SCG. Better mechanical quality was observed in the case of bio-briquette samples from SCG mixed with wood sawdust due to similar particle size, rather than with wood shavings. It indicates that the addition of SCG into the feedstock mixture influenced the final chemical and mechanical quality of bio-briquette samples in a very expressive way. The amount (mass ratio) of SCG in the feedstock mixtures should be stated carefully; a lower mass ratio of spent coffee ground than 1:1 (50% of SCG) or 1:3 (25% of SCG) should be used in case of mixing with wood biomass. In conclusion, the addition of SCG improved the heating abilities of produced bio-briquette samples, however, decrease their mechanical quality. Such a negative result can be easily improved by the creation of feedstock mixtures with a different mass ratio of SCG.

Author Contributions: Conceptualization, A.B. and M.B.; methodology, A.B. and M.B.; validation, A.B. and M.B.; formal analysis, A.B. and H.R.; investigation, A.B. and M.B.; resources, A.B., H.R. and D.H.; data curation, A.B.; writing—original draft preparation, A.B.; writing—review and editing, H.R., M.B., A.H., U.H., D.A.I., D.H.; visualization, H.R.; supervision, M.B.; project administration, H.R. and D.H.; funding acquisition, H.R. and D.H.". All authors have read and agreed to the published version of the manuscript.

Funding: This research was funded by European Union (EU), managing authority of the Czech Operational Programme Research, Development and Education within the project "Supporting the development of international mobility of research staff at CULS Prague", grant number CZ.02.2.69/0.0/0.0/16_027/0008366. Further, research was supported by the Internal Grant Agency of the Czech University Life Sciences Prague, grant number 20173005 (31140/1313/3108) and by Internal Grant Agency of the Faculty of Engineering, Czech University of Life Sciences Prague, grant number 2019:31140/1312/3103.

Conflicts of Interest: The authors declare no conflict of interest.

References

1. Belitz, H.D.; Grosch, W.; Schieberle, P. Coffee, tea, cocoa. *Food Chem.* **2009**, 938–951. [CrossRef]
2. Etienne, H. Somatic embryogenesis protocol: Coffee (*Coffea arabica* L. and *C. canephora* P.). In *Protocol for Somatic Embryogenesis in Woody Plant*; Springer: Dordrecht, The Netherlands, 2005; pp. 167–168. [CrossRef]
3. Grigg, D. The worlds of tea and coffee: Patterns of consumption. *GeoJournal* **2002**, *57*, 283–294. [CrossRef]
4. Nabais, J.M.V.; Nunes, P.; Carrott, P.J.M.; Carrott, M.R.; García, A.M.; Díez, M.A.D. Production of activated carbons from coffee endocarp by CO2 and steam activation. *Fuel Process. Technol.* **2008**, *89*, 262–268. [CrossRef]
5. Mussatto, S.I.; Dragone, G.; Roberto, I.C. Brewer's spent grain: Generation, characteristics and potential applications. *J. Cereal Sci.* **2006**, *43*, 1–14. [CrossRef]
6. FAO. Food and Agriculture Organization of the United Nations. 2019. Available online: http://www.fao.org/faostat/en/?#data (accessed on 5 March 2019).
7. ICO. International Coffee Organization. 2010. Available online: http://www.ico.org/historical/1990%20onwards/PDF/1a-total-production.pdf (accessed on 5 March 2019).
8. Acevedo, F.; Rubilar, M.; Scheuermann, E.; Cancino, B.; Uquiche, E.; Garcés, M.; Inostroza, K.; Shene, C. Bioactive compounds of spent coffee grounds, a coffee industrial residue. In Proceedings of the III Symposium on 212 A.E. ATABANI ET AL. Agricultural and Agroindustrial Waste Management, Sao Pedro, Brazil, 12–14 March 2013.
9. Corro, G.; Pal, U.; Cebada, S. Enhanced biogas production from coffee pulp through deligninocellulosic photocatalytic pretreatment. *Energy Sci. Eng.* **2014**, *2*, 177–187. [CrossRef]
10. Machado, E.S.M. Reaproveitamento de Resíduos da Indústria do Café Como Matéria-Prima Para a Produção de Etanol. Master's Thesis, Department of Biological Engineering, University of Minho, Braga, Portugal, 2009.
11. Karmee, S.K. A spent coffee grounds based biorefinery for the production of biofuels, biopolymers, antioxidants and biocomposites. *Waste Manag.* **2018**, *72*, 240–254. [CrossRef] [PubMed]

12. Gardy, J.; Rehan, M.; Hassanpour, A.; Lai, X.; Nizami, A.S. Advances in nano-catalysts based biodiesel production from non-food feedstocks. *J. Environ. Manag.* **2019**, *249*, 109316. [CrossRef]
13. Sendzikiene, E.; Makareviciene, V.; Janulis, P.; Kitrys, S. Kinetics of free fatty acids esterification with methanol in the production of biodiesel fuel. *Eur. J. Lipid Sci. Technol.* **2004**, *106*, 831–836. [CrossRef]
14. Kondamudi, N.; Mohapatra, S.K.; Misra, M. Spent coffee grounds as a versatile source of green energy. *J. Agric. Food Chem.* **2008**, *56*, 11757–11760. [CrossRef]
15. Zabaniotou, A.; Kamaterou, P. Food waste valorization advocating Circular Bioeconomy—A critical review of potentialities and perspectives of spent coffee grounds biorefinery. *J. Clean. Prod.* **2019**, *211*, 1553–1566. [CrossRef]
16. Ballesteros, L.; Teixeira, J.; Mussato, S. Chemical, functional, and structural properties of spent coffee grounds and coffee silverskin. *Food Bioprocess Technol.* **2014**, *7*, 3493–3503. [CrossRef]
17. Claude, B. Étude bibliographique: Utilisation dês sous-produits du café. *Café Cacao Thé* **1979**, *23*, 146–152.
18. Givens, D.I.; Barber, W.P. In vivo evaluation of spent coffee grounds as a ruminant feed. *Agric. Wastes* **1986**, *18*, 69–72. [CrossRef]
19. Cruz, G.M. Resíduos de cultura e indústria. *Inf. Agropecuário* **1983**, *9*, 32–37.
20. Kang, S.B.; Oh, H.Y.; Kim, J.J.; Choi, K.S. Characteristics of spent coffee ground as a fuel and combustion test in a small boiler (6.5 kW). *Renew. Energy* **2017**, *113*, 1208–1214. [CrossRef]
21. Silva, M.A.; Nebra, S.A.; Machado Silva, M.J.; Sanchez, C.G. The use of biomass residues in the Brazilian soluble coffee industry. *Biomass Bioenergy* **1998**, *14*, 457–467. [CrossRef]
22. ABNT. *Associação Brasileira de Normas Técnicas, Resíduos Sólidos—Classificação—NBR 10.004*; ABNT: Rio de Janeiro, Brazil, 1987. Available online: https://analiticaqmcresiduos.paginas.ufsc.br/files/2014/07/Nbr-10004-2004-Classificacao-De-Residuos-Solidos.pdf (accessed on 5 August 2019).
23. Haile, M. Integrated volarization of spent coffee grounds to biofuels. *Biofuel Res. J.* **2014**, *2*, 65–69. [CrossRef]
24. Allesina, G.; Pedrazzi, S.; Allegretti, F.; Tartarini, P. Spent coffee grounds as heat source for coffee roasting plants: Experimental validation and case study. *Appl. Therm. Eng.* **2017**, *126*, 730–736. [CrossRef]
25. Okot, D.K.; Bilsborrow, P.E.; Phan, A.N. Effects of operating parameters on maize COB briquette quality. *Biomass Bioenergy* **2018**, *112*, 61–72. [CrossRef]
26. Seifi, M.R. The moisture content effect on some Physical and Mechanical Properties of Corn. *J. Agric. Sci.* **2010**, *2*, 125–134. [CrossRef]
27. Altuntaş, E.; Yıldız, M. Effect of moisture content on some physical and mechanical properties of faba bean (*Vicia faba* L.) Grains. *J. Food Eng.* **2007**, *78*, 174–183. [CrossRef]
28. Yahya, A.; Hamdan, K.; Ishola, T.A.; Suryanto, H. Physical and mechanical properties of *Jatropha curcas* L. fruits from different planting densities. *J. Appl. Sci.* **2013**, *13*, 1004–1012. [CrossRef]
29. Rubio, B.; Izquierdo, M.T.; Segura, E. Effect of binder addition on the mechanical and physicochemical properties of low rank coal char briquettes. *Carbon* **1999**, *37*, 1833–1841. [CrossRef]
30. Hu, J.; Yu, F.; Lu, Y. Application of Fischer-Tropsch synthesis in biomass to liquid conversion. *Catalysts* **2012**, *2*, 303–326. [CrossRef]
31. Din, Z.D.; Zainal, Z.A. Biomass integrated gasification-SOFC systems: Technology overview. *Renew. Sustain. Energy Rev.* **2016**, *53*, 1356–1376. [CrossRef]
32. Sołowiej, P.; Neugebauer, M. Impact of coffee grounds addition on the calorific value of the selected biological materials. *Agric. Eng.* **2016**, *20*, 177–183. [CrossRef]
33. Ciesielczuk, T.; Karwaczyńska, U.; Sporek, M. The possibility of disposing of spent coffee ground with energy recycling. *J. Ecol. Eng.* **2015**, *16*, 133–138. [CrossRef]
34. Somnuk, K.; Eawlex, P.; Prateepchaikul, G. Optimization of coffee oil extraction from spent coffee grounds using four solvents and prototype-scale extraction using circulation process. *Agric. Nat. Resour.* **2017**, *51*, 181–189. [CrossRef]
35. Caetano, N.; Silva, V.; Mata, T.M. Valorization of coffee grounds for biodiesel production. *Chem. Eng. Trans.* **2012**, *26*, 267–272.
36. Zuorro, A.; Lavecchia, R. Spent coffee grounds as a valuable source of phenolic compounds and bioenergy. *J. Clean. Prod.* **2012**, *34*, 49–56. [CrossRef]
37. Pujol, D.; Liu, C.; Gominho, J.; Olivella, M.À.; Fiol, N.; Villaescusa, I.; Pereira, H. The chemical composition of exhausted coffee waste. *Ind. Crop. Prod.* **2013**, *50*, 423–429. [CrossRef]

38. Atabani, A.E.; Mercimeka, S.M.; Arvindnarayan, S.; Shobana, S.; Kumar, G.; Cadir, M.; Al-Muhatseb, A.H. Valorization of spent coffee grounds recycling as a potential alternative fuel resource in Turkey: An experimental study. *J. Air Waste Manag. Assoc.* **2018**, *68*, 196–214. [CrossRef] [PubMed]
39. Tsai, W.T.; Liu, S.C.; Hsieh, C.H. Preparation and fuel properties of biochars from the pyrolysis of exhausted coffee residue. *J. Anal. Appl. Pyrol.* **2012**, *93*, 63–67. [CrossRef]
40. ISO. *Solid Biofuels—Determination of Mechanical Durability of Pellets and Briquettes—Part 2: Briquettes*; ISO: Geneva, Switzerland, 2015.
41. Cubero-Abarca, R.; Moya, R.; Valaret, J.; Filho, M.T. Use of coffee (*Coffea arabica*) pulp for the production of briquettes and pellets for heat generation. *Ciência e Agrotecnologia* **2014**, *38*, 461–470. [CrossRef]
42. Brunerová, A.; Roubík, H.; Brožek, M. Bamboo fiber and sugarcane skin as a bio-briquette fuel. *Energies* **2018**, *11*, 2186. [CrossRef]
43. Karunanithy, C.; Wang, Y.; Muthukumarappan, K.; Pugalendhi, S. Physiochemical characterization of briquettes made from different feedstocks. *Biotechnol. Res. Int.* **2012**, *2012*, 165202. [CrossRef] [PubMed]
44. Adapa, P.; Tabil, L.; Schoenau, G. Compaction characteristics of barley, canola, oat and wheat straw. *Biosyst. Eng.* **2009**, *104*, 335–344. [CrossRef]
45. Chiteculo, V.; Brunerová, A.; Surový, P.; Brožek, M. Management of Brazilian hardwood species (Jatoba and Garapa) wood waste biomass utilization for energy production purposes. *Agron. Res.* **2018**, *16*, 365–376. [CrossRef]
46. Brunerová, A.; Pecen, J.; Brožek, M.; Ivanova, T. Mechanical durability of briquettes from digestate in different storage conditions. *Agron. Res.* **2016**, *14*, 327–336.
47. Eissa, A.H.A.; Alghannam, A.R.O. Quality characteristics for agriculture residues to produce briquette. *World Acad. Sci. Eng. Technol.* **2013**, *78*, 166–171.
48. Repsa, E.; Kronbergs, E.; Pudans, E. Durability of compacted energy crop biomass. *Eng. Rural Dev.* **2014**, *13*, 436–439.
49. Tumuluru, J.S.; Tabil, L.G.; Song, Y.; Iroba, K.L.; Meda, V. Impact of process conditions on the density and durability of wheat, oat, canola, and barley straw briquettes. *Bioenergy Res.* **2015**, *8*, 388–401. [CrossRef]
50. Brožek, M. The effect of moisture of the raw material on the properties briquettes for energy use. *Acta Universitatis Agriculturae et Silviculturae Mendelianae Brunensis* **2016**, *64*, 1453–1458. [CrossRef]
51. Brunerová, A.; Müller, M.; Brožek, M. Potential of wild growing Japanese knotweed (*Reynoutria japonica*) for briquette production. In Proceedings of the 16th International Scientific Conference Engineering for Rural Development, Jelgava, Latvia, 24–26 May 2017; pp. 561–568.
52. Brožek, M. Evaluation of selected properties of briquettes from recovered paper and board. *Res. Agric. Eng.* **2015**, *61*, 66–71. [CrossRef]
53. Brunerová, A.; Brožek, M.; Müller, M. Utilization of waste biomass from post—harvest lines in the form of briquettes for energy production. *Agron. Res.* **2017**, *15*, 344–358.

© 2019 by the authors. Licensee MDPI, Basel, Switzerland. This article is an open access article distributed under the terms and conditions of the Creative Commons Attribution (CC BY) license (http://creativecommons.org/licenses/by/4.0/).

Article

Methane Yield Potential of Miscanthus (*Miscanthus* × *giganteus* (Greef et Deuter)) Established under Maize (*Zea mays* L.)

Moritz von Cossel [1],*, Anja Mangold [1], Yasir Iqbal [2] and Iris Lewandowski [1]

[1] Department of Biobased Products and Energy Crops (340b), Institute of Crop Science, University of Hohenheim, Fruwirthstr. 23, 70599 Stuttgart, Germany; anja.mangold@uni-hohenheim.de (A.M.); iris_lewandowski@uni-hohenheim.de (I.L.)
[2] College of Bioscience and Biotechnology, Hunan Agricultural University, Changsha 410128, China; yasir.iqbal1986@gmail.com
* Correspondence: moritz.cossel@uni-hohenheim.de; Tel.: +49-711-459-23557

Received: 11 November 2019; Accepted: 2 December 2019; Published: 9 December 2019

Abstract: This study reports on the effects of two rhizome-based establishment procedures 'miscanthus under maize' (MUM) and 'reference' (REF) on the methane yield per hectare (MYH) of miscanthus in a field trial in southwest Germany. The dry matter yield (DMY) of aboveground biomass was determined each year in autumn over four years (2016–2019). A biogas batch experiment and a fiber analysis were conducted using plant samples from 2016–2018. Overall, MUM outperformed REF due to a high MYH of maize in 2016 (7211 m^3_N CH_4 ha^{-1}). The MYH of miscanthus in MUM was significantly lower compared to REF in 2016 and 2017 due to a lower DMY. Earlier maturation of miscanthus in MUM caused higher ash and lignin contents compared with REF. However, the mean substrate-specific methane yield of miscanthus was similar across the treatments (281.2 and 276.2 l_N kg^{-1} volatile solid^{-1}). Non-significant differences in MYH 2018 (1624 and 1957 m^3_N CH_4 ha^{-1}) and in DMY 2019 (15.6 and 21.7 Mg ha^{-1}) between MUM and REF indicate, that MUM recovered from biotic and abiotic stress during 2016. Consequently, MUM could be a promising approach to close the methane yield gap of miscanthus cultivation in the first year of establishment.

Keywords: biogas; biomass; cropping system; establishment; intercropping; low-input; maize; miscanthus; methane yield; perennial crop

---

1. Introduction

Miscanthus (*Miscanthus* spp.) is a fast growing perennial C4-grass [1], which has the potential to deliver high biomass yields and to grow on marginal agricultural land [2–5]. A wide range of miscanthus genotypes have been screened for different marginality factors such as salinity [6] and erosion [7]. Miscanthus biomass has quality characteristics that allow it to be used to manifold ways: as a combustion fuel, [8–10], bioethanol [11–13], bedding material [10,14,15], building material [16–20] and in biogas production [21,22]. For example, low inorganic constituents and high lignin content is preferred for combustion [23], whereas low lignin is required for efficient biogas [24] as well as ethanol production [25]. Miscanthus can also be a feedstock to processes including pyrolysis that can produce hydrocarbon fuels such as gasoline, diesel and jet fuel [26]. The variety of available genotypes, which have been developed over the years offer the possibility of selecting genotypes with optimal quality characteristics for a specific end use [27]. This study focuses on the use of miscanthus biomass for biogas production as it is considered one of the foremost promising bioenergy pathways [28–33].

Currently, some of the major impeding factors for miscanthus production across Europe are (i) high initial establishment costs, (ii) a lack of harvestable biomass in the first year [34–37] and (iii) a comparatively long crop establishment period [10,38]. Initial establishment costs largely depend on the establishment procedure. Over the years, different establishment procedures such as rhizome plantation [39–41], micro-propagation [42], direct seed sowing [37,39] or the use of plantlets obtained from stems or rhizomes have been tested to optimize the establishment procedure [42]. The adoption of a certain method is largely dependent on initial cost and its compatibility with the existing farming system especially in terms of farm machinery. Currently, direct plantation of rhizomes, which is inexpensive, is the mostly widely practiced establishment procedure [42]. However, it does not fit well with existing agricultural mechanization and requires specific machinery [10]. The adapted plantation method not only influences the initial cost but also impacts crop development especially during crop establishment period [34,37,42–44]. For example, vegetatively (via rhizomes) propagated miscanthus developed better canopies during the establishment period compared to rhizome based plantation [45].

Over time, the development of new machinery and new planting techniques may facilitate miscanthus cultivation and contribute towards reducing the initial establishment cost [42]. However, the absence of harvestable biomass during the first year complemented by rather high initial establishment cost aggravates the issue of economic viability of the crop during establishment period, which is one reason why farmers are reluctant to cultivate miscanthus. Consequently, there is need to identify innovative solutions for an optimized establishment of miscanthus which will make the crop more economically viable especially during the establishment period.

This study explores the potential effects of a recently developed miscanthus establishment procedure 'miscanthus under maize' (*Zea mays* L.; MUM) [34] on both methane yield per hectare (MYH) and fiber composition of miscanthus during the establishment period. It is expected that there will be a trade-off between the achievement of high MYH of the intercropped plant stand (maize and miscanthus) in MUM in the first year and the achievement of high MYH of miscanthus from the second year onwards. This assumption is based on higher biotic (intercropping competition) and abiotic stress (e.g., drought) in the first year of establishment of miscanthus, which can significantly influence its morphological development and thus its suitability as a biogas substrate in the following years [46,47].

2. Materials and Methods

This section reported on where the plant material was collected, how the plant samples were prepared and analyzed and how the results were evaluated. The major focus was on the fiber analyses and the biogas batch-experiments. Here, however, only basic information about the origin of the plant material was presented. For detailed information on the field trial, such as soil type, plant material and cultivation technique, please refer to Von Cossel et al. [34].

2.1. Origin of Plant Material

The plant material was taken from a field trial with randomized block design (three replicates per treatment) located in Hohenheim (southwest Germany). The field trial was established in 2016 (Figure 1) and has run continuously until the present. In this field trial, two miscanthus establishment procedures were tested: sole establishment (REF) and MUM. For miscanthus, rhizome-based plantlets of *Miscanthus* × *giganteus* (Greef et Deuter) were used.

Figure 1. Plant stand of miscanthus (*Miscanthus* × *giganteus* Greef et Deuter) (1) established under maize (*Zea mays* L.) (2) in July 2016.

2.2. Fibre Analyses, Determination of C- and N-Content and Biogas Batch Experiment

The plant material was dried to constant weight at 60 °C to determine the dry matter content (DMC), which was used to calculate the DMY (Equation (1)). Afterwards, the samples were milled using a cutting mill (SM 200, Retsch, Haan, Germany) with a 1 mm sieve for further analysis. No other pre-treatments, e.g., enzymatic hydrolysis, were applied in the conversion process. The contents of ash, lignin, cellulose, hemicellulose, nitrogen (N) and carbon (C) were analyzed for all samples as follows: The ash content was estimated according to Kiesel and Lewandowski [46]. The contents of lignin, cellulose and hemicellulose were analyzed according to VDLUFA Method Book III, methods 6.5.1, 6.5.2 and 6.5.3 [48]. The contents of N and C were measured according to DIN ISO 5725 using the elemental analyzer 'Vario Max CNS' (Elementar Analysensysteme GmbH, Stuttgart, Germany).

Both biogas batchtest and fiber analysis were conducted according to Von Cossel et al. [49] and Kiesel and Lewandowski [46]. For the batchtest, 200 mg of organic dry matter of the plant samples was mixed with 30 g inoculum (4% DMC, origins from a biogas plant) in 100 ml air-tight bottles and kept at 39 °C for 35 days, a standard procedure according to VDI guideline 4630 [46,49,50]. Within this period, all digestible fractions of the plant samples, such as hemicellulose and cellulose, are to a large extend degraded by microorganisms and converted into biogas, which consists predominantly of CH_4 and CO_2 [46,51]. For each sample, there were four replicates within the batchtest. Gas was collected on the third, the 10th, the 22nd and final day of the batchtest (day 35). The gas production was measured via pressure increase using a hand-held pressure measuring devices for external pressure sensors (HND-P pressure meter, Kobold Messring GmbH, Hofheim, Germany). The frequency of these measurements decreased towards the end of the batchtest, because the biogas production also decreased. Therefore, the pressure increase was measured on a daily basis until day 7, every second day until day 17, and every third day until the end of the batchtest. In total, the pressure increase was measured 19 times during the batchtest. For each of these measurements, the surrounding air pressure was also documented to standardize the values (norm conditions: 0 °C and 1013 hPa). The accumulated substrate-specific biogas yield (SBY) was set in relation to the biogas production of the control (inoculum without plant material) and the daily air pressure of the room in which the batchtest was conducted. The methane content (MC) of the collected biogas was determined using a thermal conductivity detector at a detection temperature of 120 °C (GC-2014 gas chromatograph, Shimadzu, Kyoto). The substrate-specific methane yield (SMY) was calculated following Equation (1):

$$SMY = SBY \times MC. \tag{1}$$

2.3. Dry Matter Yield Determination

The agronomic details are presented and discussed in Von Cossel et al. [34]. In addition to the dry matter yield (DMY) presented in [34], in this study the DMY (green harvest) from the vegetation period 2019 was also determined. Therefore, the DMY was calculated using the fresh matter yield (FMY) and the DMC as follows:

$$DMY = FMY \times DMC. \quad (2)$$

Furthermore, the leaf:stem ratio of miscanthus (10 shoots per field replicate) was measured in 2018 and 2019. However, only plant material from the years 2016–2018 was used for the substrate analyses described in the following section.

2.4. Statistical Analyses

The biogas batch experiment was statistically analyzed as described in Von Cossel et al. [49], whereas outliers were omitted given a coefficient of variation of >5%. The F-tests for the effects of establishment under maize on both SMY and MYH were conducted as described in Von Cossel et al. [49]. The model is shown in Equation (3):

$$y_{ijk} = \mu + b_k + (b\varphi)_{jk} + \tau_i + \varphi_j + (\tau\varphi)_{ij} + e_{ijk} \quad (3)$$

where b_k and $(b\varphi)_{jk}$ are the fixed across year and year-specific effect of the kth the pre-treatment, and μ is the intercept. e_{ijk} is the error of observation y_{ijk} with establishment procedure-specific variance. φ_j, τ_i and $(\tau\varphi)_{ij}$ are the fixed effects for the jth year, the ith establishment procedure and their interaction effects. The influence of factors was tested via a global F test. If differences were found, a multiple t-test was performed to create a letter display [52]. The assumptions of normality and homogeneous error variance were checked graphically. The best model was selected via the Akaike information criterion (AIC) [53]. All analysis run using the PROC MIXED procedure of the SAS® Proprietary Software 9.4 TS level 1M5 (SAS Institute Inc., Cary, NC, USA). For the correlation matrix and SMY prediction, PROC CORR and PROC REG (SAS ® Proprietary Software 9.4 TS level 1M5, see above) were used (see Appendix A, Table A1). Both degrees of freedom and standard errors were approximated using the Kenward–Roger method [54].

3. Results and Discussion

One of the most important results of the field trial underlying this study was the successful establishment in both establishment procedures REF and MUM. This means that in both REF and MUM all plants survived the winter periods during 2016–2019. Across years and treatments, the morphological and physiological characteristics of all observations (Table A2) were in line with current literature [10,55].

3.1. Dry Matter Yield

In both systems significant increase in dry matter yield were observed between one and four after establishments, whereas the total DMY of MUM (including the proportion of total DMY of maize in 2016) was significantly higher than that of REF (Figure 2).

The proportion of total DMY of miscanthus in MUM was significantly lower than in REF in 2016 and 2017 [34]. In the later years 2018 and 2019, however, there were no significant differences between REF and MUM [34]. This is in line with a finding from a recent study on intercropping miscanthus and legumes, in which similar effects were reported [47]. However, the potentially higher yielding variant REF was water limited in 2018 due to summer drought [34]. Therefore, a significantly higher dry matter yield could have been expected for REF than for MUM under normal precipitation conditions (>700 mm yr^{-1}) in 2018 (Figure 2). The underlying agronomic aspects of this observation are further described and discussed in detail in Von Cossel et al. [34]. Another aspect that could be of great importance in the context of the expansion of miscanthus cultivation in the future is the susceptibility of miscanthus to the Barley Yellow Dwarf Virus (BYDV). BYDV can be transmitted to miscanthus by the corn leaf aphid (*Rhopalosiphum maidis* Fitch) [56]. According to Hugget et al. [56], an expansion of the cultivation of miscanthus could lead to a further spread of the BYDV, which would have to be taken into account in the plant protection management of winter cereals. This has already been observed in France in the course of the spread of maize (also a host crop for the BYDV) [56]. However, BYDV will spread less strongly in miscanthus plant stands harvested in autumn (for biogas production) than in miscanthus plant stands harvested in winter (winter harvest for combustion and other utilization pathways) [56]. This is because miscanthus, which is only harvested in winter, can serve as an intermediate host for the corn leaf aphid before they can infest the winter cereals [56]. However, in the following sections, aspects of DMY formation and the expansion of miscanthus cultivation are not further discussed, as the present study focuses on the effects of establishment procedures on the biogas substrate properties of miscanthus.

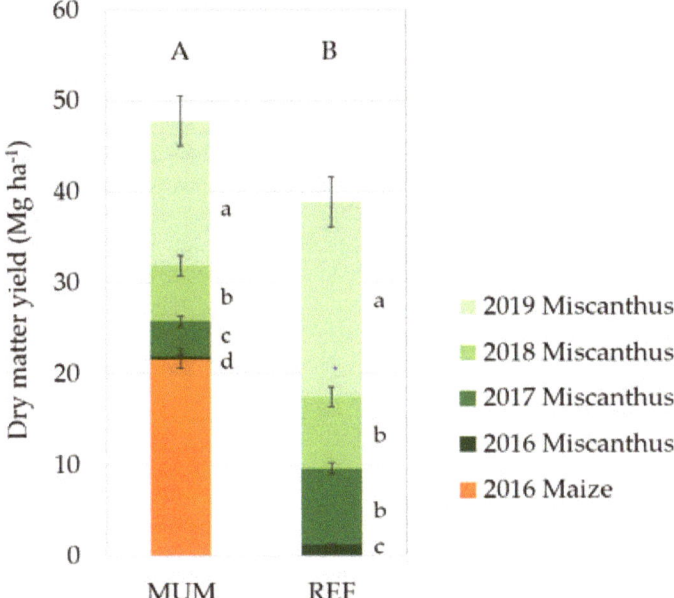

Figure 2. Stacked annual dry matter yields (DMY) of miscanthus (*Miscanthus* × *giganteus* Greef et Deuter) established under maize (*Zea mays* L.; MUM) and alone (REF) during 2016 and 2019. For MUM, the proportion of total annual DMY of maize in 2016 was also included. Different upper case letters denote for significant differences between the four-year accumulated DMY of MUM and REF; different lower case letters denote for significant differences between annual DMY within treatments.

3.2. Methane Yield Potential

It was shown, that the accumulated MYH of a miscanthus biomass production system could be significantly improved during the first years of establishment through establishing miscanthus under maize. This was mainly caused by the high maize MYH in 2016 (Table A3). The MYH of MUM during the years 2016 and 2017 was significantly lower than REF due to an earlier maturation of the miscanthus plants in MUM and better development of miscanthus in REF. This could be explained by a slower development of the miscanthus plantlets in MUM starting from 2016 presumably due to both biotic (intercropping conditions) and abiotic (drought) stress conditions. Von Cossel et al. [34] point out that it must be carefully examined whether MUM starts catching up with REF table onwards. 2018 was a challenging year for both MUM and REF due to low precipitation. For REF, the water limitation was even more severe because the biophysical yield gap (the difference between potentially realizable and actual yield) is higher in REF than in MUM. Abiotic stress, especially drought is critical for defining the biogas substrate quality (BSQ) of miscanthus [46]. Therefore, it can be assumed that the low SMY of both treatments in 2018 (Table 1) was due to changes in biomass composition—especially an increase in lignin content as a response to drought stress. This hypothesis can be supported by the findings of numerous other studies, where lignin content increased under drought stress [57,58]. Furthermore, the results from the stepwise regression analysis show that lignin is the most important regressor for SMY-prediction models (Tables 2 and A1). This is basically in line with the literature [51,59], and a validation of Model 1 (with lignin as sole regressor) based on an external dataset [51] supported the high correlation between lignin content and SMY of miscanthus biomass (Table 2). However, it should still be considered that the plant samples were milled for the biogas batch experiment, which increases the methane yield compared to coarsely chopped material [46]. Therefore, it is generally recommended to evaluate the effects of MUM on the specific methane yield of miscanthus under practical conditions at large scale.

Here it was important to highlight that cellulose content was also highest during the year facing drought stress (2018), whereas hemicellulose content did not change to the same degree. Previous studies showed that (i) the efficiency of bioconversion was significantly influenced by the degree of cellulose crystallinity [60,61] and (ii) hemicellulose content was negatively correlated with cellulose crystallinity [62,63]. In addition, the correlation matrix supports this assumption: both cellulose and hemicellulose significantly correlated with the SMY (Table 3). From this it could be concluded that along with lignin content the ratio of hemicellulose to cellulose was crucial for an efficient bioconversion of miscanthus biomass. The ratio of hemicellulose to cellulose could be optimized to some extent through crop management practices such as adjusting harvesting time. For early green harvest, it was reported that (i) the contents of hemicellulose were higher, and (ii) the contents of cellulose and lignin were lower compared with late green harvest [46,64–66]. Furthermore, at early harvest a high N content is expected in the harvested biomass [8,67], which favors substrate digestion. This was also evident from the correlation matrix, where a highly significant positive correlation between SMY and N content is recorded (Table 3). Due to the earlier maturation, miscanthus biomass of MUM showed significantly lower N contents than REF in 2016. However, starting from 2017 the N contents between both variants were equal. The same applies for the C:N ratio. This indicates that the establishment of miscanthus under maize might not have a lasting effect on N content or the C:N ratio of miscanthus.

It has been reported that a high C:N ratio inhibits the digestion of biomass through production of volatile fatty acids [68–71]. Therefore, early harvest can contribute towards improving the hemicellulose to cellulose and C:N ratio as well, which will subsequently facilitate the bioconversion of biomass. However, it must be considered that the input demand of miscanthus is higher under green harvest regime crop because of poor relocation of nutrients back to rhizomes [67]. This in turn subsequently influences the environmental performance of miscanthus [3,72]. In the case of biogas production, to some extent it could be compensated by recycling nutrients through the application of digested material. Regarding the establishment procedure, this implies that earlier maturation in the first year of establishment increases the C:N ratio to the detriment of BSQ. On one hand, the DM content of miscanthus in the first year of establishment was negligible (Figure 2) and on the other hand earlier maturation had a positive effect on the back-shifting of nutrients (albeit with comparatively low quantitative relevance).

Table 1. Year-specific estimates for qualitative and quantitative traits of the miscanthus biomass in the two establishment systems "miscanthus under maize (MUM)" and "sole establishment of miscanthus (REF)". Different upper case letters denote for significant ($p < 0.05$) differences between establishment procedures within years, lower case letters for significant differences between years within establishment procedures.

Qualitative Parameter	Unit	MUM			REF		
		2016	2017	2018	2016	2017	2018
Substrate-specific methane yield	l_N kg^{-1} volatile solid^{-1}	290.8 ± 1.5 Ba	282.8 ± 1.0 Ab	269.9 ± 5.5 Ab	298.6 ± 1.5 Aa	274.3 ± 1.0 Bb	255.7 ± 5.5 Ac
Lignin	% of dry matter	7.4 ± 0.3 Ab	8.2 ± 0.1 Bb	11.1 ± 0.4 Aa	6.4 ± 0.3 Ac	9.0 ± 0.1 Ab	11.2 ± 0.4 Aa
Cellulose	% of dry matter	36.8 ± 0.3 Ac	40.0 ± 0.7 Ab	49.4 ± 0.8 Aa	32.8 ± 0.3 Bc	41.2 ± 0.7 Ab	48.9 ± 0.8 Aa
Hemicellulose	% of dry matter	28.5 ± 0.4 Ba	26.1 ± 0.4 Ab	27.4 ± 1.2 Aab	32.2 ± 0.4 Aa	27.1 ± 0.3 Ab	27.5 ± 1.2 Ab
Ash	% of dry matter	6.8 ± 0.3 Aa	6.1 ± 0.2 Aa	3.1 ± 0.1 Ab	7.3 ± 0.3 Aa	4.0 ± 0.2 Bb	2.1 ± 0.1 Bc
Carbon (C)	% of dry matter	45.7 ± 0.2 Ab	45.7 ± 0.1 Bb	47.3 ± 0.3 Aa	45.8 ± 0.2 Ab	47.1 ± 0.1 Aa	48.2 ± 0.3 Aa
Nitrogen (N)	% of dry matter	0.9 ± 0.0 Ba	0.5 ± 0.0 Ab	0.4 ± 0.1 Ab	1.3 ± 0.0 Aa	0.4 ± 0.0 Ab	0.4 ± 0.1 Ab
C:N ratio	-	2.0 ± 0.1 Ba	1.1 ± 0.1 Ab	0.7 ± 0.2 Ab	2.8 ± 0.1 Aa	0.9 ± 0.1 Ab	0.4 ± 0.2 Ab
Methane yield per hectare	m^3_N CH$_4$ ha^{-1}	74.1 ± 39.5 Bb	952.5 ± 118.9 Ba	1624.2 ± 285.3 Aa	338.3 ± 39.5 Ab	2256.5 ± 118.9 Aa	1956.5 ± 285.3 Aa

Table 2. Models for predicting the SMY of miscanthus biomass during the years 2016–2018 (* = $p < 0.05$, ** = $p < 0.01$, *** = $p < 0.001$, n.s. = not significant, n.a. = not added to the model).

Regressor	Model 1	Model 2	Model 3
Intercept	342.35 ***	329.32 ***	299.12 ***
Lignin	−7.18 ***	−3.26 **	n.a.
C:N Ratio	n.a.	1033.88 **	n.a.
Hemicellulose	n.a.	−1.24 n.s.	n.a.
Lignin × Hemicellulose	n.a.	n.a.	−0.13 ***
Coefficient of determination (R^2)	0.8261	0.9752	0.9742
Validation [a] (R^2)	0.7881 *	_ [b]	n.s.

[a] Based on miscanthus-specific observations from the supplementary dataset provided by Von Cossel et al. [51]. [b] Not applicable due to missing variables in the supplementary dataset of Von Cossel et al. [51].

Table 3. Correlation matrix (* = $p < 0.05$, ** = $p < 0.01$, *** = $p < 0.001$, n.s. = not significant) of qualitative miscanthus biomass traits.

Trait	Ash	Lignin	Cellulose	Hemicellulose	Carbon	Nitrogen
SMY	0.92 ***	−0.91 ***	−0.88 ***	0.59 *	−0.88 ***	0.92 ***
Ash		−0.95 ***	−0.93 ***	0.53 *	−0.95 ***	0.81 ***
Lignin			0.99 ***	−0.65 **	0.87 ***	−0.85 ***
Cellulose				−0.67 **	0.81 ***	−0.83 ***
Hemicellulose					−0.35 n.s.	0.82 ***
Carbon						−0.66 **

Additionally, the morphological development of miscanthus affects the SMY [73]. This was evident from results where cell wall components (lignin, cellulose and hemicellulose) varied between both stands, though the differences for some components were rather small (Tables 1 and 4). Among morphological traits, leaf:stem ratio is important because the composition of biomass varies depending on plant fraction [73]. For example, high hemicellulose, low lignin and cellulose contents were reported in leaves compared with stems [55] and which is why biomass with high leaf share was easily digestible [53,55]. In addition, the better digestibility of miscanthus biomass with high leaf share is also attributed to lignin structural differences such as lower molecular weight of leaf derived lignin compared with stem derived lignin [56]. Therefore, leaf:stem ratio is critical to determine the BSQ and subsequently bioconversion efficiency. In this study, the morphological development of miscanthus plants was also influenced by prevailing stress conditions, whereby the establishment procedure had no significant effect on the leaf:stem ratio on miscanthus from 2018 onwards (Figure A1). However, during 2018, miscanthus leaves became senescent under drought conditions, which reduced their digestibility [74]. Furthermore, leaf:stem ratio varies from species to species [46,74] and also with time of harvesting [53,56,57]. Therefore, miscanthus genotypes with a higher leaf:stem ratio than *Miscanthus × giganteus* (Greef et Deuter) also provide a better BSQ. It remains unclear whether a longer retention time (>35 d) would have significantly increased the specific methane yield of miscanthus (Figure 3), which could be inferred from the research results of Sonwai et al [75]. However, the SMY values of the present study fit well with those of existing literature [46,64,76], which is why it could be assumed that the retention time of 35 d was sufficient to compare the SMYs of miscanthus from MUM and REF. However, it has been shown that there is a clear trade-off between BSQ and biomass yield [46,77].

Table 4. Three-year average qualitative and accumulated quantitative parameters of the miscanthus plant stands established under maize (MUM) and alone (REF). Different letters denote for significant differences between MUM and REF for those parameters without significant interactions between Establishment procedure × vegetation period (Table 5).

Three-Year Average Qualitative Parameter	Unit	Establishment Procedure	
		MUM	REF
Specific methane yield	l_N kg^{-1} volatile solid^{-1}	281.2 ± 9.8	276.2 ± 9.8
Ash	% of dry matter	5.4 ± 1.3	4.4 ± 1.3
Lignin	% of dry matter	8.9 ± 1.3	8.9 ± 1.3
Cellulose	% of dry matter	41.9 ± 4.3	41.1 ± 4.3
Hemicellulose	% of dry matter	27.7 ± 1.2 b	28.8 ± 1.2 a
Carbon (C)	% of dry matter	46.2 ± 0.6 b	47.0 ± 0.6 a
Nitrogen (N)	% of dry matter	0.6 ± 0.3	0.6 ± 0.3
C:N ratio	-	1.2 ± 0.6	1.4 ± 0.6
Three-Year Accumulated Quantitative Parameter	**Unit**	**MUM**	**REF**
Dry matter yield	Mg ha^{-1}	10.2 ± 6.1	17.5 ± 6.1
Methane yield per hectare	m^3_N CH$_4$ ha^{-1}	2695.8 ± 1565.1	4506.3 ± 1565.1

Table 5. Fixed effects of 'Vegetation period', 'Establishment procedure' and their two-fold interaction on yield and quality parameters of miscanthus as biogas substrate across years (* = $p < 0.05$, ** = $p < 0.01$, *** = $p < 0.001$, n.s. = not significant). The fixed effects of 'Pre-crop 2015' and 'Pre-crop 2015 × vegetation period' were non-significant for all parameters and therefore, not added in the table.

Parameter	Effect		
	Vegetation Period	Establishment Procedure	Establishment Procedure × Vegetation Period
SMY	**	n.s.	**
Lignin	**	n.s.	*
Cellulose	**	n.s.	*
Hemicellulose	**	n.s.	n.s.
Ash	***	n.s.	**
Carbon (C)	*	**	n.s.
Nitrogen (N)	**	n.s.	*
C:N ratio	**	n.s.	*
Dry matter yield per hectare	**	*	*
Methane yield per hectare	**	*	*

So far, the differences in yield and quality parameters of MUM have been shown and discussed only regarding the miscanthus biomass. Thus, the share of maize MYH of the total MYH in 2016 must also be considered for MUM to allow for a more holistic comparison of the long-term effects of the two miscanthus establishment procedures MUM and REF. The total three-year (2016–2018) accumulated MYH of MUM—including both miscanthus and maize—accounted for about 9906 m^3_N CH$_4$ ha$^-$ (Table 1, Table A3). This was approximately two and a half times as much as was reached by REF (about 4506 m^3_N CH$_4$ ha^{-1}; Tables 1 and 4). Consequently, the establishment procedure increased the total MYH of the new establishment procedure MUM compared with REF, even though miscanthus in MUM showed a weaker morphological development of miscanthus (reduced number of shoots, smaller shoots, lower MYH, etc.) compared with REF in the second year after establishment (Tables 1 and 5; Figure 4).

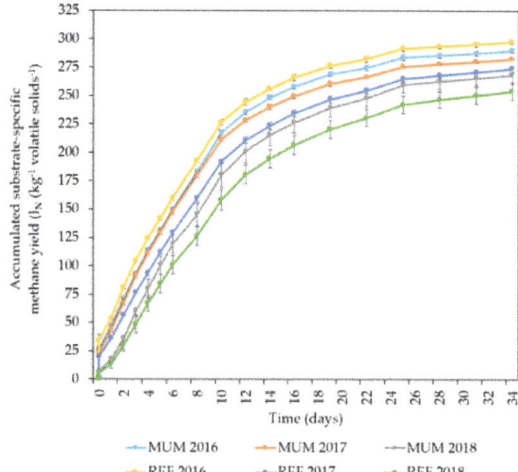

Figure 3. Development of the accumulated substrate-specific methane yields of miscanthus biomass from the two establishment procedures treatments 'miscanthus under maize' (MUM) and 'reference' (REF) during the biogas batch experiments. The error bars show the standard deviation ($n = 6$).

The establishment procedure MUM therefore provides a clear revenue advantage over the other establishment procedure REF within the first three years of establishment. Furthermore, the costs for soil tillage and herbicide measures should be virtually divided by two (maize cultivation and miscanthus cultivation). This reduces the costs for miscanthus establishment. Since the establishment costs commonly account for about a quarter of the total costs for 20 years of miscanthus cultivation [78] a reduction of the establishment costs may help to foster the implementation of miscanthus into existing farming systems.

However, the effects of both biotic (intercropping competition) and abiotic stress (e.g., drought) need to be further investigated with reference to marginal agricultural land utilization. This is because the cultivation of miscanthus should be promoted on marginal agricultural lands (rather than on favorable sites to avoid land use competition with food crop cultivation) [79–81]. The cultivation conditions on marginal agricultural lands can be challenging for miscanthus, which could worsen the recovery success of miscanthus. On the other hand, intercropping maize and miscanthus could reduce certain marginality constraints, such as wind and water erosion. Furthermore, the overall long-term performance of MUM for different end uses of miscanthus biomass should be evaluated in the future. This is because different end uses require different cultivation practices, which could affect the success of MUM in the long term. For example, a brown harvest (in winter) is usually applied for the end uses 'combustion' or 'isobutanol production'. Brown harvest regimes imply a better nutrient translocation to the rhizomes than a green harvest regimes (in autumn) [46]. A better nutrient translocation may help miscanthus plants to recover much better from the stress during the first year of establishment. Hence, the establishment of miscanthus under maize may be even more suitable for the brown harvest regime of miscanthus, and it should therefore be further investigated in the future.

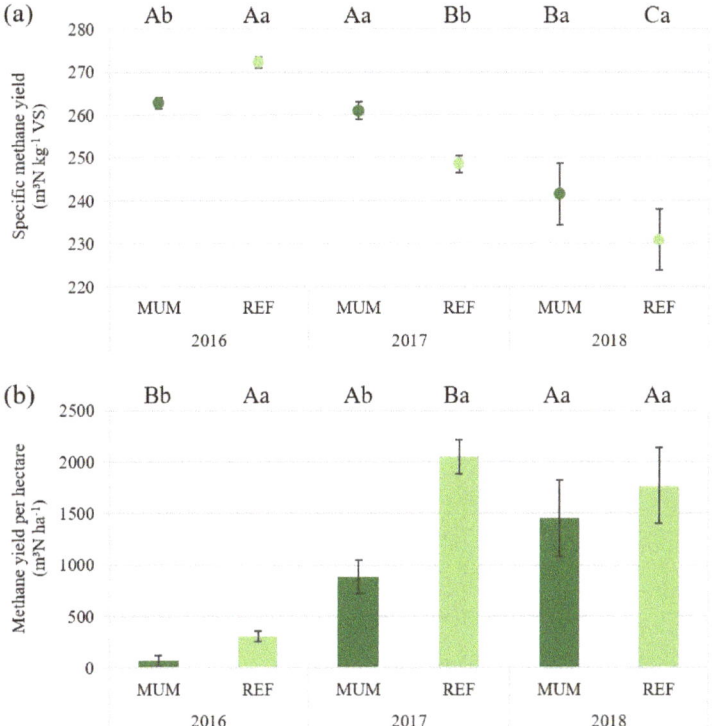

Figure 4. Overview of specific methane yields (**a**) and methane yields per hectare (MYH) of miscanthus biomass from the treatments under maize (MUM) and standard (REF) establishment. In 2016, only the proportion of total MYH of miscanthus is presented for MUM. The proportion of total MYH of maize in MUM 2016 are provided in Table A3. The error bars show the standard errors. Similar capital letters denote for non-significant ($p < 0.05$) differences within treatments between years; similar small letters refer to differences between treatments within years.

4. Conclusions

This study revealed new insights into the effects of a joint establishment of miscanthus (*Miscanthus × giganteus* Greef et Deuter) and maize (MUM) on the overall biogas yield of miscanthus for the establishment period (four years). While intercropping with maize in the first year significantly reduced the biogas yield of miscanthus in the second year after planting, no significant difference between the two establishment variants was observed in the third year after planting. In the fourth year, a non-significant difference in biomass yield indicated that miscanthus recovered from the stress of intercropping in the first year, so that no negative long-term effects on the yield level of miscanthus were to be expected due to the establishment under maize. Moreover, the high biogas yield from the maize proportion in the first year of MUM resulted in a significantly higher total biomass potential within the observation period of four years compared with the conventional establishment variant of miscanthus (REF). From this, it could be concluded that, compared to REF, MUM helped farmers to reduce the costs of miscanthus establishment by providing a first year's revenue from maize biomass.

Author Contributions: Conceptualization, M.v.C.; Data curation, M.v.C.; Formal analysis, M.v.C., A.M. and Y.I.; Funding acquisition, I.L.; Investigation, M.v.C.; Methodology, M.v.C.; Project administration, M.v.C. and I.L.; Resources, M.v.C. and A.M.; Supervision, I.L.; Validation, M.v.C. and Y.I.; Visualization, M.v.C.; Writing—original draft, M.v.C., A.M., Y.I. and I.L.; Writing—review editing, M.v.C.

Funding: This research received funding from the European Union's Horizon 2020 research and innovation program under grant agreement No 727698, and the German Federal Ministry of Education and Research (BMBF), project number: 03EK3525A. The article processing charge was funded by the University of Hohenheim.

Acknowledgments: The authors are thankful to the staff of the Agricultural Experiment Station of the University of Hohenheim for providing technical support for the field trials, and to Dagmar Metzger, Johanna Class, Friederike Selensky and Johannes Schümann for their assistance with laboratory and preparatory work. Special thanks go to Jens Hartung for his assistance with the statistical analysis.

Conflicts of Interest: The authors declare no conflict of interest. The funders had no role in the design of the study; in the collection, analyses, or interpretation of data; in the writing of the manuscript, or in the decision to publish the results.

Appendix A

Table A1. Setup of Models 1–3 for prediction of substrate-specific methane yield. For selection of regressors, stepwise selection ($p < 0.15$) was chosen.

Model	Input Regressors	Selected Regressors
1	Lignin	Lignin
2	Ash, lignin, cellulose, hemicellulose, C, N, C:N ratio	Lignin, C:N ratio, hemicellulose
3	Ash, lignin, cellulose, hemicellulose, C, N, C:N ratio, ash × ash, ash × lignin, ash × cellulose, ash × hemicellulose, ash × C, ash × N; ash × C:N ratio, lignin × lignin, lignin × cellulose, lignin × hemicellulose, lignin × C, lignin × N, lignin × C:N ratio, cellulose × cellulose, cellulose × hemicellulose, cellulose × C, cellulose × N, cellulose × C:N ratio, hemicellulose × hemicellulose, hemicellulose × C, hemicellulose × N, hemicellulose × C:N ratio, C × C, C × N, C × C:N ratio, N × N, N × C:N ratio, C:N ratio × C:N ratio	Lignin × hemicellulose

Table A2. Simple statistics of both quantitative and qualitative traits across establishment procedures.

Parameter	Unit	Mean	Standard Deviation	Minimum	Maximum	n
Number of shoots per plant	-	21.7	12.1	5.0	41.0	18
Dry matter content	% of fresh matter	0.4	0.1	0.3	0.5	18
Dry matter yield [a]	Mg ha^{-1}	8.1	7.3	0.2	26.4	24 [a]
Specific methane yield	l$_N$ kg^{-1} volatile solid^{-1}	278.7	15.1	248.1	301.0	18
Methane yield per hectare	m^3_N ha^{-1}	1200.0	854.6	53.9	2506.0	18
Methane content of biogas produced	%	55.1	0.7	54.0	56.0	18
Ash	% of dry matter	4.9	2.0	1.9	7.7	18
Lignin	% of dry matter	8.9	1.9	6.0	11.9	18
Cellulose	% of dry matter	41.5	6.3	32.7	50.4	18
Hemicellulose	% of dry matter	28.2	2.2	24.7	32.2	18
Carbon	% of dry matter	46.6	1.0	45.3	48.4	18
Nitrogen	% of dry matter	0.7	0.4	0.2	1.3	18

[a] For the dry matter yield, also data from the vegetation period 2019 was available.

Table A3. Yield and quality parameters of maize in MUM 2016.

Parameter	Unit	Value
Dry matter yield	Mg ha^{-1}	21.6 ± 1.0
Dry matter content	% of fresh matter	34.5 ± 1.1
Specific methane yield	l$_N$ kg$^-$ volatile solid^{-1}	333.2 ± 0.5
Methane yield per hectare	m$^3{}_N$ CH$_4$ ha^{-1}	7210.5 ± 348.4

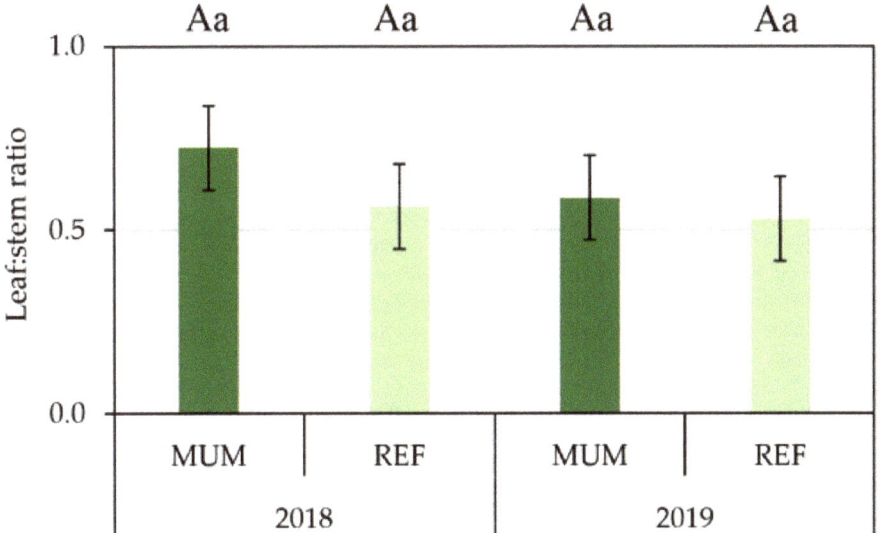

Figure A1. Leaf:stem ratio (and standard error) of miscanthus from different rhizome-based establishment procedures 'under maize' (MUM) and 'alone (REF) in 2018 and 2019 (the planting was conducted in 2016). Similar upper case letters denote for non-significant ($p < 0.15$) differences between treatments within years, lower case letters for non-significant differences between years within treatments.

References

1. Brosse, N.; Dufour, A.; Meng, X.; Sun, Q.; Ragauskas, A. Miscanthus: A fast-growing crop for biofuels and chemicals production. *Biofuels Bioprod. Biorefining* **2012**, *6*, 580–598. [CrossRef]
2. Jeżowski, S.; Mos, M.; Buckby, S.; Cerazy-Waliszewska, J.; Owczarzak, W.; Mocek, A.; Kaczmarek, Z.; McCalmont, J.P. Establishment, Growth, and Yield Potential of the Perennial Grass Miscanthus × Giganteus on Degraded Coal Mine Soils. *Front. Plant Sci.* **2017**, *8*, 726. [CrossRef] [PubMed]
3. Wagner, M.; Mangold, A.; Lask, J.; Petig, E.; Kiesel, A.; Lewandowski, I. Economic and environmental performance of miscanthus cultivated on marginal land for biogas production. *GCB Bioenergy* **2019**, *11*, 34–49. [CrossRef]
4. Von Cossel, M.; Lewandowski, I.; Elbersen, B.; Staritsky, I.; Van Eupen, M.; Iqbal, Y.; Mantel, S.; Scordia, D.; Testa, G.; Cosentino, S.L.; et al. Marginal agricultural land low-input systems for biomass production. *Energies* **2019**, *12*, 3123. [CrossRef]
5. Ramirez-Almeyda, J.; Elbersen, B.; Monti, A.; Staritsky, I.; Panoutsou, C.; Alexopoulou, E.; Schrijver, R.; Elbersen, W. Assessing the Potentials for Nonfood Crops. In *Modeling and Optimization of Biomass Supply Chains*; Elsevier: Amsterdam, The Netherlands, 2017; pp. 219–251.
6. Chen, C.-L.; van der Schoot, H.; Dehghan, S.; Kamei, C.L.A.; Meyer, H.; Schwartz, K.-U.; Visser, R.G.F.; Der Linden, V.; Gerard, C. Genetic diversity of salt tolerance in Miscanthus. *Front. Plant Sci.* **2017**, *8*, 187. [CrossRef]

7. Cosentino, S.L.; Copani, V.; Scalici, G.; Scordia, D.; Testa, G. Soil erosion mitigation by perennial species under Mediterranean environment. *Bioenergy Res.* **2015**, *8*, 1538–1547. [CrossRef]
8. Iqbal, Y.; Kiesel, A.; Wagner, M.; Nunn, C.; Kalinina, O.; Hastings, A.F.S.J.; Clifton-Brown, J.C.; Lewandowski, I. Harvest time optimization for combustion quality of different miscanthus genotypes across Europe. *Front. Plant Sci.* **2017**, *8*. [CrossRef]
9. Lewandowski, I.; Kicherer, A. Combustion quality of biomass: Practical relevance and experiments to modify the biomass quality of Miscanthus x giganteus. *Eur. J. Agron.* **1997**, *6*, 163–177. [CrossRef]
10. Anderson, E.; Arundale, R.; Maughan, M.; Oladeinde, A.; Wycislo, A.; Voigt, T. Growth and agronomy of Miscanthus x giganteus for biomass production. *Biofuels* **2011**, *2*, 71–87. [CrossRef]
11. Kim, S.J.; Um, B.H.; Im, D.J.; Lee, J.H.; Oh, K.K. Combined Ball Milling and Ethanol Organosolv Pretreatment to Improve the Enzymatic Digestibility of Three Types of Herbaceous Biomass. *Energies* **2018**, *11*, 2457. [CrossRef]
12. Kalghatgi, G.; Levinsky, H.; Colket, M. Future transportation fuels. *Prog. Energy Combust. Sci.* **2018**, *69*, 103–105. [CrossRef]
13. Von Cossel, M.; Winkler, B.; Mangold, A.; Lewandowski, I.; Elbersen, B.; Wagner, M.; Magenau, E.; Lask, I.; Staritsky, I.; Van Eupen, M.; et al. Bridging the gap between biofuels and biodiversity for a bioeconomy transition—Monetizing social-ecological effects of miscanthus cultivation for isobutanol production. *Energy Environ. Sci.*. under review.
14. Rauscher, B.; Lewandowski, I. Miscanthus horse bedding compares well to alternatives. In *Perennial Biomass Crops for a Resource-Constrained World*; Springer International Publishing: Cham, Switzerland, 2016; pp. 297–305.
15. Van Weyenberg, S.; Ulens, T.; De Reu, K.; Zwertvaegher, I.; Demeyer, P.; Pluym, L. Feasibility of Miscanthus as alternative bedding for dairy cows. *Vet. Med. (Praha)* **2015**, *60*, 121–132. [CrossRef]
16. Alzagameem, A.; Bergs, M.; Do, X.T.; Klein, S.E.; Rumpf, J.; Larkins, M.; Monakhova, Y.; Pude, R.; Schulze, M. Low-input crops as lignocellulosic feedstock for second-generation biorefineries and the potential of chemometrics in biomass quality control. *Appl. Sci. Switz.* **2019**, *9*, 2252. [CrossRef]
17. El Hage, R.; Khalaf, Y.; Lacoste, C.; Nakhl, M.; Lacroix, P.; Bergeret, A. A flame retarded chitosan binder for insulating miscanthus/recycled textile fibers reinforced biocomposites. *J. Appl. Polym. Sci.* **2019**, *136*. [CrossRef]
18. Eschenhagen, A.; Raj, M.; Rodrigo, N.; Zamora, A.; Labonne, L.; Evon, P.; Welemane, H. Investigation of Miscanthus and Sunflower Stalk Fiber-Reinforced Composites for Insulation Applications. *Adv. Civ. Eng.* **2019**, *2019*. [CrossRef]
19. Pude, R. *Bedeutung Morphologischer, Chemischer und Physikalischer Parameter Sowie ihre Interaktionen zur Beurteilung der Baustoffeignung Unterschiedlicher Miscanthus-Herkünfte*; Bad Neuenahr-Ahrweiler, Wehle: Bad Neuenahr-Ahrweiler, Germany, 2005.
20. Nichtitz, A.; Thiemann, F.; Völkering, G.; Petry, M.; Pude, R. Eignung ausgewählter mehrjähriger Biomassepflanzen für die Produktion von Hochleistungsdämmputz. *Ges Mitt. Ges. Pflanzenbauwiss.* **2016**, *28*, 180–181.
21. Lewandowski, I.; Clifton-Brown, J.; Trindade, L.M.; Van Der Linden, G.; Schwarz, K.-U.; Müller-Sämann, K.; Anisimov, A.; Chen, C.-L.; Dolstra, O.; Donnison, I.S. Progress on optimizing miscanthus biomass production for the European bioeconomy: Results of the EU FP7 project OPTIMISC. *Front. Plant Sci.* **2016**, *7*, 1620. [CrossRef]
22. Baute, K.; Van Eerd, L.L.; Robinson, D.E.; Sikkema, P.H.; Mushtaq, M.; Gilroyed, B.H. Comparing the Biomass Yield and Biogas Potential of Phragmites australis with Miscanthus x giganteus and Panicum virgatum Grown in Canada. *Energies* **2018**, *11*, 2198. [CrossRef]
23. Iqbal, Y.; Lewandowski, I. Inter-annual variation in biomass combustion quality traits over five years in fifteen Miscanthus genotypes in south Germany. *Fuel Process. Technol.* **2014**, *121*, 47–55. [CrossRef]
24. Kiesel, A.; Nunn, C.; Iqbal, Y.; Van der Weijde, T.; Wagner, M.; Özgüven, M.; Tarakanov, I.; Kalinina, O.; Trindade, L.M.; Clifton-Brown, J. Site-specific management of miscanthus genotypes for combustion and anaerobic digestion: A comparison of energy yields. *Front. Plant Sci.* **2017**, *8*. [CrossRef] [PubMed]
25. Van der Weijde, R. Targets and Tools for Optimizing Lignocellulosic Biomass Quality of Miscanthus. Ph.D. Thesis, Wageningen University, Wageningen, The Netherlands, 2016.

26. Wang, Z.; Dunn, J.B.; Wang, M.Q. *Greet Model Miscanthus Parameter Development*; Center for Transportation Research, Argonne National Laboratory: Argonne, IL, USA, 2012.
27. Van der Weijde, T.; Kiesel, A.; Iqbal, Y.; Muylle, H.; Dolstra, O.; Visser, R.G.; Lewandowski, I.; Trindade, L.M. Evaluation of Miscanthus sinensis biomass quality as feedstock for conversion into different bioenergy products. *Gcb Bioenergy* **2016**. [CrossRef]
28. Weiland, P. Biomass Digestion in Agriculture: A Successful Pathway for the Energy Production and Waste Treatment in Germany. *Eng. Life Sci.* **2006**, *6*, 302–309. [CrossRef]
29. Weiland, P. Biogas production: Current state and perspectives. *Appl. Microbiol. Biotechnol.* **2010**, *85*, 849–860. [CrossRef] [PubMed]
30. Scarlat, N.; Dallemand, J.-F.; Fahl, F. Biogas: Developments and perspectives in Europe. *Renew Energy* **2018**, *129*, 457–472. [CrossRef]
31. Daniel-Gromke, J.; Rensberg, N.; Denysenko, V.; Stinner, W.; Schmalfu, S.S.T.; Scheftelowitz, M.; Nelles, M.; Liebetrau, J. Current developments in production and utilization of biogas and biomethane in Germany. *Chem. Ing. Tech.* **2018**, *90*, 17–35. [CrossRef]
32. Divya, D.; Gopinath, L.R.; Christy, P.M. A review on current aspects and diverse prospects for enhancing biogas production in sustainable means. *Renew Sustain. Energy Rev.* **2015**, *42*, 690–699. [CrossRef]
33. Theuerl, S.; Herrmann, C.; Heiermann, M.; Grundmann, P.; Landwehr, N.; Kreidenweis, U.; Prochnow, A. The Future Agricultural Biogas Plant in Germany: A Vision. *Energies* **2019**, *12*, 396. [CrossRef]
34. Von Cossel, M.; Mangold, A.; Iqbal, Y.; Hartung, J.; Lewandowski, I.; Kiesel, A. How to Generate Yield in the First Year—A Three-Year Experiment on Miscanthus (Miscanthus × giganteus (Greef et Deuter)) Establishment under Maize (*Zea mays* L.). *Agronomy* **2019**, *9*, 237. [CrossRef]
35. Clifton-Brown, J.C.; Lewandowski, I.; Andersson, B.; Basch, G.; Christian, D.G.; Kjeldsen, J.B.; Jørgensen, U.; Mortensen, J.V.; Riche, A.B.; Schwarz, K.-U. Performance of 15 Miscanthus genotypes at five sites in Europe. *Agron. J.* **2001**, *93*, 1013–1019. [CrossRef]
36. Clifton-Brown, J.C.; Lewandowski, I. Water use efficiency and biomass partitioning of three different Miscanthus genotypes with limited and unlimited water supply. *Ann. Bot.* **2000**, *86*, 191–200. [CrossRef]
37. Ashman, C.; Awty-Carroll, D.; Mos, M.; Robson, P.; Clifton-Brown, J. Assessing seed priming, sowing date, and mulch film to improve the germination and survival of direct-sown Miscanthus sinensis in the United Kingdom. *GCB Bioenergy* **2018**, *10*, 612–627. [CrossRef] [PubMed]
38. Winkler, B.; Lemke, S.; Ritter, J.; Lewandowski, I. Integrated assessment of renewable energy potential: Approach and application in rural South Africa. *Environ. Innov. Soc. Transit.* **2017**, *24*, 17–31. [CrossRef]
39. Hastings, A.; Mos, M.; Yesufu, J.A.; McCalmont, J.; Schwarz, K.; Shafei, R.; Ashman, C.; Nunn, C.; Schuele, H.; Cosentino, S.; et al. Economic and Environmental Assessment of Seed and Rhizome Propagated Miscanthus in the UK. *Front. Plant Sci.* **2017**, *8*. [CrossRef] [PubMed]
40. Mann, J.J.; Barney, J.N.; Kyser, G.B.; Tomaso, J.M.D. Miscanthus × giganteus and Arundo donax shoot and rhizome tolerance of extreme moisture stress. *GCB Bioenergy* **2013**, *5*, 693–700. [CrossRef]
41. Olave, R.J.; Forbes, E.G.A.; Munoz, F.; Laidlaw, A.S.; Easson, D.L.; Watson, S. Performance of Miscanthus x giganteus (Greef et Deu) established with plastic mulch and grown from a range of rhizomes sizes and densities in a cool temperate climate. *Field Crops Res.* **2017**, *210*, 81–90. [CrossRef]
42. Xue, S.; Kalinina, O.; Lewandowski, I. Present and future options for Miscanthus propagation and establishment. *Renew Sustain. Energy Rev.* **2015**, *49*, 1233–1246. [CrossRef]
43. Xue, S.; Lewandowski, I.; Kalinina, O. Miscanthus establishment and management on permanent grassland in southwest Germany. *Ind. Crops Prod.* **2017**, *108*, 572–582. [CrossRef]
44. O'Loughlin, J.; Finnan, J.; McDonnell, K. Accelerating early growth in miscanthus with the application of plastic mulch film. *Biomass Bioenergy* **2017**, *100*, 52–61. [CrossRef]
45. Prcik, M.; Kotrla, M. Different planting material for establishment of the Miscanthus energy grass plantation. *J. Cent. Eur. Agric.* **2016**, *7*, 778–792. [CrossRef]
46. Kiesel, A.; Lewandowski, I. Miscanthus as biogas substrate – cutting tolerance and potential for anaerobic digestion. *GCB Bioenergy* **2017**, *9*, 153–167. [CrossRef]
47. Von Cossel, M.; Iqbal, Y.; Lewandowski, I. Improving the Ecological Performance of Miscanthus (Miscanthus × giganteus Greef et Deuter) through Intercropping with Woad (*Isatis tinctoria* L.) and Yellow Melilot (*Melilotus officinalis* L.). *Agriculture* **2019**, *9*, 194. [CrossRef]

48. Naumann, C.; Bassler, R. *VDLUFA Methodenbuch: Die Chemische Untersuchung von Futtermitteln. Band III VDLUFA-Verl*; VDLUFA-Verlag: Darmstadt, Germany, 2006.
49. Von Cossel, M.; Möhring, J.; Kiesel, A.; Lewandowski, I. Methane yield performance of amaranth (*Amaranthus hypochondriacus* L.) and its suitability for legume intercropping in comparison to maize (*Zea mays* L.). *Ind. Crops Prod.* **2017**, *103*, 107–121. [CrossRef]
50. VDI. *VDI 4630: Fermentation of Organic Materials—Characterization of the Substrate, Sampling, Collection of Material Data, Fermentation Tests*; Verein Deutscher Ingenieure e.V.-Gesellschaft Energie und Umwelt: Düsseldorf, Germany, 2016.
51. Von Cossel, M.; Möhring, J.; Kiesel, A.; Lewandowski, I. Optimization of specific methane yield prediction models for biogas crops based on lignocellulosic components using non-linear and crop-specific configurations. *Ind. Crops Prod.* **2018**, *120*, 330–342. [CrossRef]
52. Piepho, H.-P. An algorithm for a letter-based representation of all-pairwise comparisons. *J. Comput. Graph. Stat.* **2004**, *13*, 456–466. [CrossRef]
53. Wolfinger, R. Covariance structure selection in general mixed models. *Commun. Stat.Simul. Comput.* **1993**, *22*, 1079–1106. [CrossRef]
54. Kenward, M.G.; Roger, J.H. Small sample inference for fixed effects from restricted maximum likelihood. *Biometrics* **1997**, *53*, 983–997. [CrossRef]
55. Kahle, P.; Beuch, S.; Boelcke, B.; Leinweber, P.; Schulten, H.-R. Cropping of Miscanthus in Central Europe: Biomass production and influence on nutrients and soil organic matter. *Eur. J. Agron.* **2001**, *15*, 171–184. [CrossRef]
56. Huggett, D.A.J.; Leather, S.R.; Walters, K.F.A. Suitability of the biomass crop Miscanthus sinensis as a host for the aphids *Rhopalosiphum padi* (L.) and *Rhopalosiphum maidis* (F.), and its susceptibility to the plant luteovirus Barley Yellow Dwarf Virus. *Agric. For. Entomol.* **1999**, *1*, 143–149. [CrossRef]
57. Meibaum, B.; Riede, S.; Schröder, B.; Manderscheid, R.; Weigel, H.-J.; Breves, G. Elevated CO_2 and drought stress effects on the chemical composition of maize plants, their ruminal fermentation and microbial diversity in vitro. *Arch. Anim. Nutr.* **2012**, *66*, 473–489. [CrossRef]
58. Jiang, Y.; Yao, Y.; Wang, Y. Physiological response, cell wall components, and gene expression of switchgrass under short-term drought stress and recovery. *Crop Sci.* **2012**, *52*, 2718–2727. [CrossRef]
59. Dandikas, V.; Heuwinkel, H.; Lichti, F.; Drewes, J.E.; Koch, K. Correlation between Biogas Yield and Chemical Composition of Grassland Plant Species. *Energy Fuels* **2015**, *29*, 7221–7229. [CrossRef]
60. Hall, M.; Bansal, P.; Lee, J.H.; Realff, M.J.; Bommarius, A.S. Cellulose crystallinity–a key predictor of the enzymatic hydrolysis rate. *FEBS J.* **2010**, *277*, 1571–1582. [CrossRef] [PubMed]
61. Yoshida, M.; Liu, Y.; Uchida, S.; Kawarada, K.; Ukagami, Y.; Ichinose, H.; Kaneko, S.; Fukuda, K. Effects of cellulose crystallinity, hemicellulose, and lignin on the enzymatic hydrolysis of Miscanthus sinensis to monosaccharides. *Biosci. Biotechnol. Biochem.* **2008**, *72*, 805–810. [CrossRef] [PubMed]
62. Xu, N.; Zhang, W.; Ren, S.; Liu, F.; Zhao, C.; Liao, H.; Xu, Z.; Huang, J.; Li, Q.; Tu, Y.; et al. Hemicelluloses negatively affect lignocellulose crystallinity for high biomass digestibility under NaOH and H_2SO_4 pretreatments in Miscanthus. *Biotechnol. Biofuels* **2012**, *5*. [CrossRef]
63. Liitiä, T.; Maunu, S.L.; Hortling, B.; Tamminen, T.; Pekkala, O.; Varhimo, A. Cellulose crystallinity and ordering of hemicelluloses in pine and birch pulps as revealed by solid-state NMR spectroscopic methods. *Cellulose* **2003**, *10*, 307–316. [CrossRef]
64. Mangold, A.; Lewandowski, I.; Hartung, J.; Kiesel, A. Miscanthus for biogas production: Influence of harvest date and ensiling on digestibility and methane hectare yield. *GCB Bioenergy* **2019**, *11*, 50–62. [CrossRef]
65. Schmidt, A.; Lemaigre, S.; Ruf, T.; Delfosse, P.; Emmerling, C. Miscanthus as biogas feedstock: Influence of harvest time and stand age on the biochemical methane potential (BMP) of two different growing seasons. *Biomass Convers Biorefinery* **2018**, *8*, 245–254. [CrossRef]
66. Ruf, T.; Emmerling, C. Impact of premature harvest of Miscanthus x giganteus for biogas production on organic residues, microbial parameters and earthworm community in soil. *Appl. Soil Ecol.* **2017**, *114*, 74–81. [CrossRef]
67. Ruf, T.; Schmidt, A.; Delfosse, P.; Emmerling, C. Harvest date of Miscanthus x giganteus affects nutrient cycling, biomass development and soil quality. *Biomass Bioenergy* **2017**, *100*, 62–73. [CrossRef]

68. Ning, J.; Zhou, M.; Pan, X.; Li, C.; Lv, N.; Wang, T.; Cai, G.; Wang, R.; Li, J.; Zhu, G. Simultaneous biogas and biogas slurry production from co-digestion of pig manure and corn straw: Performance optimization and microbial community shift. *Bioresour. Technol.* **2019**, *282*, 37–47. [CrossRef] [PubMed]
69. Dang, Y.; Holmes, D.E.; Zhao, Z.; Woodard, T.L.; Zhang, Y.; Sun, D.; Wang, L.-Y.; Nevin, K.P.; Lovley, D.R. Enhancing anaerobic digestion of complex organic waste with carbon-based conductive materials. *Bioresour. Technol.* **2016**, *220*, 516–522. [CrossRef] [PubMed]
70. Tayyab, A.; Ahmad, Z.; Mahmood, T.; Khalid, A.; Qadeer, S.; Mahmood, S.; Andleeb, S.; Anjum, M. Anaerobic co-digestion of catering food waste utilizing Parthenium hysterophorus as co-substrate for biogas production. *Biomass Bioenergy* **2019**, *124*, 74–82. [CrossRef]
71. Xu, Z.; Zhao, M.; Miao, H.; Huang, Z.; Gao, S.; Ruan, W. In situ volatile fatty acids influence biogas generation from kitchen wastes by anaerobic digestion. *Bioresour. Technol.* **2014**, *163*, 186–192. [CrossRef] [PubMed]
72. Kiesel, A.; Wagner, M.; Lewandowski, I. Environmental Performance of Miscanthus, Switchgrass and Maize: Can C4 Perennials Increase the Sustainability of Biogas Production? *Sustainability* **2016**, *9*, 5. [CrossRef]
73. Schäfer, J.; Sattler, M.; Iqbal, Y.; Lewandowski, I.; Bunzel, M. Characterization of Miscanthus cell wall polymers. *Glob. Change Biol. Bioenergy* **2019**, *11*, 191–205. [CrossRef]
74. Mangold, A.; Lewandowski, I.; Möhring, J.; Clifton-Brown, J.; Krzyżak, J.; Mos, M.; Pogrzeba, M.; Kiesel, A. Harvest date and leaf:stem ratio determine methane hectare yield of miscanthus biomass. *GCB Bioenergy* **2019**, *11*, 21–33. [CrossRef]
75. Sonwai, A.; Pholchan, P.; Nuntaphan, A.; Juangjandee, P.; Totarat, N. Biochemical Methane Potential of Fresh and Silage 4190 Grass Under Thermophilic Conditions. *Thai Environ. Eng. J.* **2019**, *33*, 21–29.
76. Baldini, M.; da Borso, F.; Ferfuia, C.; Zuliani, F.; Danuso, F. Ensilage suitability and bio-methane yield of Arundo donax and Miscanthus×giganteus. *Ind. Crops Prod.* **2017**, *95*, 264–275. [CrossRef]
77. Wahid, R.; Nielsen, S.F.; Hernandez, V.M.; Ward, A.J.; Gislum, R.; Jørgensen, U.; Møller, H.B. Methane production potential from Miscanthus sp.: Effect of harvesting time, genotypes and plant fractions. *Biosyst. Eng.* **2015**, *133*, 71–80. [CrossRef]
78. Winkler, B.; Mangold, A.; Von Cossel, M.; Iqbal, Y.; Kiesel, A.; Lewandowski, I. Implementing miscanthus into sustainable farming systems: A review on agronomic practices, capital and labor demand. *Rev. Artic. Rev.*. under review.
79. Von Cossel, M.; Wagner, M.; Lask, J.; Magenau, E.; Bauerle, A.; Von Cossel, V.; Warrach-Sagi, K.; Elbersen, B.; Staritsky, I.; Van Eupen, M.; et al. Prospects of Bioenergy Cropping Systems for A More Social-Ecologically Sound Bioeconomy. *Agronomy* **2019**, *9*, 605. [CrossRef]
80. Yadav, P.; Priyanka, P.; Kumar, D.; Yadav, A.; Yadav, K. Bioenergy Crops: Recent Advances and Future Outlook. In *Prospects of Renewable Bioprocessing in Future Energy Systems*; Rastegari, A.A., Yadav, A.N., Gupta, A., Eds.; Biofuel and Biorefinery Technologies; Springer International Publishing: Cham, Germany, 2019; pp. 315–335. ISBN 978-3-030-14463-0.
81. Gerwin, W.; Repmann, F.; Galatsidas, S.; Vlachaki, D.; Gounaris, N.; Baumgarten, W.; Volkmann, C.; Keramitzis, D.; Kiourtsis, F.; Freese, D. Assessment and quantification of marginal lands for biomass production in Europe using soil-quality indicators. *SOIL* **2018**, *4*, 267–290. [CrossRef]

© 2019 by the authors. Licensee MDPI, Basel, Switzerland. This article is an open access article distributed under the terms and conditions of the Creative Commons Attribution (CC BY) license (http://creativecommons.org/licenses/by/4.0/).

Article

Product Inhibition of Biological Hydrogen Production in Batch Reactors

Subhashis Das [1,*], Rajnish Kaur Calay [1], Ranjana Chowdhury [2], Kaustav Nath [3] and Fasil Ejigu Eregno [1]

1. Faculty of Engineering Science and Technology, UiT-The Arctic University of Norway, 8514 Narvik, Norway; rajnish.k.calay@uit.no (R.K.C.); fasil.e.eregno@uit.no (F.E.E.)
2. Department of Chemical Engineering, Jadavpur University, Kolkata 700032, India; ranjana.juchem@gmail.com
3. Department of Biotechnology, Indian Institute of Technology, Kharagpur 721302, India; knstar.nath8@gmail.com
* Correspondence: das.subhashis@gmail.com or subhashis.das@uit.no

Received: 20 February 2020; Accepted: 10 March 2020; Published: 12 March 2020

Abstract: In this paper, the inhibitory effects of added hydrogen in reactor headspace on fermentative hydrogen production from acidogenesis of glucose by a bacterium, *Clostridium acetobutylicum*, was investigated experimentally in a batch reactor. It was observed that hydrogen itself became an acute inhibitor of hydrogen production if it accumulated excessively in the reactor headspace. A mathematical model to simulate and predict biological hydrogen production process was developed. The Monod model, which is a simple growth model, was modified to take inhibition kinetics on microbial growth into account. The modified model was then used to investigate the effect of hydrogen concentration on microbial growth and production rate of hydrogen. The inhibition was moderate as hydrogen concentration increased from 10% to 30% (*v/v*). However, a strong inhibition in microbial growth and hydrogen production rate was observed as the addition of H_2 increased from 30% to 40% (*v/v*). Practically, an extended lag in microbial growth and considerably low hydrogen production rate were detected when 50% (*v/v*) of the reactor headspace was filled with hydrogen. The maximum specific growth rate (μ_{max}), substrate saturation constant (ks), a critical hydrogen concentration at which microbial growth ceased (H_2^*) and degree of inhibition were found to be 0.976 h^{-1}, 0.63 ± 0.01 gL, 24.74 mM, and 0.4786, respectively.

Keywords: hydrogen; reactor headspace; product inhibition; kinetic modelling; *clostridium acetobutylicum*

1. Introduction

Hydrogen energy is considered one of the most promising energy storage hubs and carriers of energy harvested from renewable energy sources. Hydrogen fuel cell technology, particularly, has the potential to replace fossil fuel-based internal combustion engine mainly used in the transport sector [1]. Hydrogen is the most abundant element, but it does not exist in its molecular form and has to be produced using different technologies, such as by electrolysis from water, steam reforming, and gasification of fossil fuel. All of these technologies are energy-intensive. For example, 1 kg of hydrogen (specific energy of 40 kWh/kg) requires 50–55 kWh of electricity by electrolysis of water, which is 70–80% efficient. Therefore, exploring energy-efficient hydrogen production methods from renewable sources are necessary. Biological and thermochemical processes can convert various types of biomass such as agriculture, forest sector, and bio-waste directly into hydrogen. Biological processes for hydrogen production are more environment-friendly and consume less energy compared to thermochemical processes [2]. When using bio-waste, the production of hydrogen becomes even more cost-effective due to the utilization of low-cost waste biomass as feedstock.

The available paths of biohydrogen are typically categorized as either photo fermentation (PF) or dark fermentation (DF). PF is carried out by nonoxygenic photosynthetic bacteria, which use sunlight and biomass to produce hydrogen. DF, however, takes place under anaerobic conditions. Carbohydrate-rich biomasses, along with industrial wastes, can be used as the feedstock of DF for hydrogen production [3]. The yield of hydrogen is higher in the PF process, although there are studies [4–7] that establish that treatment capacity of organic waste and hydrogen production rate of DF is better than the PF.

Considering the potential of DF, a detailed investigation is required for scaling up the technology into industrial scale. The most critical issue that needs to be addressed is increasing the production of hydrogen, which depends on the activity of microorganisms. Production of hydrogen quantitively and qualitatively strongly depend on the metabolic pathway of microorganisms. The metabolic pathway of microorganism in DF often deviates due to the influence of certain physicochemical parameters such as substrate composition, culture pH, or concentration of byproducts of reaction medium. On the other hand, there are various research studies [7–14] that have identified the dark fermentation process parameters that influence the production of hydrogen, such as the optimal functionality of the microorganisms, hydraulic retention time, temperature, and the partial pressure of hydrogen of reaction processes. Therefore, it is possible to enrich hydrogen productivity by improving approaches to metabolic pathway control.

The primary pathway in the dark fermentation is the breakdown of carbohydrate-rich substrates to H_2 and other intermediate products such as volatile fatty acids (VFAs) and alcohols by the use of bacteria. There are a few kinds of anaerobic mesophilic or thermophilic bacteria such as genus *Clostridium*, which can produce hydrogen at a high rate, in the course of their metabolism. During fermentative hydrogen production, polysaccharides are hydrolyzed into simpler saccharides. These simpler sugars are easily taken up by hydrogen-producing bacteria (HPB) and enter the 'Embden-Meyerhof-Parnas' pathway to produce pyruvate and nicotinamide adenine dinucleotide (NADH). NADH becomes NAD^+ by donating electrons to the electron transport chain and H^+ is transported across the membrane. The major products are further formed from pyruvate and these are mainly short-chain fatty acids (e.g., acetate, butyrate, lactate) as well as alcohols (e.g., butanol, ethanol). Among these products, lactate, butanol, and ethanol have the only contribution to re-oxidized NADH. As the reactions proceed, CO_2 formed from other metabolic reactions and increases its concentration in the liquid medium. The excess CO_2 in liquid culture reacts with pyruvate using NADH to produce succinate and oxidize NADH by reducing H^+. The product formation from pyruvate is shown in Figure 1.

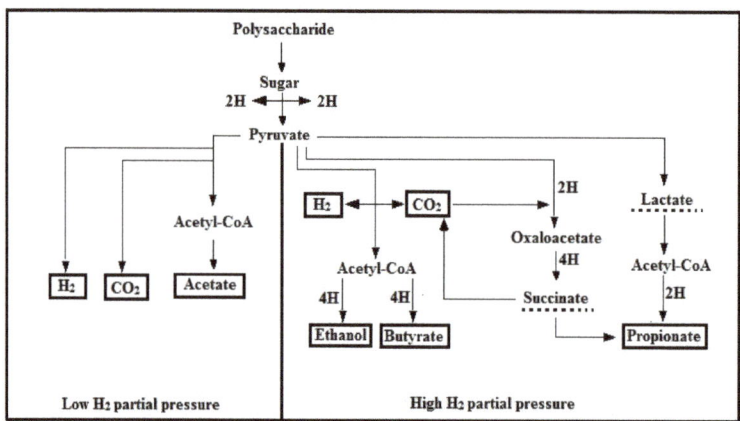

Figure 1. Metabolic pathway of H_2 formation.

On the other hand, hydrogenase, an enzyme that catalyzes the reversible oxidation of molecular hydrogen during fermentative hydrogen production, is often affected by the H_2 concentration in liquid culture. At higher H_2 partial pressure in liquid broth, the reduction of ferredoxin, which mediates electron transfer, takes place and the reduction of a proton to H_2 becomes thermodynamically less favourable, which results in a reduction of H_2 formation.

Numerous research articles have shown the inhibitory effects of hydrogen concentration in terms of partial pressure in the reactor [7–9] and these articles explain how to overcome these inhibitory effects to improve H_2 productivity. On the way to decrease partial pressure to improve H_2 productivity, several strategies were employed such as continuous gas release [6–8], larger headspace volume [9], vacuum stripping [13], or sparging with an inert gas like N_2 or CO_2 [15,16].

Different kinetic models describe the fermentation process of hydrogen production [17–19]. These models depend on physicochemical parameters and microbial environment within the reactor. Usually, these models are developed considering the effects of substrate concentration, pH, and temperature on the hydrogen production process. Kinetic models are also used to design reactors and provide proper information to adopt control strategies for hydrogen production processes. On the other hand, kinetic models are useful to describe the inhibitory effects of substrate, temperature, pH, dilution rate (in case of the continuous process), and soluble metabolites, which are generated during DF [18]. However, models that describe the effects of hydrogen accumulated in the headspace of a batch reactor on microbial growth and hydrogen production are limited. Therefore, the present study investigates the influence of hydrogen on microbial growth and evaluates how the produced hydrogen hinders the rate of production of hydrogen. The main aim of the study is to experimentally examine the adverse of produced hydrogen on microbial growth and adopted mathematical models to predict the production rate hydrogen batch reactors.

2. Materials and Methods

In order to achieve the objective, first, a series of experiments were performed in a batch reactor. A hydrogen-producing bacterium *Clostridium acetobutylicum*, which is strictly anaerobic, was selected in the present study where glucose was the sole nutrient for microbial growth. In the batch reactor operation, the headspace gas concentration, and the nutrient concentration, were varied to observe the effect of hydrogen, accumulated in the headspace on microbial growth and hydrogen production rate. Observing the nature of the batch reaction suitable microbial reaction kinetics was adopted secondly. Furthermore, the kinetic parameters were determined by using experimental data. The experimental procedure is described below.

2.1. Experiments

2.1.1. Inoculum

A pure lyophilized strain of *Clostridium Acetobutylicum* (NCIM 2337) was procured from National chemical laboratory (NCL), Pune, India. Cooked meat (CM) medium containing beef extract 45 g/L, glucose 2 g/L, peptone 20 g/L, NaCl 5 g/L was used for the growth of lyophilized bacterial culture at 37 °C for 72 h.

2.1.2. Reactor Setup for Batch Experiment

Batch experiments were conducted in 250 mL cork fitted Erlenmeyer flask having an outlet port at the bottom. A cork was fitted to a glass tube and connected to a gas measuring tube for gas sampling. The batch experimental setup is shown in Figure 2a and the whole system is presented schematically in Figure 2b.

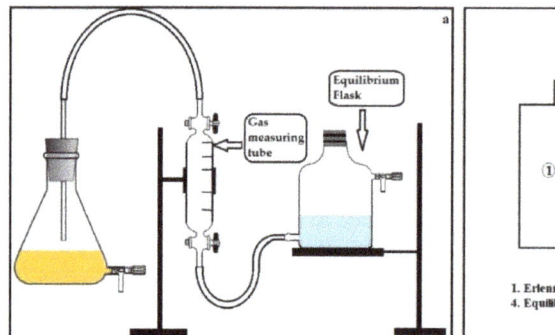 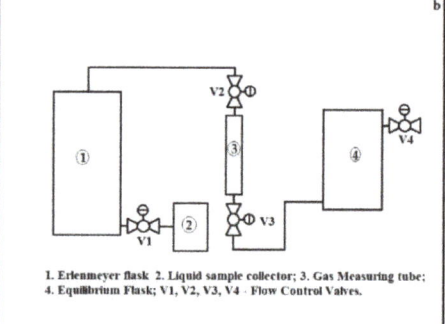

Figure 2. Experimental setup for batch test (**a**); schematic diagram of the experimental setup (**b**).

An Erlenmeyer flask was half filled with the CM medium along with 1% (v/v) of inoculum into the medium then remaining void was filled completely with the CM medium. The flask was then firmly sealed with the cork and 150 mL of argon gas was passed through the flask by a glass tube for displacing the CM media from the bottom of the flask. Thus, the flask was left with 100 mL of CM medium with bacterial culture and 150 mL headspace was occupied by a mixture of argon and hydrogen. The whole setup was then kept in the incubator at maintaining the temperature of 37 °C and pH of 7.2, which are the ideal conditions for this bacteria. Initial hydrogen concentration in the reactor headspace were varied in the range of 0% (v/v) to 50% (v/v). Initially, H_2 experiments were performed by varying glucose concentration of the modified CM medium in the range of 2 g/L to 5 g/L. At each initial substrate concentration, microbial growth pattern was observed for 30 h. An interval of 3 h samples were collected. The produced gas was accumulated in the reactor headspace and was taken out from the flask using a gas sampling tube, connected with the flask for examining the gas composition after each interval. Each experimental run was repeated three times to ensure the repeatability and the statistical accuracy of the results.

2.1.3. Sample Collection and Analysis

The biomass concentration of each sample was determined in terms of optical density with a spectrophotometer at 600 nm wavelength. Each liquid sample was centrifuged at 10,000 rpm and the supernatant was collected in order to find out the reducing sugar concentration using the dinitrosalicylic (DNS) acid reagent [20]. Next, 100 mL of gas was collected after every 3 h interval from the headspace of batch reactor using a gas sampling tube. Collected gas was then passed through an ORSAT apparatus for the removal of CO_2 gas present in the gas sample. Furthermore, the remaining gas composition was analyzed by gas chromatography. On a molecular sieve column (13×, 180 cm by 1⁄4 inch, 60–80 mesh), the gases were separated where argon was the carrier gas at 100 °C. A 406 Packard GC equipped with a thermal conductivity detector (TCD, 100 mA) was used to measure hydrogen concentration.

2.2. Kinetic Modelling

Herein, we consider the Monod model, which is the most popular and simplest model for describing the microbial reaction of microbial growth within a single substrate. The reaction kinetics are expressed as:

$$\text{Substrate(S)} \xrightarrow{\text{Cell(X)}} \text{more Cells(X)} + \text{Product (H}_2) \qquad (1)$$

the rate of reaction will be:

$$r_c = \frac{dX}{dt} = \frac{\mu_{max} S}{k_S + S} X \qquad (2)$$

where, r_c is the microbial growth rate, $(gL^{-1}h^{-1})$; X is the dry cell concentration, (gL^{-1}); t is time, (h); μ_{max} is the maximum specific growth rate of cells, (h^{-1}); S is the substrate concentration, (gL^{-1}); and k_s is the Monod constant or substrate saturation constant, (gL^{-1}).

The inhibition in microbial growth occurs typically due to the excess presence of substrate, product, or other inhibitory substance in the cell growth medium. Hans and Levenspiel [21] express the inhibition of microbial growth model as:

$$\frac{dX}{dt} = \mu_{max}\left(1 - \frac{H_2}{H_2^*}\right)^n \left(\frac{S \cdot X}{S + k_s\left(1 - \frac{X}{X^*}\right)^m}\right) \qquad (3)$$

where, H_2^* is the critical molar concentration of hydrogen at which microbial reaction ceases, (M); n is the degree of inhibition; and m is the degree of inhibition.

Evaluation of the Constants

Taking inhibition of microbial growth Equation (3) into account, Equation (2) can be expressed as a generalized Monod model:

$$\mu = \frac{\mu_{max,obs} \cdot S}{k_{s,obs} + S} \qquad (4)$$

where, μ is the pecific microbial growth rate (h^{-1}) and $_{obs}$ is the experimentally observed value.

$$\mu = \frac{1}{X} \cdot \frac{dX}{dt} \qquad (5)$$

$$\mu_{max,obs} = \mu_{max}\left(1 - \frac{H_2}{H_2^*}\right)^n \qquad (6)$$

$$k_{s,obs} = k_s\left(1 - \frac{X}{X^*}\right)^m \qquad (7)$$

By reciprocating Equation (4),

$$\frac{1}{\mu} = \frac{k_{s,obs}}{\mu_{max,obs}} \cdot \frac{1}{S} + \frac{1}{\mu_{max,obs}} \qquad (8)$$

plots of $1/\mu$ and $1/S$ can be obtained at each initially added hydrogen in reactor headspace, which is shown in the Figure 3. $\mu_{max,obs}$ and $k_{s,obs}$ at each headspace H_2 concentration can be determined by evaluating the intercepts and abscissas on Figure 3.

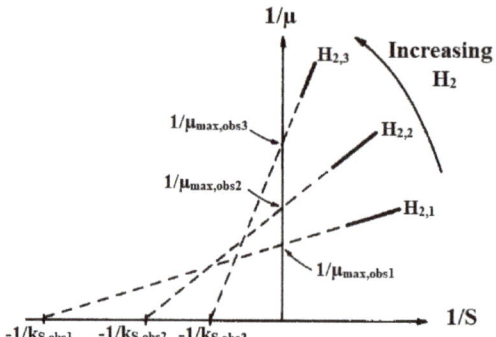

Figure 3. Evaluation procedure of $\mu_{max,obs}$ and $k_{s,obs}$ at various concentration of inhibitor; Reproduced with permission from Keehyun Han and Octave Levenspiel, Biotechnology & Bioengineering; published by John Wiley and Sons, 2004 [21].

After determining values of $\mu_{max,obs}$ and $k_{s,obs}$ at different headspace H_2 concentration, constants in Eqation (3) can be evaluated. On taking logarithms of Eqation (5) i.e.,

$$\ln(\mu_{max,obs=}) = n \cdot \ln\left(1 - \frac{H_2}{H_2^*}\right) + \ln(\mu_{max}) \tag{9}$$

a plot of $\ln(\mu_{max,obs})$ and $\ln(1-H_2/H_2^*)$ gives the values of μ_{max} and n. If the values of H_2^* is not identified from the experiments then a guessed value of H_2^* have to be has to be considered. A corrected value of H_2^* can be determined until a straight line is obtained which is shown in Figure 4.

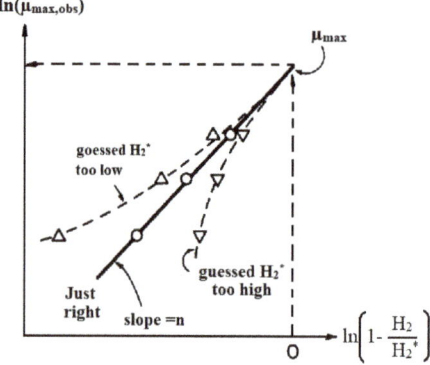

Figure 4. Evaluation procedure of μ_{max}, n and H_2^* for product inhibition; Reproduced with permission from Keehyun Han and Octave Levenspiel, Biotechnology & Bioengineering; published by John Wiley and Sons, 2004 [21].

3. Results

3.1. Effects of Added H_2 in the Reactor Headspace

Effects of hydrogen concentration accumulated in the reactor headspace on microbial growth and hydrogen production were studied by conducting experiments in batch reactor. The results were shown in Figures 5–10. In these figures, the time history of biomass concentration and produced hydrogen concentration were showed when initial hydrogen concentration in the reactor headspace was varied. From these figures, it is clear that microbial growth as well as hydrogen productivity were greatly influenced by the presence of hydrogen in the reactor headspace.

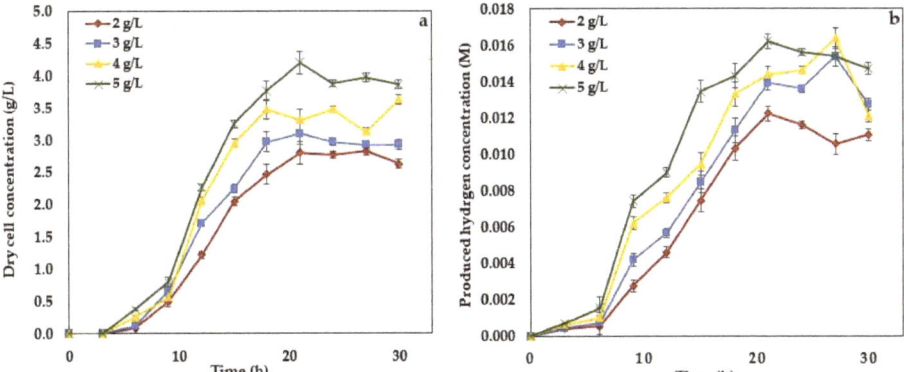

Figure 5. Experimental time histories of dry cell concentration (**a**) and hydrogen concentration (**b**) with initial 0% H_2 in reactor headspace at different substrate concentration.

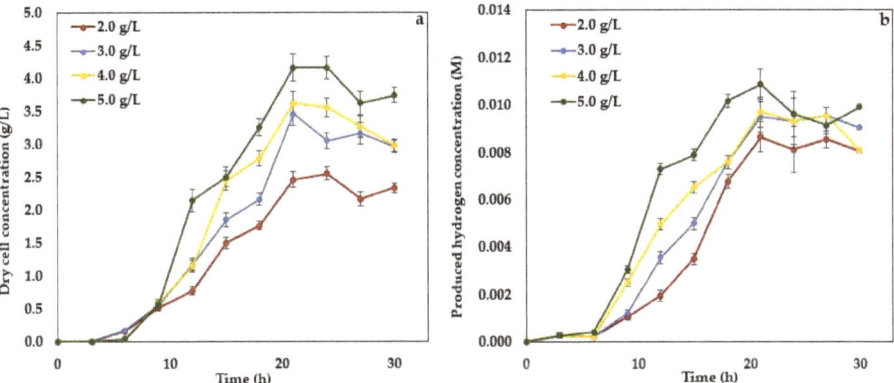

Figure 6. Experimental time histories of dry cell concentration (**a**) and hydrogen concentration (**b**) with initial 10% H_2 in reactor headspace at different substrate concentration.

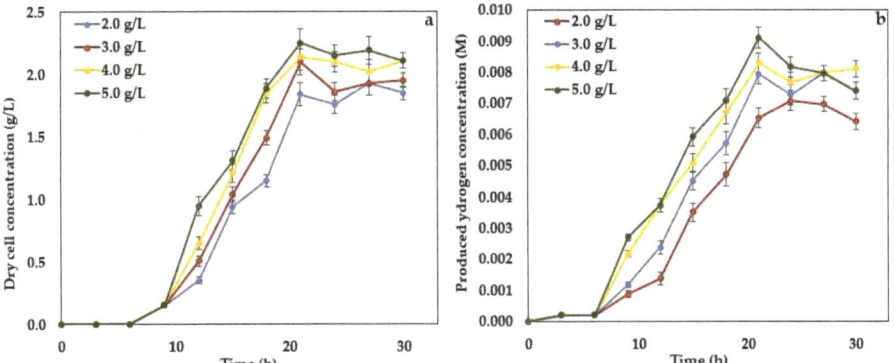

Figure 7. Experimental time histories of dry cell concentration (**a**) and hydrogen concentration (**b**) with initial 20% H_2 in reactor headspace at different substrate concentration.

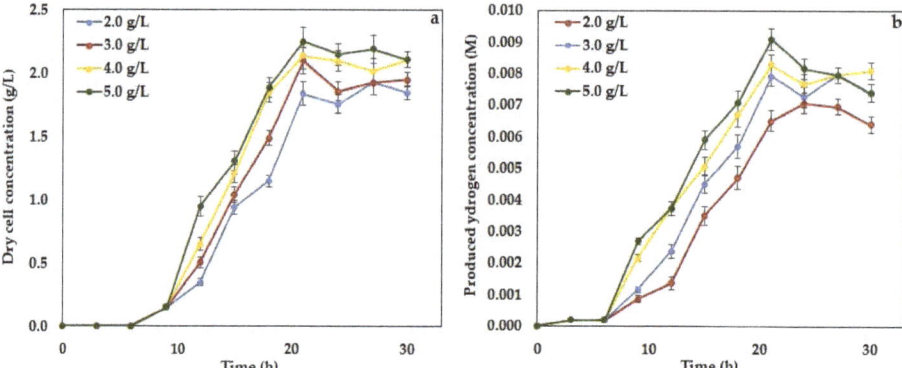

Figure 8. Experimental time histories of dry cell concentration (**a**) and produced hydrogen concentration (**b**) with initial 30% (*v/v*) H_2 in reactor headspace at different substrate concentration.

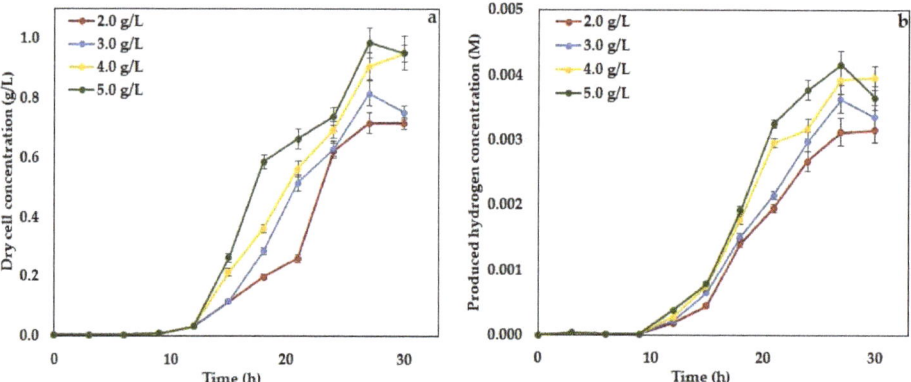

Figure 9. Experimental time histories of dry cell concentration (**a**) and hydrogen concentration (**b**) with initial 40% H_2 in reactor headspace at different substrate concentration.

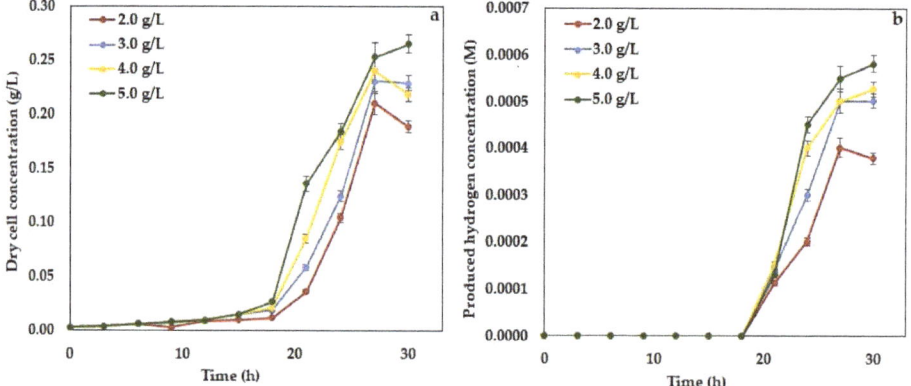

Figure 10. Experimental time histories of dry cell concentration (**a**) and hydrogen concentration (**b**) with initial 50% H_2 in reactor headspace at different substrate concentration.

Initially, experiments were started when argon gas was present in the reactor headspace and initial glucose concentration in the liquid medium was varied in the range of 2 g/L to 5 g/L. At each variation of glucose, the microbial growth pattern and hydrogen production rate were observed for 30 h. The time histories of microbial growth and hydrogen production rate were presented in Figure 5. From this figure, it is observed that microbial growth, as well as hydrogen production, was started immediately after 3 h of reaction time. There was no significant lag phase of microbial growth detected. A stationary phase was started at 21 h for every initial glucose concentration. The maximum productivity of hydrogen was 7.81 mML^{-1}h^{-1} when initial glucose concentration in the liquid medium was 5 g/L.

On the other hand, when 10% (v/v) of H$_2$ added to reactor headspace, microbial growth and hydrogen production started after 6 h of incubation time, which is shown in Figure 6. Although there was no such difference in specific growth and hydrogen production observed for different initial substrate concentration, the maximum hydrogen productivity decreased to 5.17 mML^{-1}h^{-1}, which is comparable to the 0% (v/v) added H$_2$ condition. The exponential phase of microbial growth ended at 21 h which was same as the previous condition.

In the case of 20% (v/v) added H$_2$ in the reactor headspace, the hydrogen production rate as well as biomass production rate further decreased, which can be observed from Figure 7. In this condition, propagation of hydrogen production and bacterial growth was quite similar to that of 10% (v/v) added H$_2$ condition, where microbial growth reached at its exponential phase at 6 h and it extended up to 21 h. However, in this condition, maximum hydrogen productivity decreased to 4.33 mML^{-1}h^{-1}.

A different phenomenon was observed when 30% (v/v) H$_2$ was added to the reactor headspace. In this case, exponential phase of bacterial growth started after 6 h of incubation time but it extended to 24 h where the stationary phase started. The time histories of dry cell concentration and produced hydrogen were presented in Figure 8. Monotonic decreases of microbial growth rate and hydrogen production were observed where hydrogen productivity reduced to 3.075 mML^{-1}h^{-1} when initial substrate concentration in liquid medium was 5 g/L. However, there was no such significant change in growth pattern observed for different substrate concentration in liquid medium.

An extended lag phase in microbial growth was noticed as the quantity initially added H$_2$ increased from 30% (v/v) to 40% (v/v). At this condition, the exponential phase commenced at 12 h of incubation time and extended until 27 h. A sharp degradation in microbial growth, as well as biohydrogen production, were also observed (Figure 9). There were no such effects of substrate concentration in liquid medium experience. The hydrogen productivity in this condition was estimated as 1.54 mML^{-1}h^{-1}, which is a sharp alteration compared to 30% (v/v) added H$_2$ condition.

Furthermore, when 50% (v/v) H$_2$ was added to the reactor headspace, almost no growth condition was observed which is presented in Figure 10. In this case, an extended lag phase with no microbial growth and hydrogen production was seen for 18 h. A short period of exponential phase ended at 27 h was noticed. Almost no hydrogen production condition with productivity of 0.19 mML^{-1}h^{-1} was estimated.

3.2. Inhibition Kinetics

In the present investigation, a total of 24 experimental runs were conducted at different initial H$_2$ and substrate concentration. Each initially added H$_2$ concentration and substrate concentration, and was varied from 2 g/L to 5 g/L. For each combination of H$_2$ concentration and substrate concentration, specific growth rate microorganisms were determined. By using Equation (8), plots of 1/µ and 1/S were obtained at each initially added hydrogen in reactor headspace, which is demonstrated in Figure 11. $\mu_{max,obs}$ and $k_{s,obs}$ at each headspace H$_2$ concentration were determined by evaluating the intercepts and abscissas on Figure 3. The values of $\mu_{max,obs}$ and $k_{s,obs}$ are provided in the Table 1.

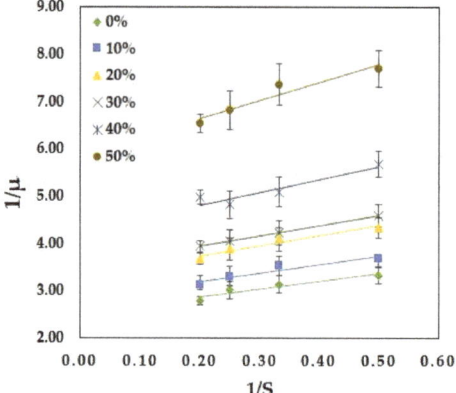

Figure 11. Determination of $\mu_{max,obs}$ and $k_{s,obs}$ at different initial concentration of H_2.

Table 1. Values of observed rate constants from experiments.

Initial H_2 Concentration in Reactor Headspace (v/v)	$\mu_{max,obs}$	$k_{s,obs}$
0%	0.6209	0.6318
10%	0.5523	0.6395
20%	0.4792	0.6285
30%	0.4643	0.6141
40%	0.3695	0.6343
50%	0.2613	0.6521

From Table 1, it can be observed that $k_{s,obs}$ does not change in any systematic manner with the change of added hydrogen in the reactor headspace. Therefore, m = 0 in Equation (7), which infers the adopted model is a noncompetitive inhibition model and $k_{s,obs}$ = ks which will be constant.

After determining values of $\mu_{max,obs}$ and $k_{s,obs}$ at different headspace H_2 concentration, constants in Equation (3) can be evaluated by plotting $\ln(\mu_{max,obs})$ vs $\ln(1-H_2/H_2^*)$ from Equation (9), which is shown in Figure 12. Figure 12 gives the values of μ_{max} and n. As H_2^* was not identified from the experiments, a guessed value of 61.5 (v/v) H_2^* (24.74 mM) was considered, which gives a straight line with R^2 = 0.9823. From Figure 12, the intercept and slope give the value of μ_{max} = 0.976 h^{-1} and n = 0.4786.

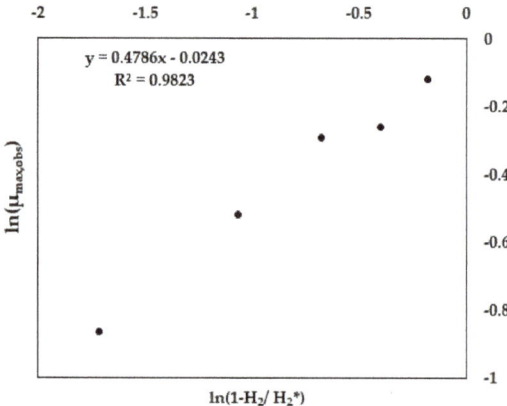

Figure 12. Determination of μ_{max}, n, and H_2^* for product inhibition.

4. Discussion

Initially, when there was no hydrogen present in the headspace and the reactor substrate concentration varied from 2 g/L to 5 g/L, the production rates were high and it significantly decreased as the hydrogen concentration increased gradually. When no hydrogen was added to the reactor, the growth phase started after 3 h of incubation time and reached the stationary phase at 21 h. For the 10% added hydrogen condition, the exponential phase started at 6 h and it went until 21 h. On the other hand, when 20% and 30% hydrogen were added, the exponential phase started at 6 h and went until 24 h. Further, when increasing hydrogen concentration by 40% of total headspace, the lag phase elongated by 12 h and the growth phase started at 15 h until 27 h. Almost no growth of bacteria was observed when 50% of the reactor headspace filled with hydrogen. The specific growth rate of biomass decreased as the hydrogen concentration increased in the reactor headspace.

This increased hydrogen concentration reduced the glucose degradation efficiency of bacteria, resulting in a lower hydrogen yield. Hydrogen yield gradually decreased along with specific the growth rate from 1.11 to 0.56 mol/mol.glucose and 0.621 ± 0.019 h^{-1} to 0.261 ± 0.021 h^{-1}, respectively. As the initial hydrogen concentration increases from 0.0 to 0.0161 M, the hydrogen productivity depletion becomes more rapid when the initial hydrogen concentration altered from 0.0161 M to 0.0201 M. Thus, the final partial pressure of hydrogen in the product gas declined and initially added hydrogen concentration increased. The effect of added hydrogen in the reactor on the specific microbial growth rate, final hydrogen concentration and hydrogen yield were calculated by dividing the total amount hydrogen produced by the amount of glucose consumed are summarized in Table 2.

Table 2. Effects of added hydrogen on specific growth rate, hydrogen production.

Added Hydrogen (M)/(v/v)	Specific Growth Rate (h^{-1})	Hydrogen Yield (mol-H$_2$/mol-glucose)	Final H$_2$ Partial Pressure (atm)
0.0/(0%)	0.621 ± 0.019	1.11 ± 0.0026	0.280 ± 0.015
4.023×10^{-3} (10%)	0.552 ± 0.028	1.01 ± 0.0020	0.229 ± 0.005
8.045×10^{-3} (20%)	0.479 ± 0.029	0.92 ± 0.0022	0.214 ± 0.008
1.207×10^{-2} (30%)	0.464 ± 0.032	0.88 ± 0.0016	0.165 ± 0.013
1.610×10^{-2} (40%)	0.369 ± 0.015	0.56 ± 0.0010	0.094 ± 0.005
2.012×10^{-2} (50%)	0.261 ± 0.021	0.21 ± 0.0010	0.0127 ± 0.003

From the current study, it is apparent to see that as hydrogen concentration increases in the reactor headspace, it restricts the mass transfer from the liquid to the gaseous phase. Thus, liquid to gas transfer becomes a rate-limiting step that dominates microbial reactions. Consequently, low microbial growth and less hydrogen production occur. On the other hand, hydrogenase, which mobilizes the reversible oxidation of molecular hydrogen, is affected by a high concentration of hydrogen and the process becomes thermodynamically unfavorable for H$_2$ formation.

From this analysis, it is clear that the initial addition of hydrogen has a significant impact on microbial growth, hydrogen yield, and hydrogen productivity. As per non-competitive inhibition is concerned, substrate concentration neither influences specific growth rate nor substrate utilization. In order to maintain the hydrogen production at an optimal level, the accumulated hydrogen in reactor headspace should not be more than 8 mMol. It can also be concluded from the present study that when 24.85 mMol hydrogen accumulated in the reactor headspace, reaction stopped and no hydrogen was produced.

5. Conclusions

Growth inhibition caused by hydrogen was examined through the acidogenesis of glucose by a bacterium, *Clostridium acetobutylicum*. Until now, the research showed that the inhibition is caused by H$_2$ present in the liquid medium, whereas kinetic models are developed to describe how that inhibition kinetically related to hydrogen production rate but not microbial growth rate. This study presented the kinetic model that describes how the microbial growth inhibited H₂ in the reactor headspace. The experiments were conducted in a batch reactor to observe the effects of hydrogen accumulated in the reactor headspace on hydrogen production from acidogenesis of glucose by a bacterium, *Clostridium acetobutylicum*. The concluding remarks can be made based on the data of the experiments and prediction of the kinetic model.

A nonlinear and non-competitive inhibition model described the inhibition kinetics of initially added hydrogen concentration on microbial growth and hydrogen production. The maximum specific growth rate (μ_{max}), substrate saturation constant (ks) critical added hydrogen concentration at which microbial growth ceased (H_2^*), and degree of inhibition were found to be 0.976 h^{-1}, 0.63 ± 0.01 g/L, 24.74 mM, and 0.4786, respectively. It was observed from the experiment that hydrogen could be an acute inhibitor if allowed to accumulate in reactor headspace. From 10% to 30% (v/v) concentration of hydrogen concentration, the microbial growth was decreased linearly. However, as more hydrogen was added in the headspace, microbial activity was inhibited exponentially, particularly after 30% (v/v), where there was a potent inhibition in microbial growth and hydrogen production rate. Practically, an extended lag phase in microbial growth and considerably low microbial growth and hydrogen production rate was detected when 50% of total reactor headspace filled with hydrogen. After 61.5% (v/v) (i.e., 24.74 mM) of hydrogen accumulated in the reactor, no microbial growth took place and production of hydrogen and microbial growth ceased. Furthermore, different process industries such as biohydrogen and biobutanol production can use the results obtained from the present study for reactor safety and adaptation of control strategies where these kinds of phenomenon occur.

Author Contributions: Conceptualization, S.D., R.C. and R.K.C.; Formal analysis, K.N.; Investigation, S.D., R.C. and R.C.; Methodology, R.C. and K.N.; Supervision, R.K.C. and R.C.; Writing—original draft, S.D.; Writing—review & editing, R.C. and F.E.E. All authors have read and agreed to the published version of the manuscript.

Funding: The publication charges for this article have been funded by a grant from the publication fund of UiT The Arctic University of Norway.

Acknowledgments: The authors wish to thank the Department of Chemical Engineering, Jadavpur University, Kolkata, India, for providing laboratory facilities for this study.

Conflicts of Interest: The authors declare no conflict of interest.

References

1. Offer, G.J.; Howey, D.; Contestabile, M.; Clague, R.; Brandon, N.P. Comparative analysis of battery electric, hydrogen fuel cell and hybrid vehicles in a future sustainable road transport system. *Energy Policy* **2010**, *38*, 24–29. [CrossRef]
2. Sivagurunathan, P.; Kumar, G.; Bakonyi, P.; Kim, S.H.; Kobayashi, T.; Xu, K.Q.; Lakner, G.; Tóth, G.; Nemestóthy, N.; Bélafi-Bakó, K. A critical review on issues and overcoming strategies for the enhancement of dark fermentative hydrogen production in continuous systems. *Int. J. Hydrog. Energy* **2016**, *41*, 3820–3836. [CrossRef]
3. Das, D.; Veziroglu, T. Advances in biological hydrogen production processes. *Int. J. Hydrog. Energy* **2008**, *33*, 6046–6057. [CrossRef]
4. Zhang, T.; Jiang, D.; Zhang, H.; Jing, Y.; Tahir, N.; Zhang, Y. Comparative study on bio-hydrogen production from corn stover: Photo-fermentation, dark-fermentation and dark-photo co-fermentation. *Int. J. Hydrog. Energy* **2020**, *45*, 3807–3814. [CrossRef]
5. Hay, J.X.W.; Wu, T.Y.; Juan, J.C.; Md. Jahim, J. Biohydrogen production through photo fermentation or dark fermentation using waste as a substrate: Overview, economics, and future prospects of hydrogen usage. *Biofuels Bioprod. Biorefin.* **2013**, *7*, 334–352. [CrossRef]
6. Eker, S.; Sarp, M. Hydrogen gas production from waste paper by dark fermentation: Effects of initial substrate and biomass concentrations. *Int. J. Hydrog. Energy* **2017**, *42*, 2562–2568. [CrossRef]
7. Levin, D.B.; Pitt, L.; Love, M. Biohydrogen production: Prospects and limitations to practical application. *Int. J. Hydrog. Energy* **2004**, *29*, 173–185. [CrossRef]
8. Guo, X.M.; Trably, E.; Latrille, E.; Carrre, H.; Steyer, J.P. Hydrogen production from agricultural waste by dark fermentation: A review. *Int. J. Hydrog. Energy* **2010**, *35*, 10660–10673. [CrossRef]

9. Oh, S.E.; Zuo, Y.; Zhang, H.; Guiltinan, M.J.; Logan, B.E.; Regan, J.M. Hydrogen production by *Clostridium acetobutylicum* ATCC 824 and megaplasmid-deficient mutant M5 evaluated using a large headspace volume technique. *Int. J. Hydrog. Energy* **2009**, *34*, 9347–9353. [CrossRef]
10. Logan, B.E.; Oh, S.E.; Kim, I.S.; Van Ginkel, S. Biological hydrogen production measured in batch anaerobic respirometers. *Environ. Sci. Technol.* **2002**, *36*, 2530–2535. [CrossRef]
11. Chang, S.; Li, J.; Liu, F.; Yu, Z. Effect of different gas releasing methods on anaerobic fermentative hydrogen production in batch cultures. *Front. Environ. Sci. Eng. China* **2012**, *6*, 901–906. [CrossRef]
12. Esquivel-Elizondo, S.; Chairez, I.; Salgado, E.; Aranda, J.S.; Baquerizo, G.; Garcia-Peña, E.I. Controlled Continuous Bio-Hydrogen Production Using Different Biogas Release Strategies. *Appl. Biochem. Biotechnol.* **2014**, *173*, 1737–1751. [CrossRef] [PubMed]
13. Foglia, D.; Wukovits, W.; Friedl, A.; De Vrije, T.; Claassen, P.A.M. Fermentative hydrogen production: Influence of application of mesophilic and thermophilic bacteria on mass and energy balances. *Chem. Eng. Trans.* **2011**, *25*, 815–820.
14. Rafieenia, R.; Pivato, A.; Schievano, A.; Lavagnolo, M.C. Dark fermentation metabolic models to study strategies for hydrogen consumers inhibition. *Bioresour. Technol.* **2018**, *267*, 445–457. [CrossRef]
15. Kim, D.-H.; Shin, H.-S.; Kim, S.-H. Enhanced H2 fermentation of organic waste by CO2 sparging. *Int. J. Hydrog. Energy* **2012**, *37*, 15563–15568. [CrossRef]
16. Nguyen, T.A.D.; Han, S.J.; Kim, J.P.; Kim, M.S.; Sim, S.J. Hydrogen production of the hyperthermophilic eubacterium, Thermotoga neapolitana under N2sparging condition. *Bioresour. Technol.* **2010**, *101*, 38–41. [CrossRef]
17. Saady, N.M.C. Homoacetogenesis during hydrogen production by mixed cultures dark fermentation: Unresolved challenge. *Int. J. Hydrog. Energy* **2013**, *38*, 13172–13191. [CrossRef]
18. Wang, J.; Wan, W. Kinetic models for fermentative hydrogen production: A review. *Int. J. Hydrog. Energy* **2009**, *34*, 3313–3323. [CrossRef]
19. Bundhoo, M.A.Z.; Mohee, R. Inhibition of dark fermentative bio-hydrogen production: A review. *Int. J. Hydrog. Energy* **2016**, *41*, 6713–6733. [CrossRef]
20. Miller, G.L. Use of Dinitrosalicylic Acid Reagent for Determination of Reducing Sugar. *Anal. Chem.* **1959**, *31*, 426–428. [CrossRef]
21. Han, K.; Levenspiel, O. Extended monod kinetics for substrate, product, and cell inhibition. *Biotechnol. Bioeng.* **1988**, *32*, 430–447. [CrossRef] [PubMed]

© 2020 by the authors. Licensee MDPI, Basel, Switzerland. This article is an open access article distributed under the terms and conditions of the Creative Commons Attribution (CC BY) license (http://creativecommons.org/licenses/by/4.0/).

Article

Coproduction of Furfural, Phenolated Lignin and Fermentable Sugars from Bamboo with One-Pot Fractionation Using Phenol-Acidic 1,4-Dioxane

Li Ji [1,2], Pengfei Li [2,3], Fuhou Lei [3], Xianliang Song [2], Jianxin Jiang [1,2],* and Kun Wang [2],*

1. Beijing Advanced Innovation Center for Tree Breeding by Molecular Design, Beijing Forestry University, Beijing 100083, China; bjfu090524116@163.com
2. Department of Chemistry and Chemical Engineering, MOE Engineering Research Center of Forestry Biomass Materials and Bioenergy, Beijing Forestry University, Beijing 100083, China; lipfgxun@126.com (P.L.); sxlswd@163.com (X.S.)
3. Guangxi Key Laboratory of Chemistry and Engineering of Forest Products, College of Chemistry and Chemical Engineering, Guangxi University for Nationalities, Nanning 530006, China; leifuhougxun@126.com
* Correspondence: jiangjx2004@hotmail.com (J.J.); wangkun@bjfu.edu.cn (K.W.); Tel.: +86-010-6233-8267 (J.J.); +86-010-6233-6007 (K.W.)

Received: 16 September 2020; Accepted: 6 October 2020; Published: 12 October 2020

Abstract: A one-pot fractionation method of Moso bamboo into hemicellulose, lignin, and cellulose streams was used to produce furfural, phenolated lignin, and fermentable sugars in the acidic 1,4-dioxane system. Xylan was depolymerized to furfural at a yield of 93.81% of the theoretical value; however, the prolonged processing time (5 h) led to a high removal ratio of glucan (37.21%) in the absence of phenol. The optimum moderate condition (80 °C for 2 h with 2.5% phenol) was determined through the high fractionation efficiency. Consequently, 77.28% of xylan and 84.83% of lignin were removed and presented in the hydrolysate, while 91.08% of glucan was reserved in the solid portion. The formation of furfural from xylan remained high, with a yield of 92.92%. The extracted lignin was phenolated with an increasing content of phenolic hydroxyl. The fractionated lignin yield was 51.88%, which suggested this could be a low-cost raw material to product the activated carbon fiber precursor. The delignified pulp was subjected to enzymatic hydrolysis and the glucose yield reached up to 99.03% of the theoretical.

Keywords: one-pot fractionation with acidic 1,4-dioxane; Moso bamboo; furfural; phenolated lignin; enzymatic hydrolysis; high-efficiency fractionation

1. Introduction

Lignocellulosic biomass is a resource with great capacity and huge development potential of becoming a biorenewable feedstock for the bioeconomy [1,2]. It could be utilized as a renewable resource for producing materials, chemicals, and fuels [3,4]. However, for these purposes, it is essential to break down the crosslinking of the plant cell walls for the selective conversion of various chemicals [5,6]. For instance, hexose and pentose sugars can be transformed from the cellulose and hemicellulose fractions, the intermediate products for ethanol, furfural, 5-hydroxymethylfurfural, levulinic acid, and other chemicals productions [7]. However, the irreversibility of condensation tends to occur during the isolation process, which hinders the potential use of lignin as a renewable resource [8]. Consequently, reducing side reactions (e.g., condensation) during early-stage biomass fractionation is essential for establishing a lignocellulose-to-chemicals value chain with competitive advantages [9,10].

An alternative strategy, called "lignin first process" (LFP), was introduced in order to maximize lignin valorization through a combination of extracting and depolymerizing lignin with a stabilized intermediate product [2]. The LFP concept has been well summarized in recent articles [11]. The lignin oil that was obtained through LFP has a low molecular weight and it illustrates a sharp contrast from traditional pulping methods (e.g., kraft, soda), which produce lignin with more condensation and less functionality [11,12]. There are still drawbacks despite the widespread usage of this process. The occurrence of repolymerization promotes a lower monomer yield and selectivity under the harsh processing conditions. The upgrading of obtained products has created a bottleneck, requiring further biorefinery for bulk chemicals (e.g., phenol and propylene) [13]. The efficiency of biomass refining can further increase if appropriate modified synthetic materials were directly obtained in the reaction system.

The principle way for lignin to overcome the constraint in LFP is likely by directly converting lignin into a polymeric matrix for material applications. It is known that activated carbon fibers (ACFs) have a highly developed pore structure with a specific surface area and it is uniformly microporous; therefore, they are superior to commercial activated carbons [14]. However, the high cost and limited supply of this material are obstacles for high quality production on a large scale. Recently, the development of products from the phenol-formaldehyde reaction has attracted more attention. As products of the lignin-to-materials value chain, there is a promising future for these products in the fields of biology, environment, and energy [15–17]. Phenolated lignin is a desired product of phenol-formaldehyde reaction that is created by controlling the reaction conditions. Phenolated lignin is a raw material for ACFs production that has a low price due to its high accessibility and renewability [16]. Phenolation treatment was an effective method for improving lignin reactivity under basic and acidic conditions. Hence, phenolated lignin constitutes a better product for macromolecular application (i.e., lignin-to-materials).

Recently, targeting the one-pot fractionation has been reported to extract and depolymerize both lignin and hemicellulose for stable target products (phenolics and polyols, respectively) [18]. Relative studies have shown that a disadvantage of LFP is the irreversible modification of the polysaccharide fraction, in addition to bulk chemical refining issues [13]. The accurate control of the reaction conditions is required to exploit hemicellulose for stable and soluble platform chemicals [3,19]. One possible solution has been demonstrated in the dioxane solvent system with acid catalytic to convert solubilized hemicellulose into high-yield furfural via dehydration [20]. No relevant literature is available on simultaneously depolymerizing lignin and hemicellulose into a polymeric matrix and furfural through one-pot fractionation. Precise tuning of process conditions is required in order to further optimize the fractionation efficiency.

Here, a creative biorefinery concept was investigated, aiming at the one-pot fractionation of raw lignocellulose into three streams: (i) hemicellulose-derived furfural, (ii) lignin-derived phenolated lignin, and (iii) cellulosic pulp. Fractionation was performed in the mixture of 1,4-dioxane and water, hydrochloric acid (HCl) as a catalyst with phenol addition. After the reaction, the three main components of lignocellulose were separated, and an efficient method of transformation from those into the corresponding products was proposed. This work highlighted the potential benefits of phenol as an increasingly efficient promoter of the preferred process of fractionation and it provided a foundation for further development of integrated biorefinery with acidic 1,4-dioxane system.

2. Materials and Methods

2.1. Materials

Moso bamboo, from the International Center of Bamboo and Rattan (Beijing, China), was air dried and screened through 40 mesh. The water content of the Moso bamboo material was approximately 10.03%. The utilized chemicals were analytical grade.

2.2. Standard Procedure for Lignocellulose Fractionation

The substance of liquid phase was composed of 1,4-dioxane, HCl and deionized water. The 1,4-dioxane concentration in liquid flow was 90%, 70%, and 50% (v/v), corresponding to 900, 700, and 500 mL addition, according to a previous study [20], and the concentration of HCl was 4.2% (v/v). The deionized water was added to reach 1000 mL of the liquid volume. In solid phase, bamboo (100 g) and phenol (0, 1.0%, 1.5%, and 2.5%, w/v) were mixed with the solution for a subsequent fractionation process. The mixture was transferred to a stainless steel tubular homogeneous reactor (Shanghai Yanzheng Experimental Instrument Co., LTD., Shanghai, China), heated between 80 °C and 140 °C, and then kept for 1–5 h. Two parallel experiments were investigated, and detailed reaction conditions are shown in Table S1.

After the reaction, the mixture was filtered for solid-liquid separation. The solid portion was washed with 2000 mL deionized water, while the liquid portion was added into H_2O drop by drop to prepare two fractionated lignin samples (in order to ensure the pH value was 2, the volume of H_2O was 320 mL). The lignin in sample II is treated with phenol while that in sample I is not.

2.3. Enzymatic Hydrolysis

Commercial enzyme cellulase of Cellic Ctec2 from Novozymes (Beijing, China) was used. The filter paper activity of Cellic Ctec2 was 160 FPU/mL, and the contained cellobiase activity was 24 CBU/mL. The dosage of Cellic Ctec2 was 18 FPU/g glucan. The hydrolysis experiments were performed at 50 °C and 150 rpm for 72 h with three parallel experiments [21]. Acetate buffer solution (pH 4.8) and 5% (w/v) substrate concentration were proposed.

2.4. Analysis Methods

The NREL method was utilized to detect the material composition [9,22]. The amount of glucose, furfural and xylose were analyzed by HPLC-Waters 2695e (Waters, Milford, MA, USA) according to previous research [21].

The resultant solution of 1 mL was sampled for phenol analysis without any further treatment other than the addition of 100 μL as-prepared internal standard (100 mg octanol in 50 mL acetone) [23]. The solution (~1.1 mL) was analyzed with a gas chromatography (GC, Agilent 7890B series), according to the Reference [23]. The injection temperature was 250 °C with column temperature program of: 80 °C (1 min), 20 °C/min to 120 °C, 50 °C/min to 280 °C, and 280 °C (2 min). The detector temperature was of 300 °C. The production yields were calculated by following Equations (1)–(7).

$$\text{Furfural yield (\% of the theoretical value)} = \frac{\text{Furfural detected} \times 132 \times 100}{\text{Xylan in substrate} \times 96} \quad (1)$$

$$\text{Glucose yield (\% of the theoretical value)} = \frac{\text{Glucose detected} \times 0.9 \times 100}{\text{The theoretical glucan value in substrate}} \quad (2)$$

$$\text{Xylose yield (\% of the theoretical value)} = \frac{\text{Xylose detected} \times 0.88 \times 100}{\text{The theoretical xylan value in substrate}} \quad (3)$$

$$\text{Solid yield (\%)} = \text{Treated dry solid mass} \times 100/\text{Untreated dry solid mass} \quad (4)$$

$$\text{Retention ratio (\%)} = \text{The corresponding content of the treated material} \times \text{Solid yield} \times 100 \quad (5)$$

$$\text{Removal ratio (\%)} = \frac{(\text{The corresponding content of the untreated material} - \text{corresponding content of the treated material} \times \text{Solid yield}) \times 100}{\text{The corresponding yield of the untreated material}} \quad (6)$$

$$\text{Fractionated lignin ratio (\%)} = \text{Mass of acid extraction of lignin} / \text{(Mass of lignin in raw material + Mass of reacted phenol)} \tag{7}$$

2.5. Characterization of the Fractionated Lignin I and II

Mw and Mn, as well as polymer dispersity index (PDI) of the lignin-derived products, were testified by GPC-Waters 2695e (Waters, Milford, MA, USA) analysis, according to Wen's method [24]. FT-IR was analyzed by Bruker-VERTEX 70 (Bruker, Germany) according to Xie's method [25]. The 2D HSQC analysis was performed according to Yang's method [26].

2.6. Statistical Analysis

The data analysis was evaluated by SPSS statistics 22.0 according to Zhou's method [27]. The p-values were examined for the significant difference that was influenced by phenol-acidic 1,4-dioxane fractionation process, and each value of p less than 0.05 indicated the difference was significant [27].

2.7. Severity Factor

The severity factor (expressed as $\log R_0$) was used to quantify the intensity of phenol-acidic 1,4-dioxane fractionation process of biomass by Equation (8) [27], in which t is the residence time (mi) and T is the temperature of the reaction condition. The reaction temperature and time duration can be used as a parameter to critically control the experiment conditions, as a stronger reaction intensity of Moso bamboo increased the mass of dissolved components. The severity of this experimental design was from 1.49 to 3.26, as shown in Table S1. These conditions were selected in order to ensure distributed severities within the equal range of reaction time and temperature.

$$\log R_0 = \log \{t \times exp\ [(T - 100)/14.75]\} \tag{8}$$

3. Results and Discussion

Figure 1 shows the overall scheme of the one-pot fractionation with acidic 1,4-dioxane. 1,4-dioxane of 900 mL, phenol (2.5%, w/v), and HCl (4.2%, v/v) catalyst were added to a 2000 mL stainless steel tubular reactor, together with 100 g of Moso bamboo. As was determined using the NREL method, Moso bamboo were composed of ash 1.80%, lignin 24.90%, glucan 46.00%, hemicellulose (a.k.a. xylan (24.00%), substitute for arabinan (0.30%), and mannan (0.20%)). After the reaction, a majority of the glucose in the solid fraction was recovered after the process of enzymatic hydrolysis, while most of the xylose was recovered as furfural in liquid hydrolysates. The solution was analyzed with GC in order to calculate the quantity of phenol in the process: 17.9 g of phenol was reacted with 25.0 g of lignin, indicating the numerical results of phenolation reaction. The total amount of fractionated lignin II containing phenolated lignin was 22.3 g, and the recovered yield of fractionated lignin II was 51.88% when the dosage of phenol was 2.5% (w/v) at 80 °C. Overall, the majority of the glucose was recovered from the enriched cellulosic pulp, a large fraction of furfural was obtained from xylan, and the phenolated lignin level was improved by increasing the phenol dosage, which resulted in a pulp fraction with a relatively higher purity.

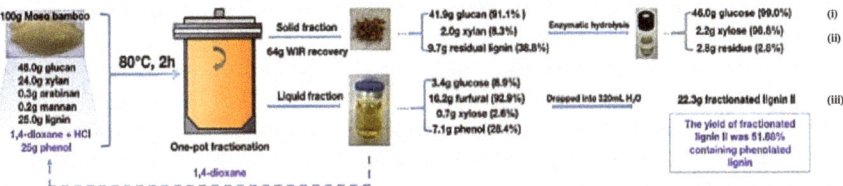

Figure 1. General scheme of the one-pot frationation process targeting (i) hemicellulose-derived furfural, (ii) lignin-derived phenolated lignin, and (iii) cellulosic pulp. WIR, water insoluble residues.

3.1. Influence of 1,4-Dioxane Dosage on Fractionation

Various solvents, including N,N-dimethylformamide, DMSO, methanol, acetone, 1,4-dioxane, and so on, would be appropriate options for lignin dissolution and separation [28]. As a common solvent for lignin, 1,4-dioxane still has some drawbacks, such as the irritation to eyes and upper respiratory tract, flammability, and higher cost [29]. Despite its weaknesses, 1,4-dioxane can be miscible with water and most organic solvents for dissolving more of lignin for value-added products.

Brown et al. reported that cyclic ethers, such as dioxane, could act as an efficient solvent for lignin extraction [30]. In the present case, 1,4-dioxane, boiling point at 100 °C, was selected as the extraction medium. This enables the separation of lignin from biomass into an acid solution and the subsequent recovery by distillation.

Figure 2 compared the distribution and composition of the biomass in the 1,4-dioxane/HCl system at various temperatures and reaction times. The retention ratio of glucan/xylan/lignin increased significantly by decreasing the 1,4-dioxane dosage from 900 to 500 mL/100 g bamboo ($p < 0.05$). The mass reduction of lignin with a 1,4-dioxane dosage of 500 mL/100 g bamboo during a 3 h reaction was 81.71%, while it was 87.80% under conditions of 900 mL/100 g bamboo with the same reaction time (Figure 2B). This phenomenon indicated that 1,4-dioxane played a key role in lignin dissolution and removal [20].

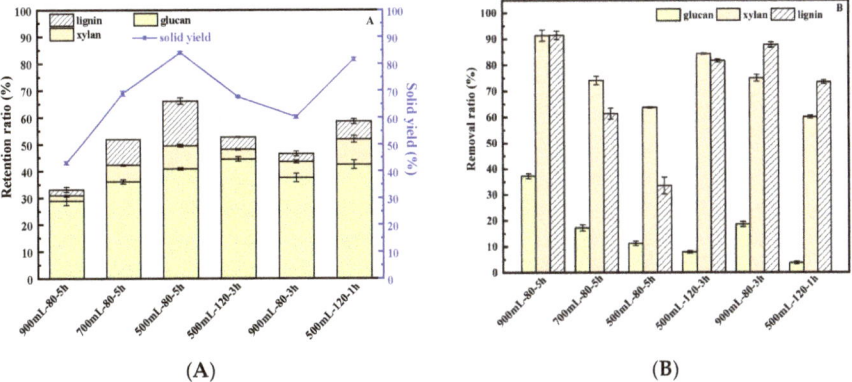

Figure 2. Effects of 1,4-dioxane (500, 700, and 900 mL/100 g bamboo), reaction temperature (80 and 120 °C), and reaction duration (1, 3, and 5 h) on (**A**) the retention of various components, and (**B**) glucan/xylan/lignin removal ratio based on the original weight of components.

Glucan remained solid in the pulp for subsequent enzymatic hydrolysis for glucose. The retention ratio of glucan increased at a higher temperature of 120 °C within a shorter reaction time of 1 h with a lower 1,4-dioxane dosage of 500 mL/100 g bamboo. This indicated that glucan degradation mainly depends on the 1,4-dioxane dosage and reaction time. As a member of Gramineae plants, bamboo presents a typical structure with brittle skeleton that contributes to the enhanced degradation of cell wall polysaccharides during the fractionation process. This explains how the content of xylan was substantially reduced ($p < 0.05$) [31]. Stein et al., suggested that the separation of cellulose, hemicellulose, and lignin occurred due to the difference in their structure and solubility properties [32].

For any biorefinery scheme, the separation of various lignocellulose fractions is crucial. Table 1 showed the effects of 1,4-dioxane (500–900 mL/100 g bamboo) under the given temperatures and time conditions. Increasing the amount of 1,4-dioxane increased the efficiency of soluble product and increased the fractionated lignin I yield to 54.00%. However, a greater decomposition of glucan was observed when the substrate was treated with 1,4-dioxane of 900 mL/100 g bamboo. No apparent improvement in lignin extraction was observed under a higher 1,4-dioxane dosage. A 1,4-dioxane dosage of 900 mL/100 g bamboo was considered to be the optimal concentration for fractionation efficiency, although it showed a weaker selectivity regarding glucan fraction reservation (Table 1).

Table 1. Comparison of 1,4-dioxane addition during the fractionation of lignin.

Entry	1,4-Dioxane Addition (mL/100 g Bamboo)	T (°C)	Time (h)	Fractionated Lignin I Yield (%)
1	900	80	5	54.00 ± 0.88
2	700	80	5	42.24 ± 0.93
3	500	80	5	23.95 ± 0.91

3.2. Influence of Phenol Addition on Fractionation

When considering the HCl-catalyzed reaction, various phenol amounts (1.0–2.5%, *w/v*) and varying temperatures (from 80–140 °C) were applied in this study in order to observe their influence on biomass fractionation. The fractionation process was performed for a shorter reaction time of 2 h when compared to the process without phenol. The weight loss of glucan substantially increased from 3.32–16.74% as reaction temperature increased from 80–100 °C, as shown in Table 2.

The pulp yield sharply decreased as temperature increased up to 140 °C, as seen in Figure S1 in the supplementary data. The retention ratio of glucan was less than 10% under the reaction condition. This was because glucan was dramatically degraded into liquid in the presence of phenol at a temperature no less than 120 °C. Because the majority of the glucan degradation mostly occurred in the initial phase of delignification, a higher temperature would lead to more glucan degradation [33]. High temperatures were not conducive to a higher fractionated lignin II yield. The retention ratio of lignin in the solid phase was slightly greater than 80%, as shown in Figure S1. Lignin was condensed under high intensity reaction conditions, and the condensation reactions occurred on the aromatic ring of lignin (C2, C5, or C6), forming new structures and remaining in the solid phase [8]. In Table 2, the weight loss of glucan under the reaction condition of 80 °C, 2 h with the addition of phenol was less than 10% (Entry 3). However, the loss of glucan reached 37% under the treatment condition of 80 °C, 5 h without phenol addition (Figure 2).

Table 2. Comparison of phenol addition in the fractionation of lignin treated with 1,4-dioxane of 900 mL/100 g bamboo at various temperatures.

Entry	Phenol Addition (%, w/v)	T (°C)	Time (h)	Removal Ratio of Lignin (%)	Yield of Fractionated Lignin II (%)	Weight Loss of Glucan (%)	Enzymatic Hydrolysis (%)	Furfural Conversion (% of Theoretical Yield)
1	1.0	80	2	78.62 ± 1.21	43.36 ± 1.39	3.32 ± 0.07	93.52 ± 1.13	89.78 ± 1.48
2	2.0	80	2	81.65 ± 0.76	46.54 ± 0.88	4.59 ± 0.11	95.14 ± 1.16	91.06 ± 1.26
3	2.5	80	2	84.83 ± 1.43	51.88 ± 0.75	8.92 ± 0.13	99.03 ± 0.97	92.92 ± 1.23
4	1.0	100	2	81.71 ± 1.28	37.92 ± 0.54	11.80 ± 0.42	96.25 ± 1.63	87.57 ± 0.79
5	1.5	100	2	85.22 ± 0.99	40.56 ± 0.92	15.50 ± 0.56	99.22 ± 0.86	88.94 ± 0.53
6	2.5	100	2	86.98 ± 0.23	42.79 ± 1.03	16.74 ± 1.01	99.26 ± 1.31	90.78 ± 1.09

Previous studies showed that the content of glucan was primarily affected by the reaction time. The internal crystalline structure of cellulose was damaged, and the degree of polymerization value of cellulose was decreased [6]. It can be observed that increasing the amount of phenol led to an increase in the soluble product yield, with 2.5% (w/v) of phenol being the optimal amount. More lignin was removed, as most was dissolved during the process, and the phenolated lignin was subsequently precipitated by placing it into water. Phenolated lignin is a more effective raw material for polymer production and has many potential uses in the fields of environment and energy [34]. The phenol reacted with the lignin and bound to its active site in order to form a higher activity product. A significantly higher extraction yield of fractionated lignin II (51.88%) was obtained ($p < 0.05$) with phenolated lignin being added.

3.3. Conversion of Hemicellulose to Furfural

Improved hydrolysis of the acid-catalyzed dioxane degraded a majority of the hemicellulose (a.k.a. xylan) into a liquid mixture for further furfural extraction. The previous results indicated that, under the acidic 1,4-dioxane experimental conditions, the removal ratio of Moso bamboo particles was, as follows: 33.48–91.45% of lignin, 60.10–91.37% of xylan, and 62.76–96.53% of glucan remained in the solid fraction (see Figure 2B).

Xylan depolymerization was observed via furfural formation in the aqueous phase. The highest furfural conversion yield of 93.81% of the theoretical value was consistent with highest furan yield in the presence of organic solvents and low water concentration (Table 3) [20]. Normally, xylose is produced during the depolymerization of xylan that is treated with a dilute acid. The dehydration of xylose treated with dilute Brønsted acid resulted in the formation of furfural, a platform molecule [35]. It can be observed that adding phenol to the one-pot fractionation system maintained the high yield of furfural conversion at approximately 90% of the theoretical as compared with the no-phenol-addition process (Tables 2 and 3). Increasing the amount of phenol led to an increase in the soluble product yield, with 2.5% (w/v) of phenol being the optimal amount. The formation of furfural from xylan in the hydrolysate had a high yield of 92.92%.

Table 3. Results of the theoretical furfural yield from Moso bamboo under various reaction conditions.

Entry	1,4-Dioxane Solvent Addition (mL/100 g Bamboo)	T (°C)	Time (h)	Furfural Conversion Yield (% of Theoretical Yield)
1	900	80	5	93.81 ± 1.05
2	700	80	5	69.03 ± 0.89
3	500	80	5	57.43 ± 1.01
4	500	120	3	80.42 ± 1.23
5	900	80	3	91.43 ± 0.77
6	500	120	1	72.46 ± 0.69

In previous studies, various organic solvents (e.g., γ-valerolactone (GVL), tetrahydrofuran (THF), and 2-methyltetrahydrofuran (2-MTHF)) were designed for furfural production from xylan and several catalysts were added in order to promote the production yield of furfural (Table 4) [36–41]. A bio-based system of 2-MTHF/H_2O was reported while using $SnCl_4$ as catalyst with the highest furfural yield of 78.1%. With a silicoaluminophosphate (SAPO) catalyst by continuous stripping and minimizing of the amount of humin output, an improved furfural yield of 63% could be achieved; however, this manufacturing technique was expensive and complicated [37]. Ionic liquids are preferable solvents for the conversion of xylan-type hemicellulose that have gained wide attention. A selective conversion of hemicellulose fraction was performed in [bmim] [HSO_4] ionic liquids to produce 36.2% furfural [40].

Table 4. Comparison of furfural production yield under various reaction conditions.

Starting Materials	Solvent	Catalyst	Conditions	Yield [a]	Reference
xylan	GVL-10% H_2O	$FeCl_3 \cdot 6H_2O$	170 °C, 100 min	68.6% [a]	Zhang et al. [36]
xylan	H_2O-THF	$AlCl_3 \cdot 6H_2O$-NaCl	140 °C, 45 min	64% [a]	Yang et al. [37]
xylan	2-MTHF-H_2O	$NaCl$-SO_4^{2-}/Sn-MMT	160 °C, 90 min	77.35% [a]	Lin et al. [38]
bagasse	H_2O/toluene	SAPO	170 °C, 8 h	63% [b]	Bhaumik et al. [39]
wheat straw	[bmim][HSO_4]	-	160 °C, 120 min	36.2% [b]	Dussan et al. [40]
corncob	H_2O	0.9 mmol H_2SO_4	160 °C, 5 h	71% [b]	Wang et al. [41]
Moso bamboo	H_2O/1,4-dioxane	HCl	80 °C, 3 h	91.47%	In this study
Moso bamboo	H_2O/phenol-acidic1,4-dioxane	HCl	80 °C, 2 h	92.92%	In this study

[a] furfural yield (accounted on C5 fraction in hemicellulose) [b] furfural yield (accounted on C5 fraction in biomass).

In this study, we provided a simpler, more competitive, and highly efficient fuel precursor for a more cost-effective furfural separation process. Generally, reactions at higher temperature with longer time and using a higher dosage of 1,4-dioxane indicate a higher yield of furfural. HCl is a type of Brønsted acid, and it was found to be the most effective catalyst for xylan conversion efficiency [42]. As seen in the tables, furfural almost reached its maximum amount considering the chemical composition of bamboo, while arabinose had minimal influence on the furfural yield.

3.4. Conversion of Lignin to Phenolated Lignin

Multiple methods have been reported on the increase of lignin reactivity, including methylolation/ hydroxymethylation, phenolation, demethylation, oxidation/reduction, and hydrolysis/hydrogenolysis. A high consumption of hazardous chemicals commonly was required in order to obtain a high conversion yield through oxidation/reduction and demethylation. It showed poor selectivity of compound formation during hydrolysis and hydrogenolysis, including the free ortho numbers and para positions that are associated with the aromatic ring. In addition, the hydroxymethylation of lignin showed a higher selectivity of maximum activation levels, but moderate yield. It has been reported that phenolated lignin has stronger mechanical behaviors than raw or hydroxymethylated lignin [43]. Furthermore, the β-O-4 bond could generate new reactive sites with aicidic catalysts [44]. Nonaka et al. reported that phenolation could be directly applied to wood pulp directly in order to obtain phenolated lignin [45].

GPC results showed that the added phenol had a positive result on the Mw of the lignin-derived products (see Table 5 and Figure S2). The Mw of lignin I and lignin II were lower than that of pretreated solid lignin [45]. As a typical organosolv, dioxane, with its appropriate solubility parameter, is an attractive isolation agent of lignin fraction [18]. Phenol, as a suppressor, was proven to prevent repolymerization of lignin [45]. This effect increased the solubility of the lignin with the reduced molecular weight. The products from the above strategies have higher monomer yields and selectivities than kraft or soda lignin. Phenolysis occurred through the attachment of phenol to lignin through the catalysis of acid, forming the substances with low molecular weight that corresponded to the results of Funaoka's research, as shown in Figure 3 [46].

Table 5. The Mw, the Mn, and polymer dispersity index (PDI) of the lignin-deriver products.

Lignin-Derived Products	Mn (g/mol)	Mw (g/mol)	PDI
lignin I	6.515×10^2 (±2.467%)	8.237×10^2 (±2.241%)	1.264 (±1.895%)
lignin II	5.611×10^2 (±3.048%)	6.936×10^2 (±2.924%)	1.236 (±4.224%)

Figure 3 presents the FT-IR spectra of the fractionated lignin I and II samples from the acid-catalyzed 1,4-dioxane-treated Moso bamboo. The bands at 1423 cm^{-1} corresponded to the aromatic skeletal vibrations, while the deformed C-H combining with the aromatic ring vibration was at 1461 cm^{-1} [25]. These two characteristic absorption peaks gradually receded with phenol addition, which was in accordance with the 2D-heteronuclear singular quantum correlation (HSQC) result. After phenolation, the peak intensity of lignin II at 1370 cm^{-1} was strengthened as compared to that of lignin I, indicating the increased content of phenolic hydroxyl. The impact of this increase was primarily due to the combination of phenol with lignin and not the hydrolysis of aryl ethers [47].

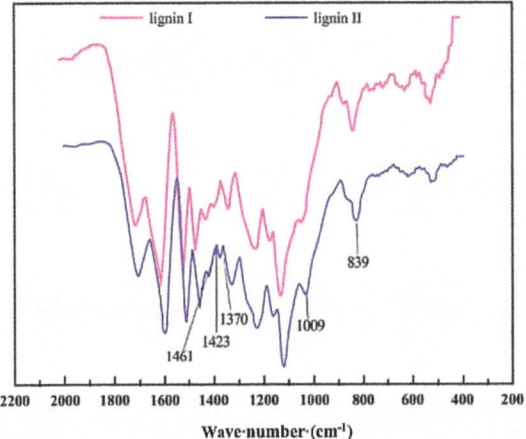

Figure 3. FT-IR spectra of Moso bamboo treated with acid-catalyzed 1,4-dioxane of 900 mL at 80 °C for 5 h and treated with 900 mL 1,4-dioxane at 80 °C with phenol (2.5%, w/v) for 2 h.

The phenolic hydroxyl group of lignin has important functions in further applications, specifically its positive influence on lignin molecule functionalization [48,49]. Higher substitution rates of lignin to phenol could be achieved through the enhanced lignin solubility and chemical reactivity related to the increased hydroxyl content [49,50]. Historically, the study of macromolecular applications has driven much of the research regarding developed lignin valorization by analyzing variations of hydroxyl content. This research contains a vast array of promising applications for carbon nanofibers [12,44]. It has been proven that the as-obtained product of phenolated lignin can be applied to macromolecular material production, in particular to the low-cost production of activated carbon fiber precursors [10,16,51], providing the opportunity to create the achievement of lignin-to-materials [9,52].

Figure 4 shows the 2D-HSQC spectra of the fractionated lignin I and II. The peak assignments/distribution were based on Ref. [26]. Figure 3 depicts the sub-units of the identified lignin. The cross signals of methoxyls ($\delta C/\delta H$ 55.9/3.73), β-O-4, and the Cβ-Hβ correlations corresponding to the β-β linkages were in the side chain of lignin. It can be seen from the spectrum that the β-O-4 substructure intensities in the spectrum of lignin were obviously lower when compared to the original Moso bamboo lignin detected in previous work [24]. The β-O-4 linkage constitutes the most abundant link of lignin. During the process, bamboo was treated with acidic dioxane both with and without phenol; the β-O-4 linkages of the bamboo lignin were broken because of the lability of alkyl aryl ethers under acidic conditions. Thus, most of the cleavage occurred on the β-O-4 bonds. Weak signals of side-chain linkages (β-O-4, β-5, β-β) represent a low Mw and a relatively high content of phenolic hydroxyl in the lignin. Yang et al. observed similar results [26]. Hence, this lignin could be a valuable phenolic material.

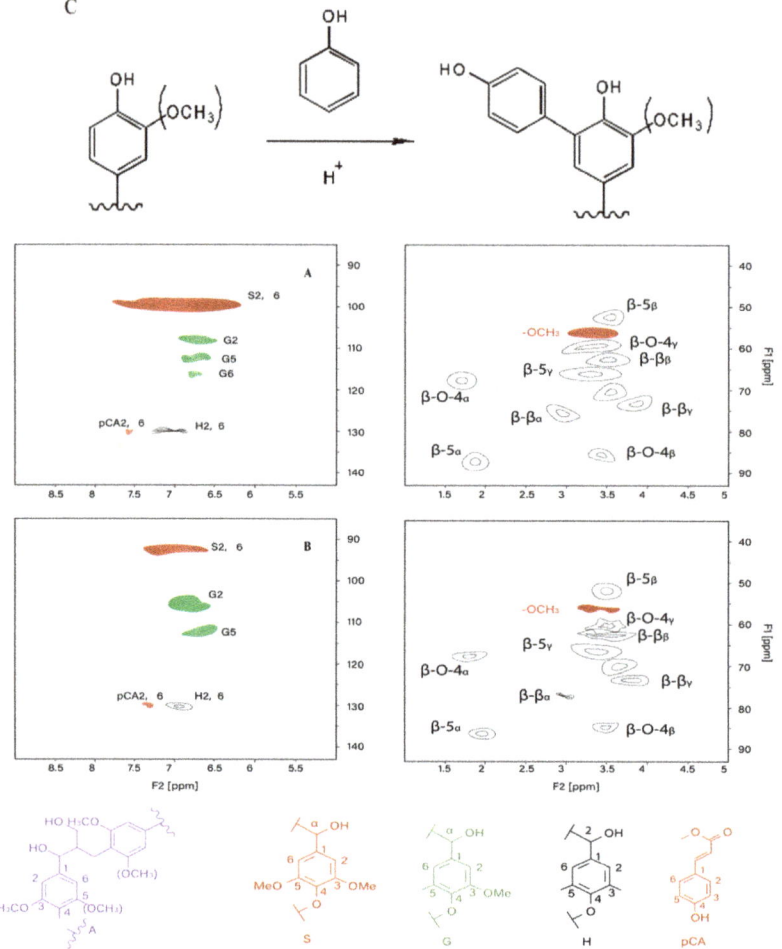

Figure 4. 2D-HSQC NMR spectra of lignin treated with 1,4-dioxane of 900 mL/100 g bamboo, acid-catalyzed at 80 °C for Moso bamboo. (**A**) Samples of the fractionated lignin I treated for 5 h without phenol. (**B**) Samples of the fractionated lignin II treated for 2 h with 2.5% phenol (*w/v*). (**C**) The reaction scheme of phenol with lignin units. The hollow area represents the reacted group.

The main 2D-HSQC spectra cross-signals of lignin in the aromatic region were appointed to the lignin units, including H, G, and S structures [26]. A significant difference was observed in the fractionated lignin II and I, treated with and without phenol respectively. The phenolation of lignin units was a crucial step toward achieving a higher phenolated lignin yield when compared to the reaction system without phenol addition [49]. The obvious disappearance of G_6 and $H_{2,6}$ linkages illustrated that phenolated lignin was produced by the reaction between phenol and lignin under acidic 1,4-dioxane with added phenol. Figure 4 shows the reaction scheme. Low intensity p-coumarate (PCA) signals were detected in the spectrum. The existence of phenol caused a frequency shift on the electron cloud of C_α.

3.5. Conversion of Carbohydrate Pulps to Fermentable Sugars

The effective saccharification of cellulose to fermentable sugars is the fundamental step required for carbohydrate pulp valorization [3,53,54]. Figure 5 shows both the glucose and xylose yields of raw Moso bamboo and acidic 1,4-dioxane treated residues. The lowest glucose yield of 36.38% was obtained after a 72-h enzymatic hydrolysis with raw material as the substrate. The efficiency of enzymatic hydrolysis was enhanced after reaction with acidic 1,4-dioxane. The glucose yield was improved with the increased dosage of 1,4-dioxane during the fractionation process. The glucose yield of 98.74% was obtained after 72 h from the substrate treated with 1,4-dioxane of 900 mL/100 g bamboo. Most of the xylose was recovered in the form of furfural in liquid hydrolysates and furfural yield was 93.81% of theoretical value (Table 3). There was still 2.10 g xylan (of 100 g Moso bamboo) remained in the solid fraction after acid-catalyzed dioxane reaction. An extra 2.33 g of xylose was later recovered from xylan after enzymatic hydrolysis; thus, the overall yield ratio was 97.75%. Evidence suggested that the xylan removal improved the enzymatic hydrolysis of cellulose [55]. It can be seen from Figure 2 that the removal ratio of xylan and lignin was close to 90%, obtaining a solid pulp of high-purity glucan with a retention ratio over 90%. The glucose yield reached 92.46% of the theoretical value for 12 h after enzymatic hydrolysis (Figure 5). This was essentially ascribed to the effective removal of lignin and xylan from the treated substrates, so that cellulase and cellobiase had more space to get into glucan and hydrolyzed glucan to glucose [56].

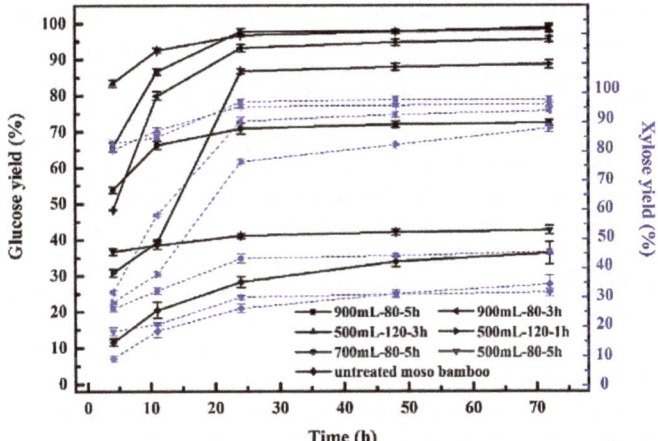

Figure 5. Effects of 1,4-dioxane (500, 700, and 900 mL/100 g bamboo), reaction temperature (80 and 120 °C), and reaction duration (1, 3, and 5 h) on the retention of various components, and glucan/xylan/lignin removal ratio based on the original weight of components.

After being treated in the presence of phenol, the remaining glucan rapidly depolymerized to glucose in the presence of enzymes (99.03% in 72 h), as seen from Table 2. Significant enhancement of enzymatic hydrolysis was achieved when most of the hemicellulose was removed and degraded to furfural under the conditions in this study. The high value was comparable to that obtained in the biodegradation of commercial Avicel [57]. The substrate that was treated in the absence of phenol showed similar enzymatic digestibility. This additional glucose production likely illustrated that phenol addition had no negative influence on enzyme binding and enzymatic digestibility.

Cellulase and xylanase are regular modular proteins with two modules—carbohydrate-binding module (CBM) and catalytic module [58]. Hence, it is necessary to know the adsorption process of cellulase and lignin in order to understand how glucan conversion is enhanced by the delignification of substrates. The influence of 1,4-dioxane dosage on alteration activities of β-glucosidase and exo-β-1,4-glucanase in the supernatants from the hydrolysis process could be observed in Figure S3 (see supplementary data). The activities of both β-glucosidase and exo-β-1,4-glucanase in the supernatant with the substrate that was treated with 1,4-dioxane of 500 mL/100 g bamboo at 120 °C for 3 h showed higher values of the overall trend compared with the enzyme activity of the sample treated with 1,4-dioxane of 700 mL/100 g bamboo at 80 °C for 5 h. This demonstrated that the decreased enzyme activity was due to a higher amount in the cellulase adsorption on lignin based on the difference in its content (81.71% vs. 61.40%) of the two samples. The increased degree of delignification reduced the effective adsorption sites for cellulase enzyme, and it made the binding force between lignin and cellulase weaker [58]. The substrate treated with 1,4-dioxane of 500 mL/100 g bamboo at 120 °C for 3 h was apparently preferably adsorbed onto the cellulase enzyme and it had a higher content of glucan with a larger surface area owing to the higher removal rate/amount of lignin and xylan. In addition, the higher content of lignin in the sample that was treated with 1,4-dioxane of 700 mL/100 g bamboo at 80 °C for 5 h included the higher percentage of β-O-4 linkage. This indicated that the inhibitory actions of water-soluble components, such as the phenolics, could partially explain the observed inhibition of the enzymes. The inhibited activity of β-glucosidases caused by the simple phenolic-compounds was increased with their contact time [55].

4. Conclusions

An integrated process for selectively fractionating bamboo into furfural, phenolated lignin, and glucan in one pot has been investigated in an acid-catalyzed system that consists of phenol at mild temperature of 80 °C. Consequently, the disentanglement of the internal bamboo structure was achieved in a single step. In the aqueous solution, the conversion yield of furfural was of 92.92% and the lignin fractions could be converted to high performance fractionated lignin with the yield of 51.88% containing phenolic material. The high-purity glucan was exhibited with the enzymatic hydrolysis yield of 99.03%, which could be comparable to that of the commercial Avicel. This method presented a high efficiency and energy saving approach for lignocellulosic biomass fractionation and application.

Supplementary Materials: http://www.mdpi.com/1996-1073/13/20/5294/s1. Method of enzyme activity analyses, Table S1: The specific reaction condition during fractionation process, Figure S1: The effects of reaction temperature (120 °C and 140 °C) and phenol additions (1.0% and 1.5%, w/v) on lignin and glucan yield of Moso bamboo. lignin A: residual lignin; lignin B: dissolved lignin (phenolated lignin included), Figure S2: Molecular weight distributions for lignin fractions lignin I and lignin II, Figure S3: The effects of enzyme activities on Moso bamboo hydrolysis. A: The β-glucosidase activity during the time period 0 to 72 h of enzymatic hydrolysis. B: The exo-β-1,4-glucanase activity during the time period 0 to 72 h.

Author Contributions: Conceptualization, L.J. and P.L.; methodology, F.L.; software, L.J.; validation, X.S., K.W.; formal analysis, X.S., K.W.; data analysis, P.L.; writing, L.J.; editing, J.J. All authors have read and agreed to the published version of the manuscript.

Funding: GXFC17-18-08-Guangxi Key Laboratory of Chemistry Engineering of Forest Products; AD18126005-Specific research project of Guangxi for research bases and talents and 2017PT13-Fundamental Research Funds for the Central Universities.

Acknowledgments: This work was supported by International Center of Bamboo and Rattan (Beijing, China) and Novozymes Investment Co., Ltd. (Beijing, China).

Conflicts of Interest: The authors declare no conflict of interest.

References

1. Welker, C.M.; Balasubramanian, V.K.; Petti, C.; Rai, K.M.; DeBolt, S.; Mendu, V. Engineering Plant Biomass Lignin Content and Composition for Biofuels and Bioproducts. *Energies* **2015**, *8*, 7654–7676. [CrossRef]
2. Nitsos, C.K.; Choli, P.T.; Matis, K.A.; Triantafyllidis, K.S. Optimization of Hydrothermal Pretreatment of Hardwood and Softwood Lignocellulosic Residues for Selective Hemicellulose Recovery and Improved Cellulose Enzymatic Hydrolysis. *ACS Sustain. Chem. Eng.* **2016**, *4*, 4529–4544. [CrossRef]
3. Kohli, K.; Prajapati, R.; Sharma, B.K. Bio-Based Chemicals from Renewable Biomass for Integrated Biorefineries. *Energies* **2019**, *12*, 233. [CrossRef]

4. Sun, Z.; Fridrich, B.; De Santi, A.; Elangovan, S.; Barta, K. Bright Side of Lignin Depolymerization: Toward New Platform Chemicals. *Chem. Rev.* **2018**, *118*, 614–678. [CrossRef] [PubMed]
5. Soto, L.R.; Byrne, E.; van Niel, E.W.; Sayed, M.; Villanueva, C.C.; Hatti-Kaul, R. Hydrogen and polyhydroxybutyrate production from wheat straw hydrolysate using Caldicellosiruptor species and Ralstonia eutropha in a coupled process. *Bioresour. Technol.* **2019**, *272*, 259–266. [CrossRef]
6. Silveira, L.; Morais, C.; da Costa Lopes, A.M.; Olekszyszen, D.N.; Bogel-Łukasik, R.; Andreaus, J.; Pereira, L. Current Pretreatment Technologies for the Development of Cellulosic Ethanol and Biorefineries. *ChemSusChem* **2015**, *8*, 3366–3390. [CrossRef]
7. Cantero, D.A.; Martínez, C.; Bermejo, M.D.; Cocero, M.J. Simultaneous and selective recovery of cellulose and hemicellulose fractions from wheat bran by supercritical water hydrolysis. *Green Chem.* **2015**, *17*, 610–618. [CrossRef]
8. Constant, S.; Wienk, H.L.J.; Frissen, A.E.; De Peinder, P.; Boelens, R.; Van Es, D.S.; Grisel, R.J.H.; Weckhuysen, B.M.; Huijgen, W.J.J.; Gosselink, R.; et al. New insights into the structure and composition of technical lignins: A comparative characterisation study. *Green Chem.* **2016**, *18*, 2651–2665. [CrossRef]
9. Yu, H.; Xu, Y.; Hou, J.; Ni, Y.; Liu, S.; Liu, Y.; Wu, C. Efficient Fractionation of Corn Stover for Biorefinery Using a Sustainable Pathway. *ACS Sustain. Chem. Eng.* **2020**, *8*, 3454–3464. [CrossRef]
10. Renders, T.; van den Bosch, S.; Koelewijn, S.F.; Schutyser, W.; Sels, B.F. Lignin-first biomass fractionation: The advent of active stabilisation strategies. *Energy Environ. Sci.* **2017**, *10*, 1551–1557. [CrossRef]
11. Galkin, M.V.; Samec, J.M. Lignin Valorization through Catalytic Lignocellulose Fractionation: A Fundamental Platform for the Future Biorefinery. *ChemSusChem* **2016**, *9*, 1544–1558. [CrossRef] [PubMed]
12. Ragauskas, A.J.; Beckham, G.T.; Biddy, M.J.; Chandra, R.; Chen, F.; Davis, M.F.; Davison, B.H.; Dixon, R.A.; Gilna, P.; Keller, M.; et al. Lignin Valorization: Improving Lignin Processing in the Biorefinery. *Science* **2014**, *344*, 1246843. [CrossRef]
13. Liao, Y.; Koelewijn, S.F.; van den Bossche, G.; Van Aelst, J.; van den Bosch, S.; Renders, T.; Navare, K.; Nicolaï, T.; Van Aelst, K.; Maesen, M.; et al. A sustainable wood biorefinery for low–carbon footprint chemicals production. *Science* **2020**, *367*, 1385–1390. [CrossRef] [PubMed]
14. You, X.; Koda, K.; Yamada, T.; Uraki, Y. Preparation of electrode for electric double layer capacitor from electrospun lignin fibers. *Holzforschung* **2015**, *69*, 1097–1106. [CrossRef]
15. Balgis, R.; Sago, S.; Anilkumar, G.M.; Ogi, T.; Okuyama, K. Self-Organized Macroporous Carbon Structure Derived from Phenolic Resin via Spray Pyrolysis for High-Performance Electrocatalyst. *ACS Appl. Mater. Interfaces* **2013**, *5*, 11944–11950. [CrossRef] [PubMed]
16. Alma, M.H.; Salan, T.; Zhao, G. Effect of different acid catalysts on the properties of activated carbon fiber precursors obtained from phenolated wheat straw. *J. Nat. Fibers* **2018**, *16*, 781–794. [CrossRef]
17. Jiang, X.; Liu, J.; Du, X.; Hu, Z.; Chang, H.-M.; Jameel, H. Phenolation to Improve Lignin Reactivity toward Thermosets Application. *ACS Sustain. Chem. Eng.* **2018**, *6*, 5504–5512. [CrossRef]
18. Renders, T.; Cooreman, E.; van den Bosch, S.; Schutyser, W.; Koelewijn, S.-F.; Vangeel, T.; Deneyer, A.; Bossche, G.V.D.; Courtin, C.M.; Sels, B.F. Catalytic lignocellulose biorefining in n-butanol/water: A one-pot approach toward phenolics, polyols, and cellulose. *Green Chem.* **2018**, *20*, 4607–4619. [CrossRef]
19. Lopes, H.J.S.; Bonturi, N.; Miranda, E.A. Rhodotorula toruloides Single Cell Oil Production Using Eucalyptus urograndis Hemicellulose Hydrolysate as a Carbon Source. *Energies* **2020**, *13*, 795. [CrossRef]
20. Shuai, L.; Amiri, M.T.; Questell-Santiago, Y.M.; Héroguel, F.; Li, Y.; Kim, H.; Meilan, R.; Chapple, C.; Ralph, J.; Luterbacher, J.S. Formaldehyde stabilization facilitates lignin monomer production during biomass depolymerization. *Science* **2016**, *354*, 329–333. [CrossRef]
21. Tang, Y.; Chandra, R.P.; Sokhansanj, S.; Saddler, J.N. Influence of steam explosion processes on the durability and enzymatic digestibility of wood pellets. *Fuel* **2018**, *211*, 87–94. [CrossRef]
22. Sluiter, A.; Hames, B.; Ruiz, R.; Scarlata, C.; Sluiter, J.; Templeton, D. Determination of structural carbohydrates and lignin in biomass. *Lab. Anal. Proced.* **2008**, *1617*, 1–16.
23. Cao, Z.X.; Ye, Y.J.; Guo, Y.X. Determination of residual phenol in rigid phenolic foam by Gas Chromatography. *Guangdong Chem. Indus.* **2012**, *8*, 27–28.
24. Wen, J.-L.; Sun, Z.; Sun, Y.-C.; Sun, R.; Xu, F.; Sun, R.C. Structural Characterization of Alkali-Extractable Lignin Fractions from Bamboo. *J. Biobased Mater. Bioenergy* **2010**, *4*, 408–425. [CrossRef]

25. Peng, C.; Chen, Q.; Guo, H.; Hu, G.; Li, C.; Wen, J.-L.; Wang, H.; Zhang, T.; Zhao, Z.K.; Sun, R.; et al. Effects of Extraction Methods on Structure and Valorization of Corn Stover Lignin by a Pd/C Catalyst. *ChemCatChem* **2017**, *9*, 1135–1143. [CrossRef]
26. Yang, S.; Zhang, Y.; Yue, W.; Wang, W.; Wang, Y.-Y.; Yuan, T.-Q.; Sun, R. Valorization of lignin and cellulose in acid-steam-exploded corn stover by a moderate alkaline ethanol post-treatment based on an integrated biorefinery concept. *Biotechnol. Biofuels* **2016**, *9*, 1–14. [CrossRef] [PubMed]
27. Zhou, Z.; You, Y.; Lei, F.-H.; Li, P.; Jiang, J.; Zhu, L. Enhancement of enzymatic hydrolysis of sugarcane bagasse by pretreatment combined green liquor and sulfite. *Fuel* **2017**, *203*, 707–714. [CrossRef]
28. Kosyakov, D.S.; Ul'Yanovskii, N.V.; Anikeenko, E.A.; Gorbova, N.S. Negative ion mode atmospheric pressure ionization methods in lignin mass spectrometry: A comparative study. *Rapid Commun. Mass Spectrom.* **2016**, *30*, 2099–2108. [CrossRef]
29. Chen, R.; Liu, C.; Johnson, N.W.; Zhang, L.; Mahendra, S.; Liu, Y.; Dong, Y.; Chen, M. Removal of 1,4-dioxane by titanium silicalite-1: Separation mechanisms and bioregeneration of sorption sites. *Chem. Eng. J.* **2019**, *371*, 193–202. [CrossRef]
30. Brown, W. Solution properties of lignin. Thermodynamic properties and molecular weight determinations. *J. Appl. Polym. Sci.* **1967**, *11*, 2381–2396. [CrossRef]
31. He, M.-X.; Wang, J.-L.; Qin, H.; Shui, Z.-X.; Zhu, Q.-L.; Wu, B.; Tan, F.-R.; Pan, K.; Hu, Q.-C.; Dai, L.-C.; et al. Bamboo: A new source of carbohydrate for biorefinery. *Carbohydr. Polym.* **2014**, *111*, 645–654. [CrossRef] [PubMed]
32. Stein, T.V.; Grande, P.M.; Kayser, H.; Sibilla, F.; Leitner, W.; De María, P.D. From biomass to feedstock: One-step fractionation of lignocellulose components by the selective organic acid-catalyzed depolymerization of hemicellulose in a biphasic system. *Green Chem.* **2011**, *13*, 1772. [CrossRef]
33. Jafari, V.; Nieminen, K.; Sixta, H.; Van Heiningen, A. Delignification and cellulose degradation kinetics models for high lignin content softwood Kraft pulp during flow-through oxygen delignification. *Cellulose* **2015**, *22*, 2055–2066. [CrossRef]
34. Tan, M.X.; Sum, Y.N.; Ying, J.Y.; Zhang, Y. A mesoporous poly-melamine-formaldehyde polymer as a solid sorbent for toxic metal removal. *Energy Environ. Sci.* **2013**, *6*, 3254–3259. [CrossRef]
35. Shi, X.; Wu, Y.; Yi, H.; Rui, G.; Li, P.; Yang, M.; Wang, G. Selective Preparation of Furfural from Xylose over Sulfonic Acid Functionalized Mesoporous Sba-15 Materials. *Energies* **2011**, *4*, 669–684. [CrossRef]
36. Zhang, L.; Yu, H.; Wang, P.; Li, Y. Production of furfural from xylose, xylan and corncob in gamma-valerolactone using $FeCl_3 \cdot 6H_2O$ as catalyst. *Bioresour. Technol.* **2014**, *151*, 355–360. [CrossRef]
37. Yang, Y.; Hu, C.; Abu-Omar, M.M. Synthesis of Furfural from Xylose, Xylan, and Biomass Using $AlCl_3 \cdot 6H_2O$ in Biphasic Media via Xylose Isomerization to Xylulose. *ChemSusChem* **2012**, *5*, 405–410. [CrossRef]
38. Lin, Q.; Li, H.; Wang, X.; Jian, L.; Ren, J.; Liu, C.; Sun, R. SO_4^{2-}/Sn-MMT Solid Acid Catalyst for Xylose and Xylan Conversion into Furfural in the Biphasic System. *Catalysts* **2017**, *7*, 118. [CrossRef]
39. Bhaumik, P.; Dhepe, P.L. Efficient, Stable, and Reusable Silicoaluminophosphate for the One-Pot Production of Furfural from Hemicellulose. *ACS Catal.* **2013**, *3*, 2299–2303. [CrossRef]
40. Dussan, K.; Girisuta, B.; Lopes, M.; Leahy, J.; Hayes, M.H.B. Effects of Soluble Lignin on the Formic Acid-Catalyzed Formation of Furfural: A Case Study for the Upgrading of Hemicellulose. *ChemSusChem* **2016**, *9*, 492–504. [CrossRef]
41. Wang, T.; Li, K.; Liu, Q.; Zhang, Q.; Qiu, S.; Long, J.; Chen, L.; Ma, L.; Zhang, Q. Aviation fuel synthesis by catalytic conversion of biomass hydrolysate in aqueous phase. *Appl. Energy* **2014**, *136*, 775–780. [CrossRef]
42. Luo, Y.; Li, Z.; Li, X.; Liu, X.; Fan, J.; Clark, J.H.; Hu, C. The production of furfural directly from hemicellulose in lignocellulosic biomass: A review. *Catal. Today* **2019**, *319*, 14–24. [CrossRef]
43. Lu, J.; Wang, M.; Zhang, X.; Heyden, A.; Wang, F. β-O-4 Bond Cleavage Mechanism for Lignin Model Compounds over Pd Catalysts Identified by Combination of First-Principles Calculations and Experiments. *ACS Catal.* **2016**, *6*, 5589–5598. [CrossRef]
44. Jiang, W.; Chu, J.; Wu, S.; Lucia, L.A. Secondary pyrolysis pathway of monomeric aromatics resulting from oxidized β-O-4 lignin dimeric model compounds. *Fuel Process. Technol.* **2017**, *168*, 11–19. [CrossRef]
45. Nonaka, H.; Kobayashi, A.; Funaoka, M. Lignin isolated from steam-exploded eucalyptus wood chips by phase separation and its affinity to Trichoderma reesei cellulase. *Bioresour. Technol.* **2013**, *140*, 431–434. [CrossRef]

46. Funaoka, M.; Matsubara, M.; Seki, N.; Fukatsu, S. Conversion of native lignin to a highly phenolic functional polymer and its separation from lignocellulosics. *Biotechnol. Bioeng.* **1995**, *46*, 545–552. [CrossRef]
47. Wen, J.-L.; Sun, R.; Xue, B.-L.; Sun, R.-C. Quantitative structural characterization of the lignins from the stem and pith of bamboo (Phyllostachys pubescens). *Holzforschung* **2013**, *67*, 613–627. [CrossRef]
48. Serrano, L.; Esakkimuthu, E.S.; Marlin, N.; Brochier-Salon, M.-C.; Mortha, G.; Bertaud, F.; Serrano, L. Fast, Easy, and Economical Quantification of Lignin Phenolic Hydroxyl Groups: Comparison with Classical Techniques. *Energy Fuels* **2018**, *32*, 5969–5977. [CrossRef]
49. Zhang, F.-D.; Jiang, X.; Lin, J.; Zhao, G.J.; Chang, H.-M.; Jameel, H. Reactivity improvement by phenolation of wheat straw lignin isolated from a biorefinery process. *New J. Chem.* **2019**, *43*, 2238–2246. [CrossRef]
50. Chen, H.; Liu, N.; Qu, X.; Joshee, N.; Liu, S. The effect of hot water pretreatment on the heavy metal adsorption capacity of acid insoluble lignin from Paulownia elongata. *J. Chem. Technol. Biotechnol.* **2017**, *93*, 1105–1112. [CrossRef]
51. Lin, J.; Shang, J.; Zhao, G. Effect of the size of spinneret on the thermal stability of chemically liquefied wood. *Wood Res.-Slovak.* **2014**, *59*, 731–738.
52. Aro, T.; Fatehi, P. Production and Application of Lignosulfonates and Sulfonated Lignin. *ChemSusChem* **2017**, *10*, 1861–1877. [CrossRef] [PubMed]
53. Hirsch, E.; Pataki, H.; Domján, J.; Farkas, A.; Vass, P.; Fehér, C.; Barta, Z.; Nagy, Z.K.; Marosi, G.J.; Csontos, I.; et al. Inline noninvasive Raman monitoring and feedback control of glucose concentration during ethanol fermentation. *Biotechnol. Prog.* **2019**, *35*, e2848. [CrossRef] [PubMed]
54. Qu, H.; Liu, B.; Li, L.; Zhou, Y. A bifunctional recoverable catalyst based on phosphotungstic acid for cellulose hydrolysis to fermentable sugars. *Fuel Process. Technol.* **2020**, *199*, 106272. [CrossRef]
55. Zhai, R.; Hu, J.; Saddler, J.N. What Are the Major Components in Steam Pretreated Lignocellulosic Biomass That Inhibit the Efficacy of Cellulase Enzyme Mixtures? *ACS Sustain. Chem. Eng.* **2016**, *4*, 3429–3436. [CrossRef]
56. Rodríguez, F.; Sánchez, A.; Parra, C. Role of Steam Explosion on Enzymatic Digestibility, Xylan Extraction, and Lignin Release of Lignocellulosic Biomass. *ACS Sustain. Chem. Eng.* **2017**, *5*, 5234–5240. [CrossRef]
57. Ramírez, I.M.; Román, M.G.; Arteaga, A.F. Rhamnolipids: Highly Compatible Surfactants for the Enzymatic Hydrolysis of Waste Frying Oils in Microemulsion Systems. *ACS Sustain. Chem. Eng.* **2017**, *5*, 6768–6775. [CrossRef]
58. Martín, C.; Volkov, P.V.; Rozhkova, A.M.; Puls, J.; Sinitsyn, A.P. Comparative study of the enzymatic convertibility of glycerol- and dilute acid-pretreated sugarcane bagasse using Penicillium and Trichoderma-based cellulase preparations. *Ind. Crop. Prod.* **2015**, *77*, 382–390. [CrossRef]

© 2020 by the authors. Licensee MDPI, Basel, Switzerland. This article is an open access article distributed under the terms and conditions of the Creative Commons Attribution (CC BY) license (http://creativecommons.org/licenses/by/4.0/).

Article

Investigation of Ash Deposition Dynamic Process in an Industrial Biomass CFB Boiler Burning High-Alkali and Low-Chlorine Fuel

Hengli Zhang, Chunjiang Yu, Zhongyang Luo * and Yu'an Li

State Key Laboratory of Clean Energy Utilization, Zhejiang University, Hangzhou 310027, China; hlzhang@zju.edu.cn (H.Z.); chunjiang@zju.edu.cn (C.Y.); lyua@zju.edu.cn (Y.L.)
* Correspondence: zyluo@zju.edu.cn; Tel.: +86-571-8795-2440

Received: 17 January 2020; Accepted: 26 February 2020; Published: 2 March 2020

Abstract: The circulating fluidized bed (CFB) boiler is a mainstream technology of biomass combustion generation in China. The high flue gas flow rate and relatively low combustion temperature of CFB make the deposition process different from that of a grate furnace. The dynamic deposition process of biomass ash needs further research, especially in industrial CFB boilers. In this study, a temperature-controlled ash deposit probe was used to sample the deposits in a 12 MW CFB boiler. Through the analysis of multiple deposit samples with different deposition times, the changes in micromorphology and chemical composition of the deposits in each deposition stage can be observed more distinctively. The initial deposits mainly consist of particles smaller than 2 μm, caused by thermophoretic deposition. The second stage is the condensation of alkali metal. Different from the condensation of KCl reported by most previous literatures, KOH is found in deposits in place of KCl. Then, it reacts with SO_2, O_2 and H_2O to form K_2SO_4. In the third stage, the higher outer layer temperature of deposits reduces the condensation rate of KOH significantly. Meanwhile, the rougher surface of deposits allowed more calcium salts in fly ash to deposit through inertial impact. Thus, the elemental composition of deposits surface shows an overall trend of K decreasing and Ca increasing.

Keywords: deposit; biomass industrial boiler; alkali metal; circulating fluidized bed

1. Introduction

Compared with fossil energy, biomass has the characteristics of zero CO_2 emission. Biomass direct combustion technology for power generation, as a relatively mature large-scale biomass utilization technology, has been widely used in China. Biomass power generation in China mainly adopts grate furnaces and circulating fluidized bed (CFB) boilers [1]. CFB boilers have a wide combustion adaptability, low combustion temperature and good gas–solid mixing characteristics in the bed, making them a very suitable combustion technology for biomass. This technology has taken up a large proportion in biomass power generation projects in recent years in China [1]. The dominant elements in biomass ash are Ca, K, Cl and S [2,3]. During combustion, K element in biomass fuel is released to the gas phase in the form of KCl, KOH and K_2SO_4 [4–10]. These potassium-containing compounds cause severe slagging due to their low melting temperature [11–16].

The study of deposition on the heating surface of biomass power plants is mostly carried out by analyzing mature deposits sampled in boiler maintenance. The innermost layer of deposits typically contains iron oxide, KCl, and K_2SO_4. The intermediate layer contains melt KCl and other ash particle [1,12,13,17–19]. In some deposits, an in situ reaction between KCl and the captured ash particles will occur, leading to the release of chlorine-containing gases. But, through the analysis of mature deposits, the dynamic deposition process is difficult to study clearly and in detail. On the one hand, the long-term deposition process makes the deposit undergo a chemical reaction and changes its morphology and chemical composition, compared to its initial state when the deposit formed. On the other hand, the mature deposit derived from biomass ash is tight and hard, making it difficult to stratify in order to analyze and study the deposition process, especially the extremely thin deposit layer at the initial deposition stage.

To avoid the disadvantages of mature deposits, an ash deposition probe has been used to "real-time" sample the deposits in some of the literature. Some research works focused on the deposit formation rate and studied the influences of the probe surface temperature, flue gas temperature, fuel type (alkali metal content), probe exposure time and additives on the deposit formation rate [20–22]. Others used the deposition probe to get short-term deposits to analyze the deposition mechanism [23]. In addition, an ash deposition probe was also used to measure corrosion rate caused by alkali metal deposits at different flue gas temperatures and probe surface temperatures on three types of superheater steel [24–26]. These literatures focused more on the deposit formation rate and the corrosion rate. Although some investigation of the deposition mechanism using an ash deposition probe has been carried out, the analysis was based on one deposit sample, resulting in an unclear dynamic deposition process.

In this work, the deposit build-up process on an ash deposition probe at different deposition times is studied. Through the analysis of multiple deposits samples with different deposition times, the dynamic deposition process was shown more clearly and in detail. In addition, this work is carried out in a full-scale 12 MW biomass CFB boiler, enriching the previous studies in the field of biomass CFB boiler ash deposition, especially industrial CFB boilers.

2. Materials and Methods

2.1. MW Biomass CFB Boiler

In order to study the mechanism of deposition on a high-temperature superheater of a CFB boiler burning biomass with high-alkali metal content, the experiment was carried out in a CFB boiler at a biomass power plant (2 × 12 MW) in Jiangsu province, China. The boiler operates at a medium temperature and pressure (main steam parameters are 450 °C, 3.82 MPa.), and has the capacity of 75 tons/h. Deposition occurred during the operation process.

The 12 MW medium temperature and medium pressure CFB boiler of a biomass power plant adopts the low-temperature fluidized combustion scheme of Zhejiang University. A sketch is given in Figure 1. Through combustion organization and boiler design, the temperature of different regions in the furnace is well controlled, effectively inhibiting the alkali metal problem in the combustion process of high-alkali biomass fuel. This scheme makes use of the characteristics of high combustion activity and low burnout temperature of biomass semi-coke to control the temperature of the boiler dense phase area and return loop at 750 °C or lower, thus eliminating the bed material agglomeration or slagging in the dense phase area and return material loop. Through the arrangement of the heating surface and the control of the circulation ratio, the central volatile combustion area temperature is controlled under 820 °C, minimizing the migration of alkali metals into the gas phase, thus inhibiting deposition and high-temperature corrosion on the heating surface. This scheme has been successfully applied in a large number of engineering practices, significantly inhibiting various alkali metal problems on the basis of ensuring combustion efficiency.

Nevertheless, ash deposition on the heat transfer surface of ZJU CFB boiler has been also found, which is a hard and thick deposit on the superheater.

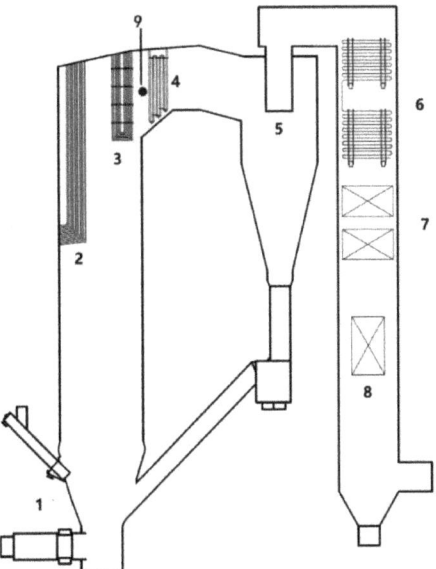

Figure 1. The 12 MW biomass circulating fluidized bed (CFB) boiler: 1. feeder; 2. water wall; 3. platen superheater; 4. high-temperature superheater; 5. cyclone separator; 6. low-temperature superheater; 7. economizer; 8. air preheater; 9. location of deposition probe.

2.2. Fuel

The blended biomass fuel consisted of rice husks (30%), bark (30%), construction plywood wastes (30%) and other biomass fuels such as lees of wine (10%). The industrial analysis, heating value and elemental analysis of the blended fuel are shown in Table 1. K was analyzed by atomic absorption spectroscopy (AAS) after microwave digestion, while Cl and SO_4^{2-} were analyzed by ion chromatography. Compared to typical straw biomass, this blended biomass fuel had a similar K content and much lower Cl content, which may have affected the migration of alkali metal (K) during combustion and deposition characteristic.

Table 1. Physicochemical properties of blended biomass fuel.

Item	Symbol	Blended Biomass Fuel
Proximate analysis	moisture (%, ad)	7.66
	ash (%, ad)	5.04
	volatile (%, ad)	69.4
	fixed carbon (%, ad)	17.9
Ultimate analysis	C (%, ad)	43.78
	H (%, ad)	4.59
	N (%, ad)	1.35
	S (%, ad)	0.09
	O (%, ad)	37.49
Heating value	Q (kJ/kg, ad)	18312
Inorganic constituent (dry basis)	K (%)	0.41
	SO_4^{2-} (%)	0.45
	Cl (%)	0.09

X-ray diffraction (XRD) and energy spectrum (EDS) results of fly ash are shown in Figure 2 and Table 2. The high contents of K and SO_4^{2-} are also reflected in the composition of fly ash. The main components of fly ash are identified as $CaSO_4$, $CaCO_3$, SiO_2 and $K_2Ca(SO_4)_2 \cdot H_2O$. SiO_2 comes from biomass or bed material. Due to the low content of other components and the complexity of ash composition, no other obvious diffraction peaks are found in the XRD diagram.

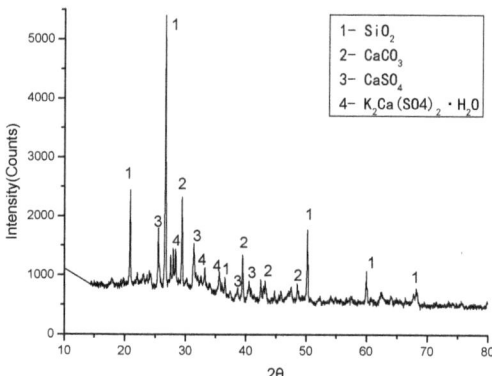

Figure 2. XRD result of fly ash.

Table 2. Elemental composition of fly ash.

	C	O	Na	Mg	Al	Si	P	S	Cl	K	Ca	Fe
Fly ash	16.61	30.91	1.30	2.05	5.38	9.72	0.72	2.40	0.87	7.37	19.26	3.40

2.3. Collection and Analysis of Deposits

A temperature-controlled deposit probe was used to sample deposits inside the boiler in this study, set in the high temperature superheater region through a reserved hole prepared for the boiler soot blowing system, shown in Figure 3. Through the analysis of the deposit surface microstructure and elemental distributions with different sampling time intervals, this study studied the dynamic process of deposition.

The normal operation of the boiler combustion temperature is between 750 and 800 °C, while the flue gas temperature in the high temperature superheater inlet is about 700 °C. A sketch of the ash deposit probe is shown in Figure 4. The whole system included a fan, an electric control valve, a digital display controller, a thermocouple and a stainless-steel probe. The probe was about 2.5 m in length with an outer diameter of 38 mm. A thin endless stainless-steel belt was attached to the probe, which was used to sample the ash deposits. In order to keep the same surface conditions as the high temperature superheater, the probe was cooled by air to keep a stable temperature of the sampling belt surface, with a K-type thermocouple testing its temperature. Considering the main steam temperature of the boiler (450 °C) and the flue gas temperature in the high temperature-superheater region (700 °C), the surface temperature of the high-temperature superheater was estimated to be around 500 °C. So, the stable endless stainless-steel belt surface temperature was set at 500 °C.

The sampling tests lasted 1, 2, 5, 15, 24, and 48 h. In each test, we used a new sampling belt to attach to the probe. After the deposits were captured on the sampling belt, the sample, along with the sampling belt, were analyzed directly by SEM-EDS. The micromorphology of ash deposits was observed with SEM. Simultaneously, the element contents of the microstructure observed were measured by EDS. Considering that EDS analysis has a relatively large testing error for element O and C, all the EDS results of deposits given below removed the data of O and C, then were normalized again. Through the experiment and analysis mentioned above, dynamic deposition experiment can conduct the dynamic collection of ash accumulation at each stage of deposition, and observe the changes in morphology and chemical composition more intuitively at the initial stage of deposition.

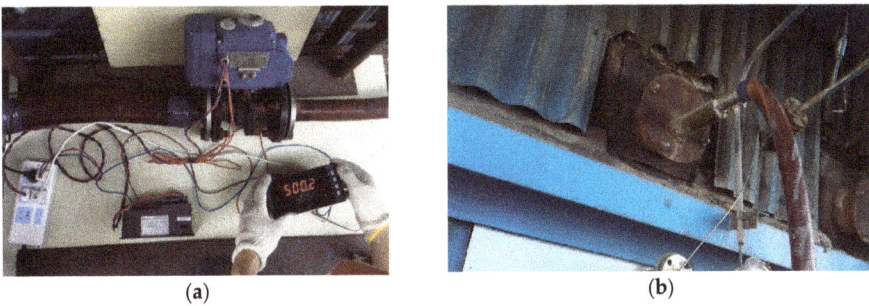

Figure 3. Temperature-controlled deposit probe system. (**a**) Temperature controlling system; (**b**) The deposit probe.

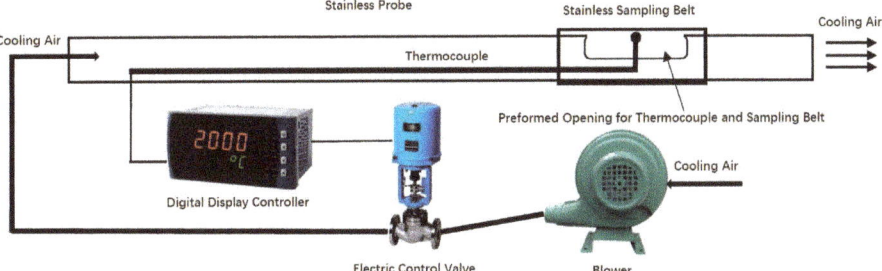

Figure 4. Sketch of temperature-controlled ash deposit probe.

3. Results and Discussion

Photos of the deposit sampling belt are shown in Figure 5. Within the first 5 h of deposition, the surface is shiny and dark gray. After 15 h, the khaki deposits began to accumulate. The deposit sampling belt is heated to 500 °C in muffle furnace in air as a control sample (Figure 6). The surface is smooth, except some fine particles, which are supposed to be caused by dust in the muffle furnace.

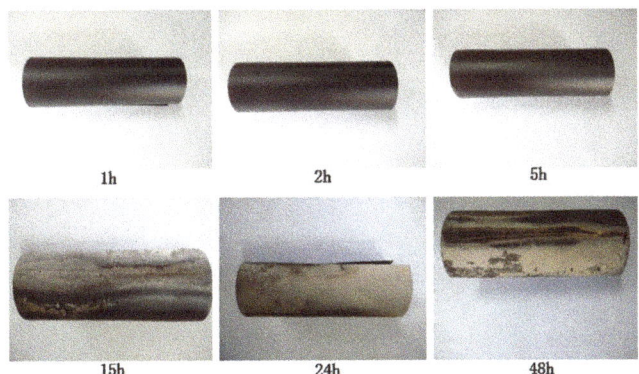

Figure 5. Sampling belt at different deposition times.

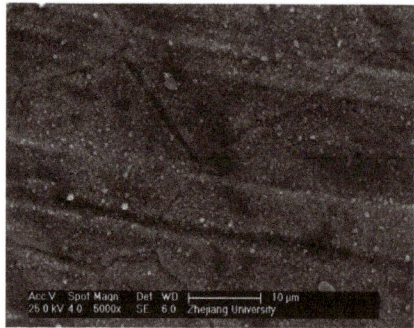

Figure 6. The surface of original sampling belt.

3.1. Initial Deposition

In the period of initial deposition (samples of 1 h deposits and 2 h deposits), the main contents of metal elements are Ca and Mg. On the contrary, the content of K is very low, with only 3.34% (1 h) and 2.44% (2 h), respectively, as shown in Table 3, which is very different to the viewpoint that initial deposition is caused by alkali salt, as argued in some studies [12,13,17–19]. The surface microstructure under SEM is shown in Figure 7. The deposits consist of particles less than 1 μm in diameter. According to deposition theory, the deposition of particles in this size is dominated by thermoplastic forces, moving from the high-temperature towards the low-temperature area, captured by the heating surface.

(a)　　　　　　　　　　　　　　　　(b)

Figure 7. Deposits after 1 and 2 h under 5000 times magnified SEM field. (**a**) Deposits after 1 h; (**b**) Deposits after 2 h.

Table 3. Elemental composition of 1, 2 h deposits and fly ash particles less than 2μm.

Element	Na	Mg	Al	Si	P	S	Cl	K	Ca
Deposits of 1 h	4.35	15.46	4.56	12.16	5.37	19.93	2.85	3.34	32.02
Deposits of 2 h	1.83	16.18	4.58	11.84	5.26	18.25	1.83	2.44	37.79
Fly ash < 2 μm	2.27	18.01	4.37	12.11	5.13	19.09	1.51	2.09	35.41

3.2. Migration and Deposition of Alkali Metal

The deposits after 5 and 15 h show a significantly different microstructure from the previous two deposit samples. The deposit particles become larger, as shown in Figure 8. The EDS analysis (Table 4) of the whole SEM field shows that the main elements are Ca, S, Si and K. Compared to the deposits after 1 and 2h, the content of K begins to rise, but the remarkable presence of Cl is still not detected, proving that alkali does not exist in form of KCl. The large smooth crystalline particles in Spot 1 and 2 are measured by EDS separately, and the results are also shown in Table 4. Their main elements are K, Ca, S. Converting mass to atomic ratio, the atomic number of K, Ca and S conforms to the relation (K/2 + Ca) = S. It is speculated that the smooth crystalline particles in the field of SEM are a mixture or compound salt of $CaSO_4$ and K_2SO_4. Similar crystalline particles in other SEM fields were also selected for EDS analysis, and similar results were obtained.

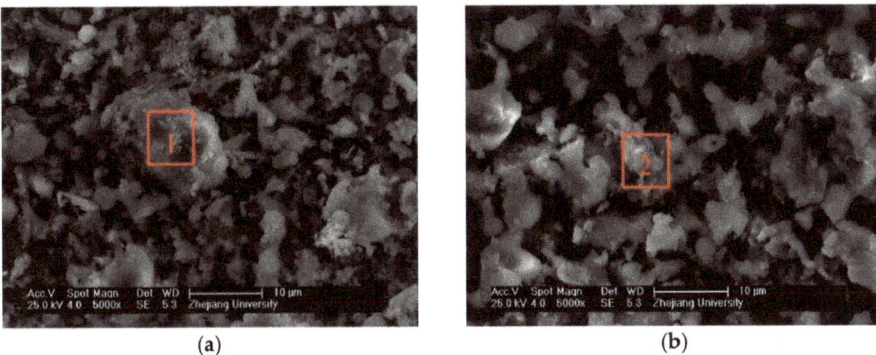

Figure 8. Deposits after 5 and 15 h under 5000 times magnified SEM field. (a) Deposits after 5 h; (b) Deposits after 15 h.

Table 4. Elemental composition of 5 h and 15 h deposits.

Element	Na	Mg	Al	Si	P	S	Cl	K	Ca
5 h	3.97	5.80	4.06	15.14	1.11	26.32	0.00	22.84	20.75
15 h	4.66	3.87	3.34	9.78	1.46	25.20	0.25	34.36	17.07
Spot 1	1.18	1.08	1.25	2.84	1.08	34.91	0.57	27.59	29.46
Spot 2	1.92	1.11	1.21	2.97	0.56	34.85	0.00	31.25	26.10

According to the K migration mechanism, during the process of combustion, K will release into the gas phase in the form of KCl, KOH and K_2SO_4, depending on the combustion temperature and fuel characteristics. In this study, with the condition of combustion temperature (800 °C), K_2SO_4 exists in a solid state in fly ash after combustion and will not release into the gas phase. However, the enrichment of potassium salt (K_2SO_4) in deposits is much higher than potassium content in fly ash, which is almost impossible directly formed as it is from the fly ash particles' inertial impact. Meanwhile, the potassium salt in deposits of 5 and 15 h have a pure crystalline structure and distribute continuously in the whole visual field of SEM, also indicating that the salt may experience condensation, melting or a chemical reaction rather than direct deposition from fly ash solid particles.

A large number of studies have indicated that the main source of element K in heating surface deposits is gas phase K in biomass combustion [4,5,8–10]. In this study, with the condition of combustion temperature (800 °C) and low Cl content in fuel, K probably enters the gas phase mainly in the form of KOH in place of KCl and K_2SO_4. Also, in the daily operation of this power plant, the moisture of the blend is up to 30%, enhancing the dissociation reaction to form KOH during combustion with the help of water vapor [8]. Further testing of potassium species in deposits is required. Due to the detection limit of complex mixtures of XRD analysis, minority constituents of the deposits cannot be clearly detected. So, the water-soluble parts of the deposits were selected to be further tested. Deposits after 15 h were dissolved in water and filtered the insoluble matter. Then, the solution was heated and evaporated in order to crystallize to conduct the XRD analysis. The result is shown in Figure 9. Without the interference of the complex insoluble part, the XRD result shows more details. Except for K_2SO_4 and $CaSO_4$, which have already been found in SEM-EDS analysis, weak diffraction peaks of

KOH and Na_2SO_4 are also found in the water-soluble part of deposits. This indicates that the source of K_2SO_4 in deposits is probably KOH.

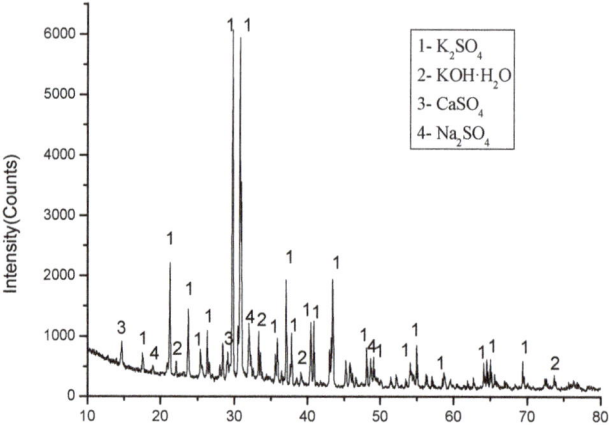

Figure 9. XRD result of water-soluble part of the 15 h deposits.

At this deposition stage, the main deposition process is as follows: during biomass combustion, under a relatively low combustion temperature, low Cl and high moisture content in biomass fuel, element K releases into the gas phase mainly in the form of KOH. Cooled by the high-temperature superheater, KOH condenses on the deposit's surface, and then reacts with SO_2, O_2 and H_2O, turning KOH into K_2SO_4. Both condensation and the reaction cause K_2SO_4 to form a pure crystalline structure and distribute continuously.

It is worth noting that the condensation of KOH and thermophoretic deposition of fine particles in fly ash happen at the same time, all caused by the temperature gradient. In deposition times of 1 and 2 h, thermophoretic deposition rate is higher than the condensation of KOH, becoming the dominant deposition method and leading to high Mg and Ca content and low K content. On the contrary, in deposition times of 5 and 15 h, condensation rate of KOH may be sped up by the rough deposit surface, which provides a condensation nucleus for gas phase KOH. As a consequence, compared to 1 and 2 h, the relative content of K in the 5 and 15 h deposition increased, while the content of Ca decreased, leading to different deposit surface microstructures.

3.3. Development of Deposition

When the deposition time reaches 24 and 48 h, the deposits becomes more and more dense. A continuous crystalline structure in Figure 10 (Spot 3 and 4) is still the main deposit component. In terms of K, Ca and S atomic number ratio, this structure is still a mixture or compound salt of $CaSO_4$ and K_2SO_4. EDS analysis (Table 5) of the whole SEM visual field shows that K, Ca and S element are not in conformity with the relation of (K/2 + Ca = S), but (K/2 + Ca) > S. This suggests that there are other potassium or calcium slats apart from K_2SO_4 and $CaSO_4$. The deposit surface is very rough, providing conditions for the formation of deposits by the inertial impact of fly ash. Combining the XRD result of fly ash shown in Figure 2, $CaCO_3$ is likely to be present in these deposits.

Comparing the K and Ca content in 15, 24 and 48 h deposits, a significant decrease occurs in K content, while Ca shows an increasing trend. This is caused by the change in condensation rate of KOH and inertial impact deposition rate. With the development of deposits, the deposition layer becomes thicker, and the surface temperature increases. The temperature gradient between flue gas and deposits surface decreases, reducing the condensation rate of KOH. In contrast, the rough surface allowed more Ca salts in fly ash to deposit through inertial impact. Thus, the deposits show an overall trend of K decreasing and Ca increasing.

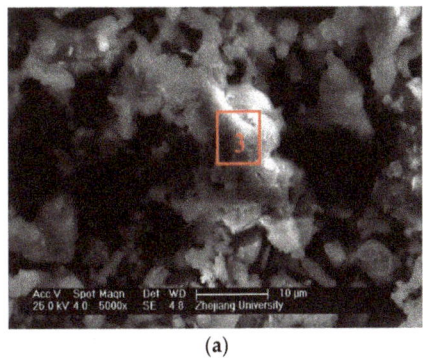 

(a) (b)

Figure 10. Deposits after 24 and 48 h under 5000 times magnified SEM field. (**a**) Deposits after 24 h; (**b**) Deposits after 48 h.

Table 5. Elemental composition of 24 h and 48 h deposits.

Element	Na	Mg	Al	Si	P	S	Cl	K	Ca
24 h	1.53	3.72	1.81	7.49	1.03	28.72	0.00	26.18	29.54
48 h	1.25	5.48	1.97	9.47	1.15	25.49	0.00	20.58	34.62
Spot 3	0.45	1.61	0.46	1.17	0.41	37.28	0.00	28.99	29.62
Spot 4	0.75	1.00	0.42	1.69	0.42	37.14	0.07	28.19	30.29

4. Conclusions

The dynamic process of ash deposition on the high-temperature superheater of a 12 MW biomass CFB boiler was studied. A temperature-controlled deposit sampling system was used to sample deposit samples with different deposition times. Combustion temperature and inorganic elements in biomass fuel significantly impact the formation of deposits. The deposition process of burning low Cl content biomass fuel is different from that of high Cl content biomass in CFB boiler. Almost no KCl was found in deposits. K exists in the form of K_2SO_4, which is derived by the reaction of KOH, O_2, H_2O and SO_2. The deposition process can be identified as three stages:

1. Initial deposition, In the initial stage, the temperature gradient leads to the deposition of fine particles from the flue gas through thermophoretic deposition.
2. Condensation of KOH, Under the condition of a relatively low combustion temperature, low Cl content and high moisture content, K in biomass fuel will release into the gas phase mainly in the form of KOH. In the second stage, after the initial deposition, the surface becomes rough, leading to an acceleration of gas phase KOH condensation on the deposits surface. Then, KOH reacts with H_2O, O_2 and SO_2 in flue gas to form the enrichment of K_2SO_4 in deposits.
3. Inertial impact of fly ash, In the third stage, the rougher surface allows fly ash to deposit through inertial impact. At the same time, with the thickening of the deposition layer, the outer layer temperature increases, leading to a significant reduction in KOH condensation rate. Thus, the elemental composition of the deposit's surface shows an overall trend of K decreasing and Ca increasing.

Author Contributions: Conceptualization, C.Y. and Z.L.; Investigation, H.Z. and Y.L.; Methodology, H.Z. and C.Y.; Supervision, Z.L.; Writing – original draft, H.Z.; Writing – review & editing, H.Z. and Z.L. All authors have read and agreed to the published version of the manuscript.

Funding: This research was funded by International Cooperation Foundation for China-USA, NSFC-NSF 51661125012.

Conflicts of Interest: The authors declare no conflict of interest.

References

1. Li, L.; Yu, C.; Huang, F.; Bai, J.; Fang, M.; Luo, Z. Study on the deposits derived from a biomass circulating fluidized-bed boiler. *Energy Fuels* **2012**, *26*, 6008–6014. [CrossRef]
2. Zajac, G.; Szyszlak-Barglowicz, J.; Golebiowski, W.; Szczepanik, M. Chemical characteristics of biomass ashes. *Energies* **2018**, *11*, 2885. [CrossRef]
3. Vassilev, S.V.; Baxter, D.; Andersen, L.K.; Vassileva, C.G. An overview of the composition and application of biomass ash. Part 1. Phase-mineral and chemical composition and classification. *Fuel* **2013**, *105*, 40–76. [CrossRef]
4. Chen, C.; Yu, C.; Zhang, H.; Zhai, X.; Luo, Z. Investigation on K and Cl release and migration in micro-spatial distribution during rice straw pyrolysis. *Fuel* **2016**, *167*, 180–187. [CrossRef]
5. Tchapda, A.H.; Pisupati, S.V. A review of thermal co-conversion of coal and biomass/waste. *Energies* **2014**, *7*, 1098–1148. [CrossRef]
6. Jensen, P.A.; Frandsen, F.J.; Dam-Johansen, K.; Sander, B. Experimental investigation of the transformation. and release to gas phase of potassium and chlorine during straw pyrolysis. *Energy Fuels* **2000**, *14*, 1280–1285. [CrossRef]
7. Bjorkman, E.; Stromberg, B. Release of chlorine from biomass at pyrolysis and gasification conditions. *Energy Fuels* **1997**, *11*, 1026–1032. [CrossRef]
8. Knudsen, J.N.; Jensen, P.A.; Dam-Johansen, K. Transformation and release to the gas phase of Cl, K, and S during combustion of annual biomass. *Energy Fuels* **2004**, *18*, 1385–1399. [CrossRef]
9. Wang, G.; Jensen, P.A.; Wu, H.; Frandsen, F.J.; Sander, B.; Glarborg, P. Potassium capture by Kaolin, Part 1: KOH. *Energy Fuels* **2018**, *32*, 1851–1862. [CrossRef]
10. Wang, G.; Jensen, P.A.; Wu, H.; Frandsen, F.J.; Sander, B.; Glarborg, P. Potassium capture by Kaolin, Part 2: K_2CO_3, KCl, and K_2SO_4. *Energy Fuels* **2018**, *32*, 3566–3578. [CrossRef]
11. Jeong, T.; Sh, L.; Kim, J.; Lee, B.; Jeon, C. Experimental investigation of ash deposit behavior during co-combustion of bituminous coal with wood pellets and empty fruit bunches. *Energies* **2019**, *12*, 2087. [CrossRef]
12. Jin, X.; Ye, J.; Deng, L.; Che, D. Condensation behaviors of potassium during biomass combustion. *Energy Fuels* **2017**, *31*, 2951–2958. [CrossRef]
13. Niu, Y.; Zhu, Y.; Tan, H.; Hui, S.; Jing, Z.; Xu, W. Investigations on biomass slagging in utility boiler: Criterion numbers and slagging growth mechanisms. *Fuel Process. Technol.* **2014**, *128*, 499–508. [CrossRef]
14. Miles, T.R.; Miles, T.R.; Baxter, L.L.; Bryers, R.W.; Jenkins, B.M.; Oden, L.L. Boiler deposits from firing biomass fuels. *Biomass Bioenerg.* **1996**, *10*, 125–138. [CrossRef]
15. Sandberg, J.; Karlsson, C.; Fdhila, R.B. A 7 year long measurement period investigating the correlation of corrosion, deposit and fuel in a biomass fired circulated fluidized bed boiler. *Appl. Energy* **2011**, *88*, 99–110. [CrossRef]
16. Baxter, L.L.; Miles, T.R.; Miles, T.R.; Jenkins, B.M.; Milne, T.; Dayton, D.; Bryers, R.W.; Oden, L.L. The behavior of inorganic material in biomass-fired power boilers: Field and laboratory experiences. *Fuel Process. Technol.* **1998**, *54*, 47–78. [CrossRef]
17. Jensen, P.A.; Frandsen, F.J.; Hansen, J.; Dam-Johansen, K.; Henriksen, N.; Horlyck, S. SEM investigation of superheater deposits from biomass-fired boilers. *Energy Fuels* **2004**, *18*, 378–384. [CrossRef]
18. Valmari, T.; Lind, T.M.; Kauppinen, E.I.; Sfiris, G.; Nilsson, K.; Maenhaut, W. Field study on ash behavior during circulating fluidized-bed combustion of biomass. 2. Ash deposition and alkali vapor condensation. *Energy Fuels* **1999**, *13*, 390–395. [CrossRef]
19. Zhou, H.; Jensen, P.A.; Frandsen, F.J. Dynamic mechanistic model of superheater deposit growth and shedding in a biomass fired grate boiler. *Fuel* **2007**, *86*, 1519–1533. [CrossRef]
20. Tobiasen, L.; Skytte, R.; Pedersen, L.S.; Pedersen, S.T.; Lindberg, M.A. Deposit characteristic after injection of additives to a Danish straw-fired suspension boiler. *Fuel Process. Technol.* **2007**, *88*, 1108–1117. [CrossRef]

21. Zbogar, A.; Jensen, P.A.; Frandsen, F.J.; Hansen, J.; Glarborg, P. Experimental investigation of ash deposit shedding in a straw-fired boiler. *Energy Fuels* **2006**, *20*, 512–519. [CrossRef]
22. Bashir, M.S.; Jensen, P.A.; Frandsen, F.; Wedel, S.; Dam-Johansen, K.; Wadenback, J.; Pedersen, S.T. Suspension-firing of biomass. Part 1: Full-scale measurements of ash deposit build-up. *Energy Fuels* **2012**, *26*, 2317–2330. [CrossRef]
23. Madhiyanon, T.; Sathitruangsak, P.; Sungworagarn, S.; Pipatmanomai, S.; Tia, S. A pilot-scale investigation of ash and deposition formation during oil-palm empty-fruit-bunch (EFB) combustion. *Fuel Process. Technol.* **2012**, *96*, 250–264. [CrossRef]
24. Retschitzegger, S.; Gruber, T.; Brunner, T.; Obernberger, I. Short term online corrosion measurements in biomass fired boilers. Part 1: Application of a newly developed mass loss probe. *Fuel Process. Technol.* **2015**, *137*, 148–156. [CrossRef]
25. Retschitzegger, S.; Gruber, T.; Brunner, T.; Obernberger, I. Short term online corrosion measurements in biomass fired boilers. Part 2: Investigation of the corrosion behavior of three selected superheater steels for two biomass fuels. *Fuel Process. Technol.* **2016**, *142*, 59–70. [CrossRef]
26. Hansen, L.A.; Nielsen, H.P.; Frandsen, F.J.; Dam-Johansen, K.; Horlyck, S.; Karlsson, A. Influence of deposit formation on corrosion at a straw-fired boiler. *Fuel Process. Technol.* **2000**, *64*, 189–209. [CrossRef]

© 2020 by the authors. Licensee MDPI, Basel, Switzerland. This article is an open access article distributed under the terms and conditions of the Creative Commons Attribution (CC BY) license (http://creativecommons.org/licenses/by/4.0/).

Article

Thermal Degradation Kinetics and FT-IR Analysis on the Pyrolysis of *Pinus pseudostrobus*, *Pinus leiophylla* and *Pinus montezumae* as Forest Waste in Western Mexico

José Juan Alvarado Flores [1,*], José Guadalupe Rutiaga Quiñones [1], María Liliana Ávalos Rodríguez [2], Jorge Víctor Alcaraz Vera [3], Jaime Espino Valencia [4], Santiago José Guevara Martínez [4], Francisco Márquez Montesino [5] and Antonio Alfaro Rosas [1]

[1] Faculty of Wood Engineering and Technology, University Michoacana of San Nicolas of Hidalgo, Edif. D, University Cd, Morelia C.P. 58060, Michoacán, Mexico; jrutiaga@yahoo.com.mx (J.G.R.Q.); anthonyalfa@hotmail.com (A.A.R.)
[2] Center for Research in Environmental Geography, National Autonomous University of Mexico, Morelia C.P. 58190, Michoacán, Mexico; lic.ambientalista@gmail.com
[3] Institute of Economic and Business Research, University Michoacana of San Nicolas of Hidalgo, University Cd, Morelia C.P. 58060, Michoacán, Mexico; talcarazv@hotmail.com
[4] Faculty of Chemical Engineering, University Michoacana of San Nicolas of Hidalgo, Edif. V-1, University Cd, Morelia C.P. 58060, Michoacán, Mexico; jespinova@yahoo.com.mx (J.E.V.); santiago_guemtz@hotmail.com (S.J.G.M.)
[5] Centre for the Study of Energy and Sustainable Technologies, University of Pinar del Rio, Martí 270 Final, C.P. Pinar del Rio 20100, Cuba; marquezmontesino1992@gmail.com
* Correspondence: doctor.ambientalista@gmail.com

Received: 24 December 2019; Accepted: 7 February 2020; Published: 21 February 2020

Abstract: For the first time, a study has been carried out on the pyrolysis of wood residues from *Pinus pseudostrobus*, *Pinus leiophylla* and *Pinus montezumae*, from an area in Western México using TGA analysis to determine the main kinetic parameters (Ea and Z) at different heating rates in a N_2 atmosphere. The samples were heated from 25 °C to 800 °C with six different heating rates 5–30 °C min^{-1}. The Ea, was calculated using different widely known mathematical models such as Friedman, Flynn-Wall-Ozawa and Kissinger-Akahira-Sunose. The Ea value of 126.58, 123.22 and 112.72 kJ/mol (*P. pseudostrobus*), 146.15, 143.24 and 132.76 kJ/mol (*P. leiophylla*) and 148.12, 151.8 and 141.25 kJ/mol (*P. montezumae*) respectively, was found for each method. A variation in Ea with respect to conversion was observed with the three models used, revealing that pyrolysis of pines progresses through more complex, multi-stage kinetics. FT-IR spectroscopy was conducted to determine the functional groups present in the three species of conifers. This research will allow future decisions to be made, and possibly, to carry out this process in a biomass reactor and therefore the production of H_2 for the generation of energy through a fuel cell.

Keywords: *Pinus pseudostrobus*; *Pinus leiophylla*; *Pinus montezumae*; pyrolysis kinetics; TGA-DTG; Friedman-OFW-KAS models; FT-IR

1. Introduction

Around the world, the amount of publications regarding renewable energies and biofuels accounts for a large percentage of the total of approximately 56% [1,2]. In human history, the use of biomass has been required to meet the energy needs. From the 20th century onwards, due to increased energy demand and the excessive use of various fuels, the incorporation of biomass in the fuel mix has been seriously considered, especially in developing countries [3]. Today, biomass is considered one of the main sources of energy, as well as one of the new alternatives that have been implemented to try to reduce the amount of emissions of CO_2, SO_x, NO_x and particulate matter produced during energy production processes [4]. It is worth mentioning that, in the case of CO_2, the use of oil-based fuels, accounts for more than 70% of human related emissions of this gas worldwide, and the rest is attributed to changes in land use. Pollution in some countries is alarming, as is the case in the United States where about 97% of all transport energy is currently derived from oil [5]. Transport energy consumes 63% of all oil used in this country. Because fossil fuels are not renewable and the United States has a need for foreign energy, there is an excellent opportunity to develop renewable energy sources. In this sense, and considering the serious consequences of the greenhouse effect and the future depletion of fossil fuels, there is the possibility of using biomass to produce energy [6]. Given the current scenario in which prosperity is directly related to the capacity to use energy, production resources must be taken into account, especially those that are well distributed in the territory, which are renewable, environmentally appropriate and contribute to reducing CO_2 emissions. Lignocellulosic waste fulfils this purpose [7,8]. It is important to mention that plant biomass is distributed throughout much of the world (except for polar ice caps and extremely dry areas) and grows in different forms (herbaceous plants, shrubs, trees, algae, etc.). In addition, the development of agricultural techniques has considerably increased soil fertility, leading to a greater use of agricultural land for the last 20 to 30 years.

This will allow the generation of energy in certain devices of the latest technology, such as fuel cells, with special emphasis on solid oxides (SOFC), which can generate electricity from the use of gases (H_2, CH_4) from the combustion of biomass or from agricultural waste, industrial and even urban waste (landfills). In this case, the process is produced from obtaining methane gas, pollutant, emitted by organic waste or biomass combustion, which at the time of entering the cell produces electricity. It is important to emphasize that the development of this technology opens in Mexico the access to a wider energy market and includes electricity generation in rural areas, in addition, where the temperature is very low in winter, these systems can work in dual form: heat and power. In winter it would work as a heating and electric power generator, and in summer as an electric generator.

Biomass can be thermally transformed through various thermal processes such as liquefaction, gasification and pyrolysis. Pyrolysis is a well-established route for thermal processing of biomass. The pyrolysis process dates back to the Egyptians, when tar was produced for ship caulking and certain embalming agents [9]. Pyrolysis leads to the conversion of biomass into non-condensable, condensable gases and higher molecular weight compounds such as coal [10]. It is a reality that the lignocellulosic waste pyrolysis process represents a very promising for the near future from the production of various chemical compounds such as bio-oil [11,12]. In thermal processes for the transformation of biomass, pyrolysis is usually the first stage [4]. Currently, there is an extensive bibliography that analyses certain kinetic mechanisms in lignocellulosic biomass. Today, the wood industry is interested in finding a more economical way to dispose of the various waste products from forestry and logging activities. These products, which have been ignored in the past, can now replace oil [13].

Therefore, in order to standardize pyrolytic processes on an industrial scale for the generation of biofuels from these forest residues, the application of modern technologies for obtaining energy carriers and displacing the fuels obtained from fossil materials and consequently the reduction of polluting agents, it is necessary to know in depth what happens in pyrolysis. One of the main aspects to know is the chemical kinetics, because this will be fundamental base for the design of the reaction zone of the process. In the development of pyrolysis, it is necessary to consider the appropriate temperature levels and heating speed. One of the pathways that has gained great diffusion in thermal decomposition analysis of biomass is the study of the decomposition process by thermogravimetric analysis (TGA) [14–16]. TGA has been formally defined as a group of techniques in which a property of the sample is controlled against time or temperature, while the temperature of the sample is programmed in a specific atmosphere [17]. The TGA in addition to being applied to plant and animal studies [18], but also for the thermal decomposition of other materials such as medical waste [19], car wrecks [20], PCB waste or sewage sludge [21]. There are also various thermoanalytic techniques classified according to the physical property subject to measurement (see Table 1).

Table 1. Classification of thermoanalytical techniques [22].

Property	Technique	Parameter Measured	Acronym
Mass.	Thermogravimetric analysis.	Sample mass.	TGA
	Derivative thermogravimetry.	First derivative of mass.	DTG
Temperature.	Differential thermal analysis.	Temperature difference between sample and inert reference material.	DTA
	Derivative differential thermal analysis.	First derivative of DTA curve.	
Heat.	Differential scanning calorimetry.	Heat supplied to sample or reference.	DSC
Pressure.	Thermomanometry.	Pressure.	
Dimensions.	Thermodilatometry.	Coefficient of linear or volumetric expansion.	
Mechanical properties.	Thermomechanical analysis.		TMA
Electrical properties.	Thermoelectrical analysis.	Electrical resistance.	TEA
Magnetic properties.	Thermomagnetic analysis.		
Acoustic properties.	Thermoacoustic analysis.	Acoustic waves.	TAA
Optical properties.	Thermoptical analysis.		TOA

By using gases such as nitrogen, argon or helium, a TGA analysis can be performed, where the amount of mass lost with respect to a programmed temperature is determined [23]. In this sense and based on various mathematical models, it is possible to obtain very valuable information about the composition of the material, reaction orders, the various stages of thermal transformation, as well as their kinetic behaviour parameters, which is essential in the knowledge of the kinetic behaviour of lignocellulosic materials. This information is basic when designing, building and operating an industrial scale reactor for pyrolysis of the material being studied or for the energetic exploitation of products that can be generated such as hydrogen, for energetic purposes, for example, in fuel cells.

For the determination of the above-mentioned kinetic parameters, non-isothermal methods can be used, which require various heating speeds, although various thermal transformation processes can be affected by changes in the heating rate, thus causing other reactions, which makes analysis by DTG more difficult. [24]. Due to the wide variation in the parameters of the Arrhenius equation [25], today, there are several mathematical models for calculating Arrhenius variables. These so-called "model-free" methods are based on an iso-conversive basis, where an assumption is made that the rate of progression of a reaction is constant and therefore the speed of the reaction depends only on the temperature of the reaction. Considering the activation energy (Ea) as the main variable, iso-conversion methods do not require prior knowledge of the reaction mechanism in thermal conversion of the biomass under study, i.e., it is not necessary to choose a reaction model [26]. Non-isothermal iso-conversional methods can follow a differential approach such as Friedman [27], and non-differential (integral) methods Flynn-Wall-Ozawa (FWO) [28] and Kissinger-Akahira-Sunose (KAS) [29,30]. Considering an attractive alternative to oil, with a zero impact of CO_2, its energy capacity and amount of waste approximately 1300 m^3/year (sawdust and shavings) of *Pinus pseudostrobus*, *Pinus leiophylla* and *Pinus Montezumae* as the most commercial and important forest species of the industrial-wood locality of San Juan Nuevo Parangaricutiro (Purépecha zone of the state of Michoacán, Mexico); the aim of this research and for the first time, the mathematical models of Friedman, FWO and KAS are used to determine the most important variables of the kinetic process (Ea and Z) in the inert atmosphere of the thermal degradation of the three selected species of pine. It is important to mention that no articles have been found that refer to the kinetic analysis of the thermogravimetric process for these forests' species pine.

On the other hand, that despite the fact that in recent years the main components of lignocellulosic materials have been analysed through infrared spectroscopy (IR), it is necessary to study their primary composition in greater depth. Today, the analysis of the main components of plant biomass such as cellulose, hemicellulose, lignin and other polymers is well studied and their chemical changes with infrared spectroscopy through Fourier Transform Infrared (FT-IR). In addition to kinetic analysis, in this paper, the results of FT-IR are presented to the three species of pine.

2. Methods

2.1. Sample Preparation and Chemical Composition Analysis

Random samples, of known origin, were taken from different workshops of San Juan Nuevo Parangaricutiro, Michoacán, Mexico, taking precautions to avoid contamination with other types of wood and other substances such as solvents, ensuring that only *P. pseudostrobus, P. leiophylla* and *P. Montezumae* waste represented each sample. Pyrolysis is the term commonly used for a high temperature treatment. The analysis of this type of treatment should include: drying, devolatilization and mainly in the events that occur in the formation of coal, which is one of the main objectives of this article. Each of the samples were placed in containers in a dry place at room temperature (25 °C) for 2 days to eliminate the superficial humidity with which the sample arrives. Afterwards, the fine grain milling was carried out. Once dried, a sieve was made in order to obtain samples with a particle size of approximately 200 µm. Finally, the biomass was taken to an oven to dry at 115°C monitoring until a constant weight was obtained.

In previous investigations, the chemical composition of *P. pseudostrobus, P. leiophylla* and *P. Montezumae*, located in the aforementioned area of Michoacán state has already been determined [31]. In this case, the ash analysis was carried out on an X-ray spectrometer, coupled to a SEM (Jeol JSM-6400) [32]. The minerals were calculated (UNE-EN 14775) [33]. It should be noted that due to the high volume (1300 m^3/year) of waste produced and the energy potential it represents for the community in question, research of these woods has continued.

2.2. Thermogravimetric Analysis TGA-DTG

A Simultaneous Thermal Analyzer STA 6000 thermogravimetric analyzer (PerkinElmer, city, state abbrev if USA, country) was taken after sample preparation and with a uniform particle size to perform a gradual mass degradation at different heating rates (β = 5–30 °C/min). The non-isothermal model was carried out with each heating rate β (six analyses), from 25 °C to 120 °C, left in isotherm for 12 min and then the system was brought to a temperature of 800 °C at the same heating rate to continue with one more isotherm for 30 min, then cooling to room temperature. In the thermobalance of the TGA equipment, a ceramic crucible was used where between 6.5 and 7.5 mg were placed of sample, in order to reduce the effects of mass transfer and heat transfer, because the presence of temperature gradients in the bed of the material and the particles cause that the biomass does not react homogeneously and there are differences in the sequence of reactions, due to the conditions of transport of the primary products of the reactions to the outside of the particles and through the bed of the material. These transport processes are largely responsible for the secondary reactions, which are generated from the primary products of biomass pyrolysis [4]. All thermal degradations were carried out in an inert atmosphere of high purity (99.99%) nitrogen (N_2) as reaction gas with a flow rate of 50 mL/min. Before each experiment, N_2 was purged for 45 min at a flow rate of 100 mL/min.

2.3. Kinetic Modeling

The kinetics of free models are based on iso-conversion methods and are mainly used for obtaining and evaluating the activation energy, which is a function of the degree of conversion of a chemical reaction. Such methods are widely recommended [34]. Because thermographs, TGA results, contain partially superimposed peaks, mathematical models are generally used for deconvolution [35]. It has been proven that the values obtained depend not only on factors such as atmosphere, gas flow, sample mass and heating rate, but also on the mathematical treatment of the data. To evaluate such data at different heating rates, this paper describes the pyrolysis process from three iso-conversional kinetic models, one differential corresponding to Friedman [26], and two integrals, one from Flynn-Wall-Ozawa (FWO) [36] and the other from Kissinger-Akahira-Sunose (KAS) [28]. With these methods, the kinetic parameters that characterize the thermal degradation process of biomass can be obtained. The data from the TGA curves were used to determine the kinetic parameters.

The general process of pyrolysis of biomass can be represented as [37]:

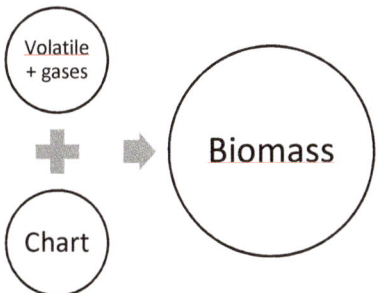

The overall kinetics of the biomass pyrolysis reaction can be described as follows:

$$\frac{d\alpha}{dt} = k(T)f(\alpha) \tag{1}$$

Practically all the proposed kinetic models employ a law that obeys Arrhenius fundamental velocity expression, so Equation (1) can be expressed as:

$$\frac{d\alpha}{dt} = \frac{A}{\beta}\exp\left(-\frac{E_a}{RT}\right)f(\alpha) \tag{2}$$

where $f(\alpha)$ is a conversion function which, as can be seen in Table 2, represents the reaction model used and depends on the control mechanism [38–40]; $d\alpha/dt$ is the speed of the isothermal process; T is the absolute temperature (K), α is the degree of conversion, A is the pre-exponential factor or frequency factor (min^{-1}), which is the frequency of molecular collisions, regardless of their energy level [41]. E_a, (activation energy) is the maximum energy required in a reaction to form a certain amount of products [42], R is the universal gas constant equal to 8.314 J/(mol K) and β is the linear heating velocity ($\beta = dT/dt$) and it's a constant. The exponential term of Equation (2) can be considered as the fraction of collisions that has sufficient kinetic energy to induce a reaction, thus the product $A\exp^{-E_a/RT}$ produces the frequency of collisions that are successful [43]. It is important to mention that derived from the exponential term of the Arrhenius equation, there is a dependence on temperature, it should also be noted that the variable A (pre-exponential factor) also depends on temperature behavior [44]. The degree of conversion (α) or also called process coefficient, can be defined as the mass fraction that has been decomposed or the mass fraction of volatile compounds and is expressed as $\alpha = (m_0 - m)/(m_0 - m_\infty)$ where m is the mass of substrate present at any time t, m_0 is the initial mass of the substrate and m_∞ is the final mass of solids (unreacted residue) that remains after the reaction.

The analysis of the thermal degradation of biomass is carried out with the application of a series of diverse kinetic models, which can be applied to its degradation process in addition to the dynamic analyses of thermal degradation. In this sense, iso-conversion methods assume a fixed value of α, thus only the temperature is a decisive factor for the speed of reaction. Thus, it is possible to calculate the E_a, considering α, independently of the reaction model $f(\alpha)$. Iso-conversion methods can be differential or integral for the treatment of differential thermal analysis data.

Table 2. Functions of the most common reaction mechanisms for gas solid reactions [34].

Symbol	Mechanism	g(α)	f(α)
D_1	Diffusion One-way transport	α^2	$1/2\alpha$
D_2	Two-way transport	$\alpha + (1-\alpha)\ln(1-\alpha)$	$[-\ln(1-\alpha)]^{-1}$
D_3	Three-way transport	$[1-(1-\alpha)^{1/3}]^2$	$(3/2)(1-\alpha)^{2/3}[1-(1-\alpha)^{1/3}]^{-1}$
G-B	Ginstling-Brounshtein equation	$1-(2/3)\alpha-(1-\alpha)^{2/3}$	$(3/2)[(1-\alpha)^{-1/3}-1]^{-1}$
Zh	Zhuravlev equation	$[(1-\alpha)^{-1/3}-1]^2$	$(3/2)(1-\alpha)^{4/3}[(1-\alpha)^{-1/3}-1]^{-1}$
A_2	Random nucleation and nuclei growth Bi-dimensional	$[-\ln(1-\alpha)]^{1/2}$	$2(1-\alpha)[-\ln(1-\alpha)]^{1/2}$
A_3	Tree-dimensional	$[-\ln(1-\alpha)]^{1/3}$	$3(1-\alpha)[-\ln(1-\alpha)]^{2/3}$
P-T_1	Prout-Tompkins ($m = 0.5$)	$\ln[(1+\alpha^{1/2})/(1-\alpha^{1/2})]$	$(1-\alpha)\,\alpha^{1/2}$
P-T_2	Prout-Tompkins ($m = 1$)	$\ln[\alpha/(1-\alpha)]$	$(1-\alpha)\,\alpha$
F_1	Chemical reaction First-order	$-\ln(1-\alpha)$	$1-\alpha$
F_2	Second-order	$(1-\alpha)^{-1}-1$	$(1-\alpha)^2$
R_1	Limiting surface reaction between both phase One dimension.	α	1
R_2	Two dimension	$1-(1-\alpha)^{1/2}$	$2(1-\alpha)^{1/2}$
R_3	Three dimension	$1-(1-\alpha)^{1/3}$	$3(1-\alpha)^{2/3}$

2.3.1. Friedman Method

In the case of iso-conversional differential models, Friedman's method is probably the most general of the derived techniques. It is based on the comparison of the conversion velocity $d\alpha/dT$ for a conversion grade α determined, using different heating rates [45]. It is necessary to work with a degree of advance range α in which the linear fit is adequate. Considering the natural logarithm on both sides of Equation (2) and simplifying, the general equation of this method is as follows:

$$\ln\left(\frac{d\alpha}{dt}\right) = \ln\left[\beta\left(\frac{d\alpha}{dT}\right)\right] = \ln[Af(\alpha)] - \frac{Ea}{RT} \qquad (3)$$

By analogy of the general equation of the straight line ($y = mx + b$), through the Friedman method, it is possible to obtain the value of the activation energy by graphing $\ln(d\alpha/dt)$ vs. $1/T$, for each conversion value, α, at different heating rates. The resulting graph will show several lines that, according to Equation (3), will have a slope (m) equal to $-Ea/R$ [46].

2.3.2. Flynn-Wall-Ozawa Method (FWO)

The FWO method is one of the most common and widely accepted methods in the scientific community for calculating thermokinetic parameters from experimental data. This method uses a correlation between the heating rate of the sample, the activation energy and the temperature inverse [47]. It is an integral and iso-conversional technique that assumes that Ea is constant in every reaction process considering time from $t = 0$ until t_∞, where t_∞ is the conversion time of α [48].

Integrating Equation (3) with respect to the variables α and T results:

$$g(\alpha) = \int_0^\alpha \frac{d\alpha}{f(\alpha)} = \frac{A}{\beta}\int_0^{T\alpha} \exp\left(\frac{-Ea}{RT}\right)dT \qquad (4)$$

where T_∞ is equal to the conversion temperature α. Considering the value of Ea/RT equal to x, Equation (4) is transformed into:

$$g(\alpha) = \frac{AE_a}{\beta R}\int_\alpha^\infty \frac{exp^{-x}}{x^2} = \frac{AE_a}{\beta R}p(x) \qquad (5)$$

where $p(x)$ represents the integrand on the right side of Equation (4) and is known as the temperature integral. This integral does not have an exact analytical solution [49], however, as noted below, it can be approximated through an empirical interpolation formula proposed by Doyle [50]:

$$\log p(x) \cong -2.315 - 0.4567x, \text{ for a range } x\colon 20 \le x \ge 60 \qquad (6)$$

Considering this approximation to the right member of Equation (4) and applying the natural logarithm on both sides of the equation the final form of the OFW model is obtained:

$$log\beta = log\left(A\frac{E_a}{Rg(\alpha)}\right) - 2.315 - 0.4567\frac{Ea}{RT} \quad (7)$$

According to Equation (7) and when graphing the $log\beta$ versus $1/T$ at different heating rates, parallel straight lines are obtained for each degree of conversion α. The value of the apparent activation energy is calculated across the slope of those lines. Such slope is proportional to the expression $-0.4567E_a/R$. The value of $logA$ is given by the intercept ($log\beta$) of each line with the vertical axis of the graph.

2.3.3. Kissinger-Akahira-Sunose Method (KAS)

The KAS method is a non-isothermal iso-conversional technique which, like the previous method, is widely used. The KAS method uses the Arrhenius equation using a differential method. This method does not require knowledge of the exact thermal degradation mechanism [51]. The KAS method is derived from Equation (2), which is integrated from specific conditions ($x = 0$, $T = T_0$), to get the following expression:

$$g(x) \int_0^x \frac{dx}{f(x)} = \frac{A}{B}\int_{T_0}^T \exp\left(-\frac{E}{RT}\right)dT \equiv \frac{AE}{\beta R}p\left(\frac{E}{RT}\right) \quad (8)$$

As can be seen from Equation (8), the frequency factor A, the function $f(x)$ and the activation energy Ea, are temperature-dependent T, however Ea and A are independent of x. By re-integrating Equation (8), an expression is obtained as a function of natural logarithms:

$$ln(g(\alpha)) = ln\left(\frac{AE}{R}\right) - ln\beta + ln\left(p\frac{E}{RT}\right) \quad (9)$$

Considering the approximation of Coats-Redfern [52] where:

$$p\left(\frac{E_a}{RT}\right) \cong \left[\frac{exp\left(-\frac{E_a}{RT}\right)}{\left(\frac{E_a}{RT}\right)^2}\right] \quad (10)$$

Combining Equations (9) and (10) and simplifying gives the main expression of the KAS method:

$$ln\frac{\beta}{T^2} = ln\left[\frac{AR}{E_a g(\alpha)}\right] - \frac{E_a}{RT} \quad (11)$$

From Equation (11) the apparent activation energy can be obtained by graphing $ln(\beta/T^2)$ versus $1/T$ where the value of the slope of the line obtained is equal to $-E_a/R$ for a constant value of the degree of conversion, α.

2.3.4. Frequency Factor (Z)

Because the vast majority of thermal analyses are carried out at a constant heating rate, a more useful approach for the Arrhenius integral has been implemented under experimental conditions of a linear temperature program and to extend these results of the unequivocal determination of the Arrhenius kinetic parameters by establishing the iso-conversional method whose final expression is [53]:

$$Z = \frac{\beta * (E_a + 2 * R * T_\alpha) * e^{\frac{E_a}{RT_\alpha}}}{R * T_\alpha^2} \quad (12)$$

where: Z, is the frequency factor (min^{-1}), β, is the heating speed (°C/min), Ea, is the activation energy (kJ/mol), Tα is the temperature (K), where the maximum conversion is reached (α).

2.4. Fourier Transform Infrared Analysis (FT-IR)

The infrared spectroscopy technique will identify the main gaseous products produced during the pyrolysis of the three forest species. The The FT-IR values of the lignocellulosic materials were obtained using the PerkinElmer ATR 400 model resolution 4 cm^{-1}. The biomass was prepared with methods well established in the literature for IR analysis. All the spectra were acquired (16 scans/sample) in the range of 650–4000 cm^{-1} with 4 cm^{-1} resolution.

3. Results and Discussion

3.1. Chemical Analysis

Table 3 shows the results of the elementary analysis of P. pseudostrobus, P. leiophylla and P. montezumae previously obtained by some of the authors who collaborated in this research [30]. A low ash content (0.13–0.23%) can be observed in all three species of pine, which is very significant and attractive in the thermal degradation processes of lignocellulosic biomass [22]. A high degree of ash in the fuel can damage combustion equipment and users when they have to do cleaning work [54]. With regard to the mineral composition of the ash, the elements, calcium, potassium, magnesium, silicon and aluminium, were the elements that were presented in greater quantity. It should be noted that the content of potassium (12.23–21.1%) and sodium (2.17–5.74%), help reduce the melting point of the ash [55,56]. A high content of magnesium (Mg), allows an increase in the melting point of the ash. It should also be mentioned that the high amount of calcium (25.48–42.46%) reduces the amount of ash, although it can increase its melting point [22]. Silicon, is fixed in silicates. In this case the amount of silicon (4.57–17.46%) helps to significantly reduce the melting point of the ash [57].

Table 3. Chemical analysis of P. pseudostrobus, P. leiophylla and P. montezumae wood sawdust [27].

Parameter	P. pseudostrobus	P. leiophylla	P. montezumae
Ash (%)	0.19 (±0.1)	0.23 (±0.06)	0.13 (±0.01)
Elemental ash composition of forest residues the three conifers (%)			
Ca	25.48 (±0.5)	42.46 (±0.98)	42.44 (±0.54)
K	12.23 (±0.77)	13.16 (±0.78)	21.1 (±0.66)
Mg	10.82 (±0.44)	21.64 (±0.31)	13.08 (±0.47)
P	9.35 (±0.54)	5.37 (±0.20)	4.06 (±0.16)
S	1.70 (±0.21)	3.8 (±0.18)	2.59 (±0.09)
Na	2.17 (±0.37)	5.74 (±0.61)	2.33 (±0.29)
Si	17.46 (±0.70)	4.57 (±0.43)	4.89 (±0.61)
Al	12.70 (±0.69)	3.22 (±0.51)	6.26 (±0.28)
Fe	7.01 (±0.33)	no detected	3.21 (±0.24)
Ti	0.5 (±0.1)	no detected	no detected

3.2. Thermogravimetric Analysis

Figure 1A–C, illustrate the P. Pseudostrobus, P. leiophylla and P. montezumae curves with respect to mass loss and temperature (TGA) for heating rates of 5, 10, 15, 20, 25 and 30 °C/min in nitrogen inert atmosphere. The curves show the typical appearance of pyrolysis of lignocellulosic materials and from them the thermal phases for each of the β can be located. The main reactions consist of broken glycosidic bonds with the consequent partial depolymerization of the cellulosic component of the wood.

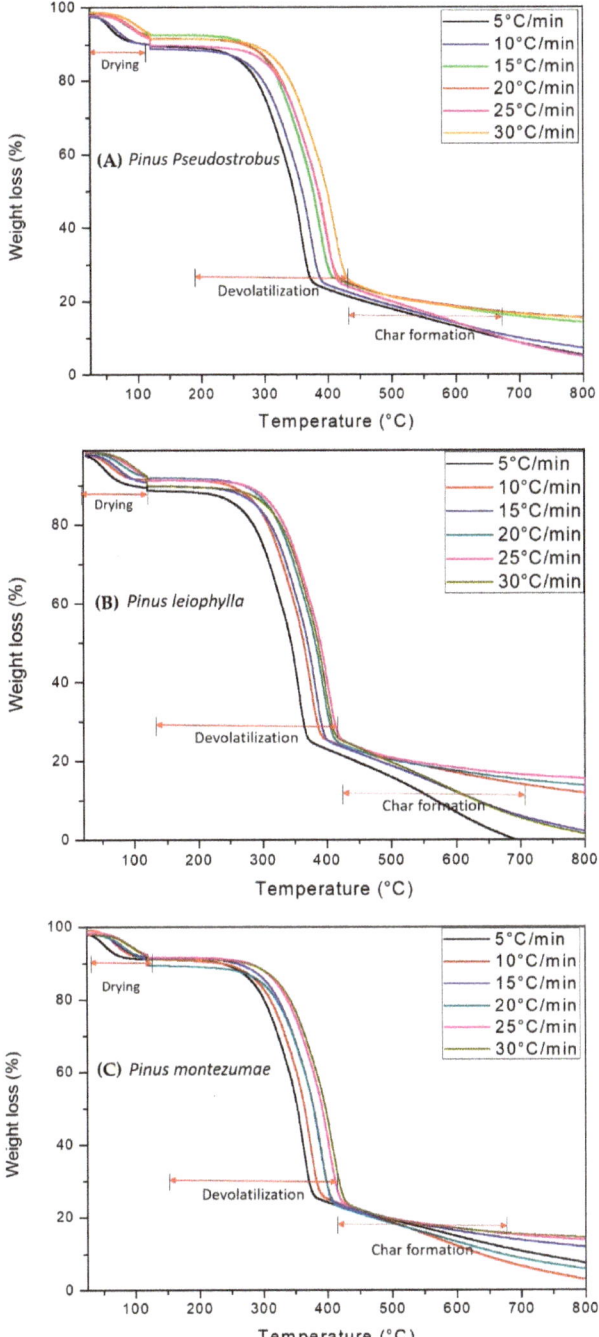

Figure 1. TGA curves and percent mass loss of *P. pseudostrobus* (**A**), *P. leiophylla* (**B**) and *P. montezumae* (**C**) under N_2 at different heating rates.

According to several researchers, hemicellulose decomposes in a range between 180–340 °C, which is less than cellulose when decomposing between 230–450 °C and the latter less than lignin, which is thermally transformed at a temperature greater than 500 °C [53,58]. Cellulose decomposition occurs in two ways. First, the bonds are divided into monomers at a lower temperature and form CO, CO_2 and carbonaceous gases. Then, at a higher temperature, liquid formation occurs. In stage three, lignin decomposes at a temperature above 500 °C and at a slower rate due to an association with a hydroxyl phenolic group. At this stage there is the presence of high molecular weight carbonaceous products.

The thermograms of Figure 1A–C, can be divided into four distinct zones, zone 1 (T < 50 °C), which corresponds to an increase in mass which is attributable to condensation of water and the formation of intermediate compounds which are subsequently decomposed. Zone 2 (T ≈ 135 °C), where moisture evaporation takes place and the release of CO, CO_2, and extractable materials. According to the literature, between 180 and 400 °C the highest devolatilization occurs, which has been designated as the zone of active pyrolysis [59]. At a temperature higher than 200 °C and up to approximately 400 °C, zone 3 is present, where the maximum degradation of hemicellulose, cellulose and lignin is reached about 80% of the total mass. During this stage most of the volatiles were released and the evolution of the secondary gases was practically completed at 400 °C, which led to the formation of carbon [60]. In the passive zone, there is no decomposition but the carbon and ashes are part of the final solid waste. It is worth mentioning that from 400 to 700 °C the condensed system grows gradually, but all peripheral atoms are bonded by chemical bonds to hydrogen atoms or hydrocarbon groups, substances that have high electrical resistivity. It is important to mention that high heat flows in the highest heating zone decrease the viscosity level in the material, while increasing the reactions that form the volatiles. This behavior has been previously described for several biofuels [61,62]. The highest mass loss is identified at the maximum peak in the thermo-differential analysis (DTG) curves. DTG is shown in Figure 2A–C. As predicted, the graphs show three main areas. First peak is observed due to the elimination of moisture and light volatile matter when heated from 50 to approximately 135 °C. The main stage of thermal decomposition is carried out in a temperature range between 200–400 °C at heating rates between 5–30 °C/min. Two peaks are observed that are evidence of hemicellulose and cellulose decomposition, while there is no indication of any peak derived from lignin decomposition [63]. The second (280–340 °C) and third peak (350–400 °C) appear when hemicellulose and cellulose compounds are transformed. In the final part, the lignin has been transformed at a lower speed, so that a maximum carbonization has been carried out. A comparison between the peaks of hemicellulose, cellulose and lignin shows that they have different height and position, which indicates the influence of the distribution of organic and inorganic compounds in the thermal degradation process of *P. pseudostrobus*, *P. leiophylla* and *P. montezumae*.

It is observed that as the heating rate increases, so does the temperature at the beginning and end of pyrolysis. The region where the moisture is volatilized does not show a greater variation with the change of the heating rate. Another important aspect is that the maximum points of the TGA curves and the minimum points of the DTG curves move towards higher temperatures. This is related to the heat transfer concept, where at a lower heating rate, there is locally greater thermal energy, which promotes that the balance with the inert atmosphere takes longer. In parallel, and in the same heating range, there is an increase in the heating rate which promotes a decomposition of the sample at a higher temperature, causing the curve in this heating zone to move in a rightward direction [64]. This phenomenon has also been observed by other researchers [65,66].

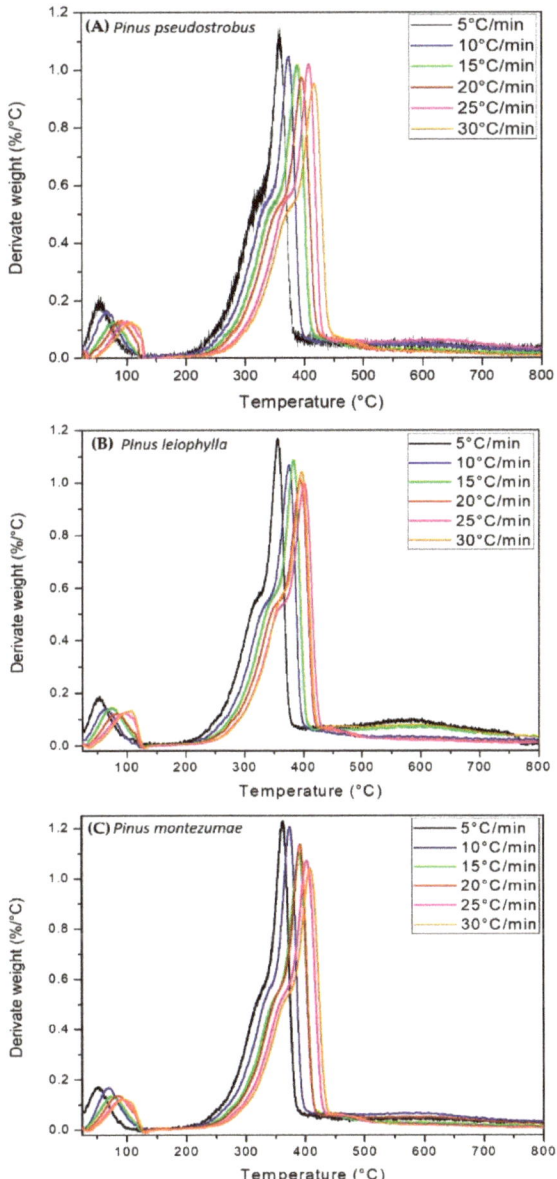

Figure 2. DTG of *P. pseudostrobus* (**A**), *P. leiophylla* (**B**) and *P. montezumae* (**C**) under N_2 at different heating rates.

3.3. Kinetic Analysis

The kinetic parameters have been studied in a distributed manner according to the possible values by careful sampling. The distribution of the kinetic parameters is reported for each model used. From the data obtained from the thermogravimetric analysis and for a given fractional conversion value (α), the three iso-conversional kinetic methods mentioned above from Friedman, FWO and KAS were used to determine the values of the main kinetic parameters such as activation energy (*Ea*) and frequency factor (Z) for each value from α, during pyrolysis of the *P. pseudostrobus*, *P. leiophylla* and *P. montezumae*. Equations (3), (7) and (11) were used for each method. Figure 3A–C, shows the conversion change, α, with respect to temperature at different heating rates for each pine species and it can be seen that activation energy is a function of fractional conversion, this is because most lignocellulosic biomass pyrolysis reactions do not represent a single-step global mechanism, on the contrary, it follows a multi-stage reaction, which means that pyrolysis of *P. pseudostrobus*, *P. leiophylla* and *P. montezumae*, is a complex process consisting of several reactions.

To determine the kinetic parameters, values have been selected from α where the calculated squares of the correlation coefficient, R^2, were greater than 0.90 for all curves at different heating rates and locating the corresponding temperature. The graphs of the Friedman methods, $\ln[d\alpha/dt]$ versus $1/T$; Flynn-Wall-Ozawa (FWO), $\ln\beta$ versus $1/T$; and Kissinger-Akahira-Sunose (KAS), $\ln(\beta/T^2)$ versus $1/T$, for different conversion values, α, are shown in Figure 4A–C, Figure 5A–C, and Figure 6A–C, for *P. pseudostrobus*, *P. leiophylla* and *P. montezumae*, respectively. The apparent activation energies, Ea, were obtained from the slopes in each model (see Table 4) and the average frequency factors from Equation (12), which can be seen in Table 5.

The Friedman, OFW and KAS models adjust to the degradation of *P. pseudostrobus*, *P. leiophylla* and *P. montezumae* since the correlation coefficient, R^2, is close to 1 in the conversion range (α) of 0.20, 0.25, 0.30, 0.35, 0.40, 0.45, 0.50, 0.55, 0.60, 0.65 and 0.70. It is important to mention that, using the available experimental data and during the adjustment of the data in each model, low correlation was observed for conversion values lower than 0.20 and higher than 0.70, which means that the R^2 values showed a low correlation value [34,67]. Iso-conversional methods (model-free) allow to estimate the activation energy as a conversion function without a previous assumption in the reaction model and allow to detect almost unequivocally the kinetics of multiple steps as a dependence of the activation energy (Ea) with respect to the conversion (α), in contrast to methods like Kissinger's, which produces a single Ea value for the whole process and the complexity of the system may not be accurately revealed [68].

The average of the activation energies (Table 4) calculated from the Friedman, OFW and KAS methods was 126.58, 123.22 and 112.72 kJ/mol respectively with regard to *P. pseudostrobus*; 146.15, 143.24 and 132.76 kJ/mol respectively for *P. leiophylla* and 148.12, 151.80 and 141.25 kJ/mol respectively for *P. montezumae*, checking the compatibility of the data obtained in the thermal transformation of biomass by TGA, being compatible with the proposed mathematical models when conversion values between 0.20 and 0.70 are used.

Similar results have been reported in other kinetic studies of thermal processes in forest residues for several pine species. For example, *Pinus insignis*, a change in Ea energy in the range of 62–206 kJ/mol was found [69]. Da silva et al., reported the kinetic mechanism of *Pinus elliottii* and calculated an average Ea of 145.24 kJ/mol [10]. Domínguez et al. reported Ea values for pyrolysis of *Pinus radiata* residues in the range of 117.7–135.5 kJ/mol [70]. As can be seen there are some differences in the Ea values obtained in this investigation, however, it should be noted that the results for the apparent activation energy reflect contributions from various stages where reactions occur that contribute to changing the speed of the overall reaction process. The previous behaviour, in processes of thermal transformation of lignocellulosic materials, presents variations both in temperature and in the reaction level or degree of reaction, frequently observing an overlapping of such parameters [71,72]. On the other hand, the activation energy depends on the pyrolysis reaction mechanism. As mentioned above, the Ea is the minimum required to carry out an alteration in the bonds of each atom involved in the reaction, therefore as the activation energy is higher, the speed of the reaction will be lower. Generally, variables such as the degree of speed of the entire reaction, as well as the system's reaction level, will be governed by this parameter.

Several authors state that the reactivity of fuels derived from the pyrolysis of biomass can be calculated from the calculated activation energy of a thermal process [73]. Fuel reactivity is of great importance when planning the design and development of a pyrolytic reactor for lignocellulosic biomass. It is worth mentioning that the current research is directed at the three main constituents of vegetal biomass, such as cellulose, hemicellulose and lignin.

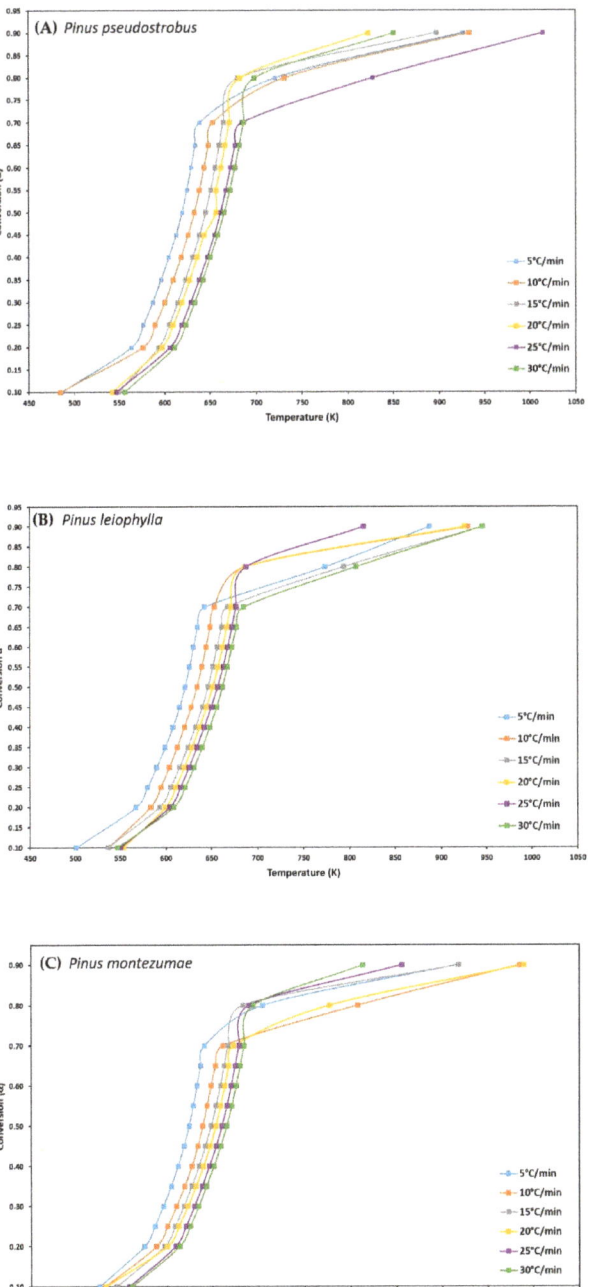

Figure 3. Isothermal residence time effect for the *P. pseudostrobus* (**A**), *P. leiophylla* (**B**) and *P. montezumae* (**C**) under N_2 at different heating rates.

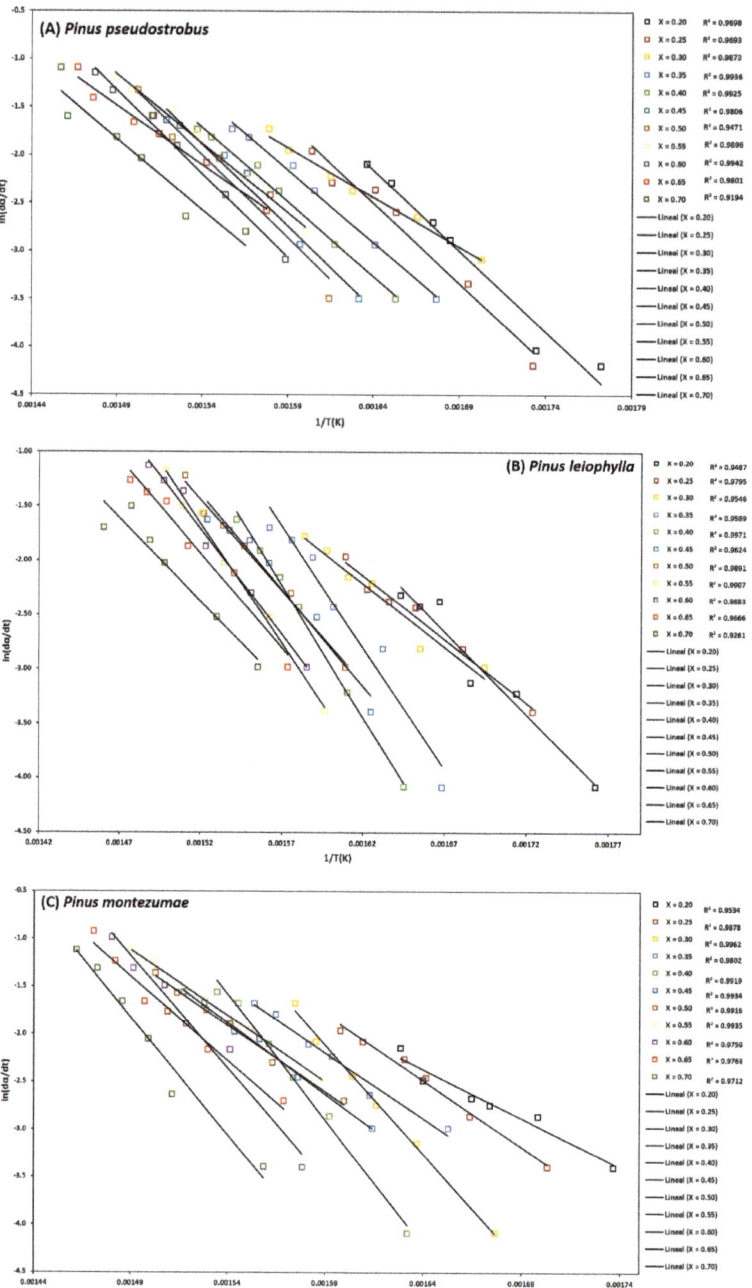

Figure 4. Determination of activation energy by the Friedman method of *Pinus pseudostrobus* (**A**), *Pinus leiophylla* (**B**) and *Pinus montezumae* (**C**).

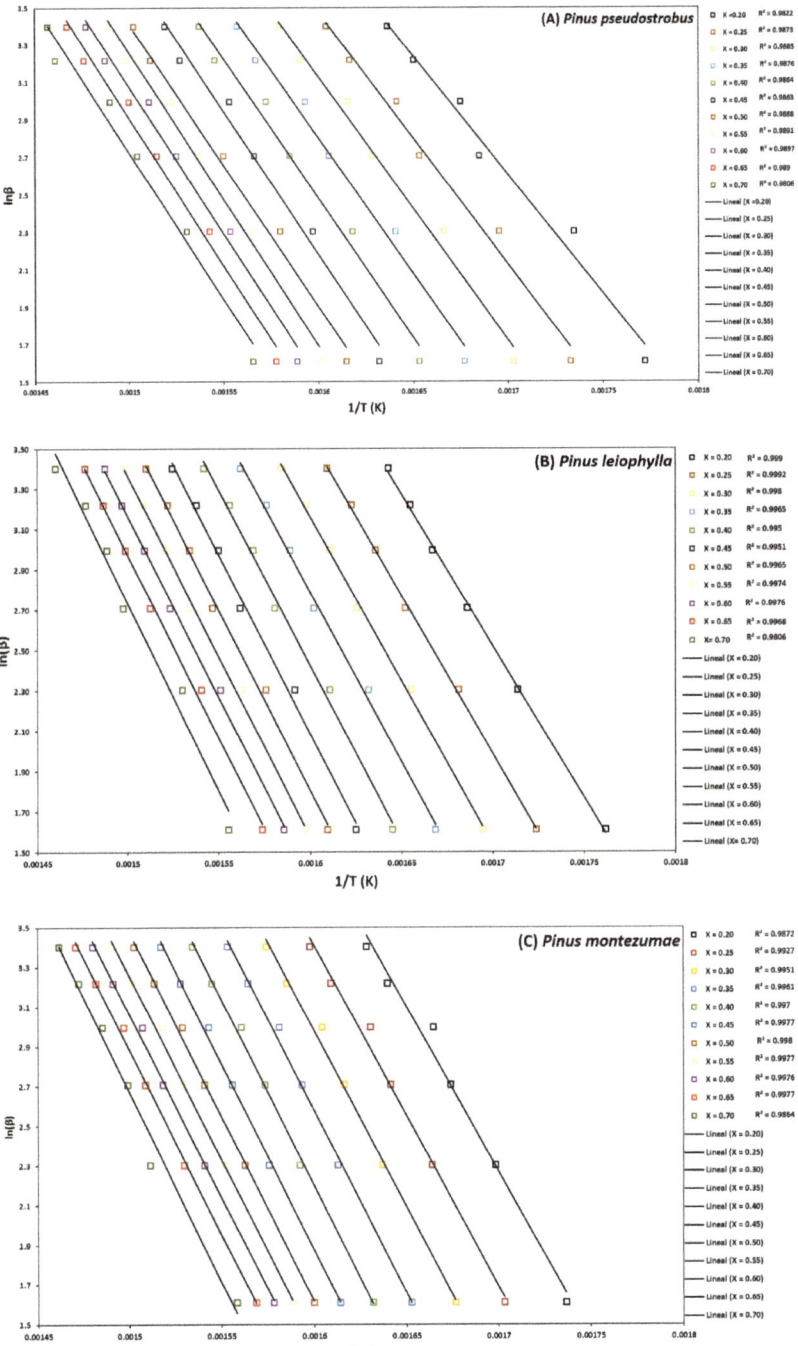

Figure 5. Determination of activation energy by the FWO method of *Pinus pseudostrobus* (**A**), *Pinus leiophylla* (**B**) and *Pinus montezumae* (**C**).

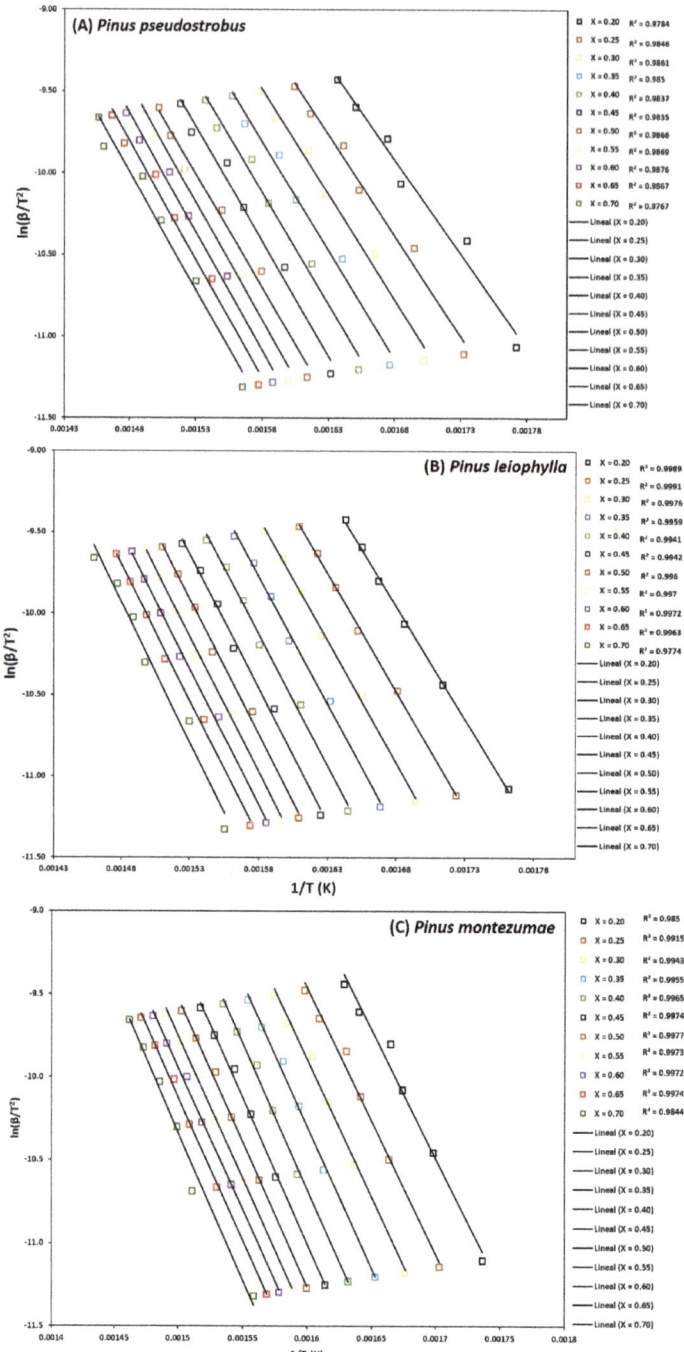

Figure 6. Determination of activation energy by the KAS method of *Pinus pseudostrobus* (**A**), *Pinus leiophylla* (**B**) and *Pinus montezumae* (**C**).

Table 4. E_a (kJ/mol) and R^2 for *P. pseudostrobus*, *P. leiophylla* and *P. montezumae* by Friedman, FWO and KAS methods, R^2 corresponding to linear fittings in Figure 4A–C, Figure 5A–C and Figure 6A–C.

	Conversion (α)	Ea, Friedman	R^2	Ea, FWO	R^2	Ea, KAS	R^2
	0.20	139.60	0.9698	105.24	0.9822	95.48	0.9784
	0.25	138.62	0.9693	112.02	0.9873	102.04	0.9846
	0.30	85.72	0.9873	116.79	0.9885	106.65	0.9861
	0.35	125.29	0.9936	120.66	0.9876	110.36	0.985
PP	0.40	128.95	0.9925	124.18	0.9864	113.75	0.9837
	0.45	141.33	0.9806	127.37	0.9863	116.80	0.9835
	0.50	143.34	0.9471	125.76	0.9888	115.082	0.9866
	0.55	119.56	0.9896	130.52	0.9891	119.75	0.9869
	0.60	144.58	0.9942	130.97	0.9897	120.11	0.9876
	0.65	102.50	0.9801	131.34	0.989	120.41	0.9867
	0.70	122.88	0.9194	130.58	0.9806	119.55	0.9767
	Average	126.58	0.9748	123.22	0.9868	112.72	0.9841
	0.20	126.63	0.9487	124.86	0.9990	115.09	0.9989
	0.25	97.06	0.9795	130.28	0.9992	120.31	0.9991
	0.30	96.81	0.9546	135.02	0.9980	124.88	0.9976
	0.35	185.14	0.9589	139.99	0.9965	129.70	0.9959
PL	0.40	201.07	0.9971	144.75	0.9950	134.32	0.9941
	0.45	146.74	0.9624	147.59	0.9951	137.03	0.9942
	0.50	139.45	0.9891	148.89	0.9965	138.23	0.9960
	0.55	182.75	0.9907	149.14	0.9974	138.40	0.9970
	0.60	161.10	0.9883	149.53	0.9976	138.71	0.9972
	0.65	143.94	0.9666	150.64	0.9968	139.73	0.9963
	0.70	126.93	0.9261	154.96	0.9806	143.94	0.9774
	Average	146.15	0.9692	143.24	0.9956	132.76	0.9948
	0.20	85.16	0.9534	138.33	0.9872	128.45	0.9850
	0.25	115.97	0.9878	142.98	0.9927	132.89	0.9915
	0.30	190.20	0.9962	147.80	0.9951	137.57	0.9943
	0.35	114.47	0.9802	152.57	0.9961	142.19	0.9955
PM	0.40	218.99	0.9919	155.63	0.9970	145.12	0.9965
	0.45	124.45	0.9934	155.93	0.9977	145.31	0.9974
	0.50	115.32	0.9916	155.09	0.9980	144.38	0.9977
	0.55	115.27	0.9935	153.93	0.9977	143.12	0.9973
	0.60	196.31	0.9759	153.69	0.9976	142.81	0.9972
	0.65	147.50	0.9763	154.74	0.9977	143.79	0.9974
	0.70	205.71	0.9712	159.129	0.9864	148.12	0.9844
	Average	148.12	0.9828	151.80	0.9948	141.25	0.9940

PP: *Pinus pseudostrobus*; PL: *Pinus leiophylla*; PM: *Pinus montezumae*.

The E_a value of such biochemical components can vary in amounts from hundreds (cellulose and hemicellulose) to tens (lignin) of kJ/mol. This means that depending on the type of lignocellulosic material being studied, variations in the activation energy will result in the whole thermal process of biomass transformation [74]. The values of the apparent activation energies for the KAS and OFW methods vary approximately from 95 to 131 kJ/mol (*P. pseudostrobus*); from 115 to 143 KJ/mol (*P. leiophylla*) and from 128 to 148 kJ/mol (*P. montezumae*), respectively, and for the Friedman method from 85 to 144 kJ/mol (*P. pseudostrobus*); from 96 to 200 kJ/mol (*P. leiophylla*) and from 85 to 200 kJ/mol (*P. montezumae*).

In this way, it can be considered that the pyrolysis process of the three selected pine species maintains different reaction mechanisms in its transformation, besides, E_a is positively a function of α. Most variations occur in the early stages of decomposition in the range of 0.20–0.30. In the later stages (α = 0.40–0.70), degradation is controlled by an almost stable activation energy due to the superposition of secondary decomposition reactions. Secondary reactions derived from various constituents of the lignocellulosic material, as well as their relationship with other compounds produced, have a reaction rate that is directly related to Ea, which will generally describe the thermal transformation. It is important to note that some activation energy values (*P. montezumae*) reappears ending the thermal decomposition process which can be influenced by the possible formation of slag that can occur at elevated temperatures. Comparing the results of the Ea through the OFW and KAS models, it can be clearly seen that the values per OFW are higher, which can be explained due to the considerations that were taken into account in the calculations for the temperature integral and the corrections during the process. With respect to the activation energies, it is worth mentioning the excellent concordance between the results obtained with deviations of less than 10% between the OFW and KAS methods. These results demonstrate that the OFW and KAS methods are highly reliable for predicting activation energy in pyrolysis processes.

Table 5. Average frequency factor Z (s^{-1}) calculation results as a function of α ($0.70 \leq \alpha \geq 0.20$) and β (°C/min) for *Pinus pseudostrobus*, *Pinus leiophylla* and *Pinus montezumae* obtained by Friedman, FWO and KAS kinetic models.

		β	5	10	15	20	25	30
PP	FRIEDMAN		6.28×10^9	6.48×10^9	4.46×10^9	4.73×10^9	3.87×10^9	3.72×10^9
	OFW	Z	1.69×10^8	1.89×10^8	1.71×10^8	1.79×10^8	1.50×10^8	1.55×10^8
	KAS		1.95×10^7	2.29×10^7	2.15×10^7	2.30×10^7	1.98×10^7	2.08×10^7
PL	FRIEDMAN		1.06×10^{14}	8.90×10^{13}	6.24×10^{13}	6.16×10^{13}	5.62×10^{13}	4.71×10^{13}
	OFW	Z	9.04×10^9	9.66×10^9	8.24×10^9	8.69×10^9	8.62×10^9	8.10×10^9
	KAS		1.07×10^9	1.19×10^9	1.06×10^9	1.13×10^9	1.14×10^9	1.09×10^9
PM	FRIEDMAN		2.65×10^{15}	1.78×10^{15}	1.59×10^{15}	1.50×10^{15}	1.22×10^{15}	1.09×10^{15}
	OFW	Z	3.88×10^{10}	3.58×10^{10}	3.64×10^{10}	3.82×10^{10}	3.45×10^{10}	3.34×10^{10}
	KAS		4.61×10^9	4.47×10^9	4.66×10^9	4.97×10^9	4.59×10^9	4.50×10^9

PP: *Pinus pseudostrobus*; PL: *Pinus leiophylla*; PM: *Pinus montezumae*. OFW: Flynn-Wall-Ozawa; KAS: Kissinger-Akahira-Sunose.

As can be seen in Table 4, the average values of the correlation coefficient (R^2) for the three methods and for each of the pine species were found to be greater than 0.96. In this case the results obtained from the three models presented an excellent correlation for a value of $\alpha = 0.20$–0.70. For Friedman, OFW and KAS methods such values of R^2 very close to the unit were 0.9748, 0.9868 and 0.9841 with regard to *P. pseudostrobus*. A variation of R^2 from 0.9692, 0.9956 and 0.9948 for *P. leiophylla* respectively and finally to *P. montezumae* the range of R^2 presented a variation of 0.9828, 0.9948 and 0.9940 in every method employed.

According to Table 5, the average frequency factor (Z) for each heating speed (β) in the Friedman, KAS and OFW methods varies from 6.28×10^9–3.72×10^9; and 1.69×10^8–1.55×10^8; and 1.95×10^7–2.08×10^7 for *P. pseudostrobus*, respectively. On average, a variation of Z was found for *P. leiophylla* from 1.06×10^{14}–4.71×10^{13}; and 9.04×10^9–8.10×10^9; and 1.07×10^9–1.09×10^9 respectively. finally, to *P. montezumae* the results of Z, were 2.65×10^{15}–1.09×10^{15}; and 3.88×10^{10}–3.34×10^{10}; and 4.61×10^9–4.50×10^9 respectively. A low Z value represents the surface reactions, however, if the surface is not involved, a low Z value means the presence of a more compressed material, and conversely, if a high Z value is present, it means that the material is more relaxed. If surface corrections can be made, high values of the frequency factor can be achieved, as long as the complexes present can be moved on the surface. It is important to consider the Z factor as an indicator of molecularity, because it is not easy to control high concentrations in solids. For low values of Z around 10^9 s^{-1} or lower, it is possible to have such behavior. However, the reactions present will be bimolecular if they are elementary [75]. It is important to mention that the frequency factor calculated from the iso-conversational methods has no physical importance; it is only considered as an adjustment parameter [34]. It should be noted that, according to the literature review, this research is the first attempt to describe a detailed thermokinetic process considering each stage of pyrolysis of *P. pseudostrobus*, *P. leiophylla* and *P. montezumae* forest residues.

Figure 7A–C show the behaviour of the degree of advance with respect to the activation energy of the *P. pseudostrobus*, *P. leiophylla* and *P. montezumae*, where the high dependence of the activation energy with respect to the degree of conversion is first observed. Due to different reaction mechanisms, it is observed that the activation energy increases proportionally to the degree of conversion in the three Friedman, FWO and KAS models, reaching a maximum value when α is in the range of 0.55 to 0.70.

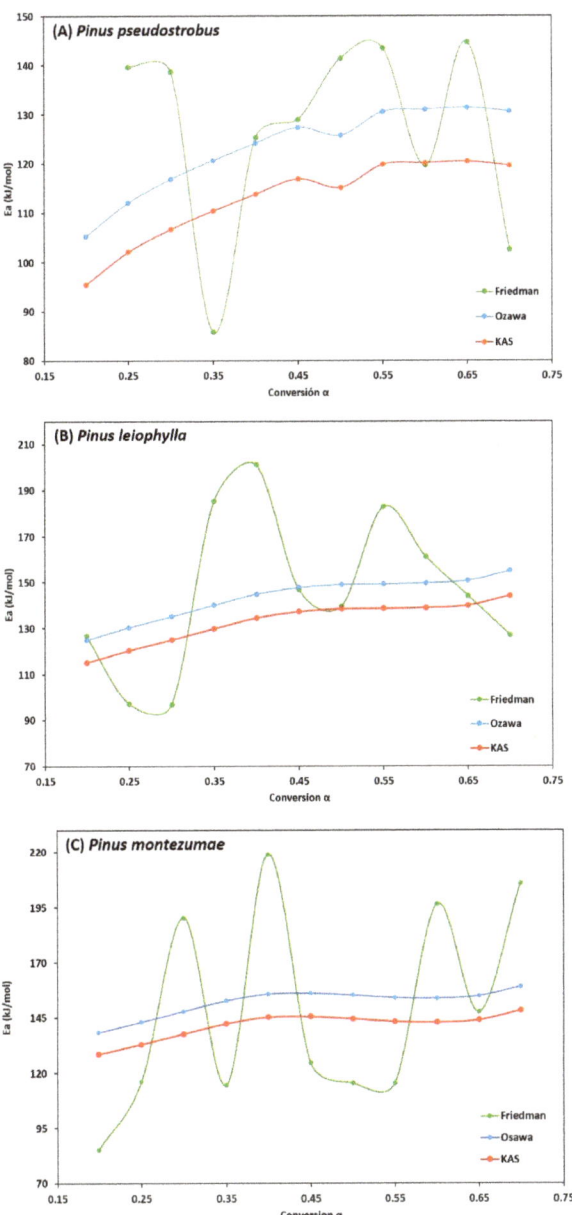

Figure 7. Behavior of *Ea* as a function of α for the three kinetic models applied of *Pinus pseudostrobus* (**A**), *Pinus leiophylla* (**B**) and *Pinus montezumae* (**C**).

It is also observed that in the FWO and KAS methods, the higher activation energy values in the pyrolysis process occur at higher heating rates, which can be explained by the shorter residence time, with the higher activation energy needed to overcome the activated complex of various chemical reactions such as depolymerization and repolymerization.

3.4. Fourier Transform Infrared Analysis (FT-IR)

The species of the wood samples, which were analyzed in this study, are *P. pseudostrobus*, *P. leiophylla* and *P. montezumae*. The FT-IR spectrum of the three pine species is shown in Figure 8.

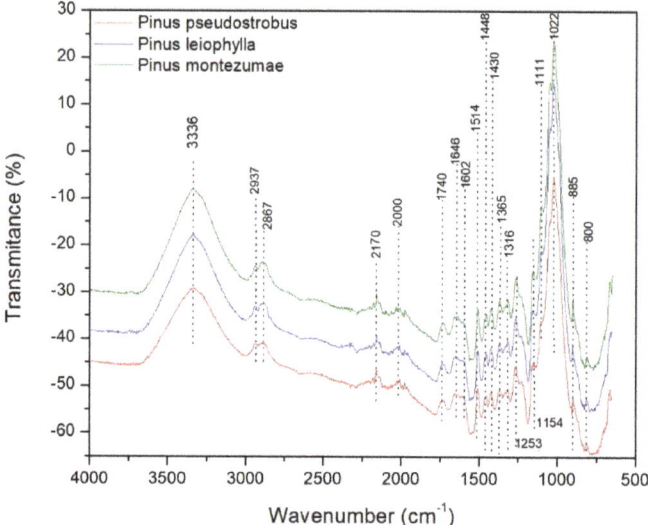

Figure 8. FT-IR of *Pinus pseudostrobus*, *Pinus leiophylla* and *Pinus montezumae*.

The following main absorption regions may be highlighted. Firstly, bands in the range 3000 to 3500 cm^{-1} are observed due to the O-H stress of the intermolecular hydrogen bonded. Such a region is defined for hydroxyl groups according to a certain frequency value, for example the indicated maximum of 3336 cm^{-1}. According to literature when an inter-intra-molecular combination of H$_2$ bonds is present, it is possible to cause an increase in the OH band of the IR [76]. In the band range between 3000 and 2700 cm^{-1} symmetrical and non-symmetrical vibrations are associated according to the presence of methylene, methoxyl C-H and methyl groups, which are constituents of hemicellulose [77]. In the band range from 2170 to 2000 cm^{-1} there is the presence of gases such as CO$_2$ and CO. When observing all the spectra, it can be observed that there is a concentration of peaks that stands out in the determined region between 800 and 1800 cm^{-1}. Such peaks represent some bands that stretch and deform to various groups and vibrational values that correspond to the constituents of the lignocellulosic material. This range represents very valuable information in lignocellulosic materials regarding changes (stretching and deformation vibrations) in the components of cellulose, hemicellulose and lignin [78]. Then, there are some absorption peaks between 1646 and 1740 cm^{-1} corresponding to the vibrations of the carbonyl groups C=O, acetyl and carboxyl groups present in hemicellulose and cellulose [77]. The band of absorption between 1514 and 1646 cm^{-1} is attributed to the vibrations of the groups C-O-C of the ring of the β-glucopyranosa that constitutes the cellulose. There are bands in the range of 1430–1602 cm^{-1} corresponding to the vibration of the structure of the aromatic rings (C = C) characteristic of cellulose and lignin. When identifying the loss of acetyl-type groups, usually there is a reduction in the band with a value of 1253 cm^{-1}, as can be seen in Figure 8, which represents the vibration in C-O bonds of Ph-C, that is an important part of both the lignin aromatic ring and the xyloglucan [79]. The band present in 1154 cm^{-1} represent the vibrational behavior of C-O-C bonds of the esters characteristic of hemicellulose and cellulose. The 1022 cm^{-1} signal to the C-O vibration of alcohols is attributed to hemicellulose, cellulose and lignin. Finally, the bands in the range 800–895 cm^{-1} are those corresponding to the vibration of the structure of the aromatic rings (C-C) characteristic of lignin degradation. It is important to mention that, the wavelength value equal to 1514 cm^{-1} corresponds to the structure of the aromatic ring of the lignin, i.e., the C-C bonds [80,81]. According to those reported by other authors, glucose ring stretching vibration is found at wave number 1111 cm^{-1} [82], which is also shown in Figure 8.

4. Conclusions

For the first time, the kinetics of the pyrolytic process of *P. pseudostrobus*, *P. leiophylla* and *P. montezumae* in an inert atmosphere has been studied by thermogravimetric analysis to determine its most representative kinetic parameters, such as activation energy and frequency factor. According to the TGA analysis, the most important quantitative phase of pyrolysis of the three pine species studied takes place in the 150 to 400 °C range. Approximately up to 250 °C a loss of 10 to 15% of mass occurs, corresponding to the first stage, i.e., loss of water and extractives. From 250 °C to 400 °C most of the volatile substances are released, mainly hemicellulose and cellulose being decomposed, with a loss of mass of about 80%. A rapid reduction of volatile compounds was observed at high temperatures (T > 400 °C) as well as coke formation.

The average activation energy determined by the Friedman, OFW and KAS methods was 126.58, 123.22 and 112.72 kJ/mol respectively with regard to *P. pseudostrobus*; 146.15, 143.24 and 132.76 kJ/mol respectively for *P. leiophylla* and 148.12, 151.80 and 141.25 kJ/mol, respectively, for *P. montezumae*, respectively, resulting in similar values in magnitude for the three methods and maintaining a tendency to increase with increasing heating velocity. For the range $\alpha = 0.20$–0.70 and according to the results, it was possible to observe an optimal correlation (average $R^2 > 0.97$) in practically all the experimental data used in the mathematical models applied in this investigation to determine the E_a, in this sense, the non-consideration of some data for the calculation and analysis of E_a, it does not affect the veracity and quality of the results.

The average frequency factor (Z) for each heating rate (β) in the Friedman, KAS and OFW methods varies from 6.28×10^9–3.72×10^9; and 1.69×10^8–1.55×10^8; and 1.95×10^7–2.08×10^7 for *P. pseudostrobus*, respectively. On average, a variation of Z was found for *P. leiophylla* from 1.06×10^{14}–4.71×10^{13}; and 9.04×10^9–8.10×10^9; and 1.07×10^9–1.09×10^9, respectively. Finally, to *P. montezumae* the results of Z, were 2.65×10^{15}–1.09×10^{15}; and 3.88×10^{10}–3.34×10^{10}; and 4.61×10^9–4.50×10^9, respectively. When performing the kinetic analysis and taking into account the error values and the determination coefficient, it is established that the FWO and KAS models are best suited to the thermal degradation of *P. pseudostrobus*, *P. leiophylla* and *P. montezumae*. The variability of the E_a values when applying these methods confirm the complexity of the degradation process. Finally, it should be mentioned that the low ash content of *P. pseudostrobus*, *P. leiophylla* and *P. montezumae* make them good candidates for the production of biochemical products such as methane and its subsequent processing for the production of hydrogen. According to FT-IR analysis, it was observed that all relative intensities in all bands were higher for *P. montezumae*, followed by *Pinus leiophylla* and finally *P. pseudostrobus*. The band is mainly distinguished in the maximum peak of 1022 cm^{-1}. The differences between the IR values in the softwoods used in this research may be related to the amount of the main biochemical components of the lignocellulosic material, i.e., cellulose, hemicellulose and lignin.

Author Contributions: Writing—original draft preparation, J.J.A.F.; formal analysis, J.J.A.F., J.G.R.Q. and F.M.M.; methodology, J.J.A.F., A.A.R., F.M.M., S.J.G.M. and J.E.V.; investigation, J.J.A.F., M.L.Á.R. and A.A.R.; conceptualization, J.J.A.F., J.G.R.Q. and J.V.A.V.; supervision, M.L.Á.R. and J.V.A.V.; software, J.J.A.F. and M.L.Á.R. All authors have read and agreed to the published version of the manuscript.

Funding: The project was supported by funds of the Scientific Research Coordination of the University Michoacana de San Nicolás de Hidalgo No. 14516.

Acknowledgments: The authors wish to thank the Faculty of Wood Technology Engineering of the University Michoacana de San Nicolás de Hidalgo for the facilities for this work. This work was supported by the Scientific Research Coordination of the UMSNH No. 14516.

Conflicts of Interest: The authors declare no conflict of interest.

References

1. Manzano, F.; Alcayde, A.; Montoya, F.; Zapata, A.; Gil, C. Scientific production of renewable energies worldwide: An overview. *Renew. Sustain. Energy Rev.* **2013**, *18*, 134–143. [CrossRef]
2. Varma, A.; Thakur, L.; Shankar, R.; Mondal, P. Pyrolysis of wood sawdust: Effects of process parameters on products yield and characterization of products. *Waste Manag.* **2019**, *89*, 224–235. [CrossRef]
3. Saldarriaga, J.; Aguado, R.; Pablos, A.; Amutio, M.; Olazar, M.; Bilbao, J. Fast characterization of biomass fuels by thermogravimetric analysis (TGA). *Fuel* **2015**, *140*, 744–751. [CrossRef]
4. Soto, N.; Machado, W.; López, D. Determinación de los parámetros cinéticos en la pirólisis del pino ciprés. *Quim. Nova* **2010**, *33*, 1500–1505. [CrossRef]
5. Davis, S.; Diegel, S.; Boundy, R. *Transportation Energy Data Book*; Oak Ridge National Laboratory: Oak Ridge, TN, USA, 2009; No. ORNL-6984.
6. Bridgwater, A. The production of biofuels and renewable chemicals by fast pyrolysis of biomass. *Int. J. Glob. Energy* **2007**, *27*, 160–203. [CrossRef]
7. Ragauskas, A.; Williams, C.; Davison, B.; Britovsek, G.; Cairney, J.; Eckert, C.; Frederick, W.; Hallett, J.; Leak, D.; Liotta, C.; et al. The path forward for biofuels and biomaterials. *Science* **2006**, *311*, 484–489. [CrossRef] [PubMed]
8. Frombo, F.; Minciardi, R.; Robba, M.; Rosso, F.; Sacile, R. Planning woody biomass logistics for energy production: A strategic decision model. *Biomass Bioenergy* **2009**, *33*, 372–383. [CrossRef]
9. Mullaney, H.; Farag, I.; La Claire, C.; Barrett, C. *Technical, Environmental and Economic Feasibility of Bio-Oil in New Hampshire's North Country*; New Hampshire Industrial Research Center: Durham, NH, USA, 2002.
10. Da Silva, J.; Alves, J.; de Araujo, W.; Andersen, S.; de Sena, R. Pyrolysis kinetic evaluation by single-step for waste wood from reforestation. *Waste Manag.* **2018**, *72*, 265–273. [CrossRef]
11. Janković, B. The pyrolysis process of wood biomass samples under isothermal experimental conditions-energy density considerations: Application of the distributed apparent activation energy model with a mixture of distribution functions. *Cellulose* **2014**, *21*, 2285–2314. [CrossRef]
12. Mohan, D.; Pittman, C.; Steele, P. Pyrolysis of wood/biomass for bio-oil: A critical review. *Energy Fuels* **2006**, *20*, 848–889. [CrossRef]
13. Soltes, E.; Elder, T. *Pyrolysis in Organic Chemicals from Biomass. IS Goldstein*; CRC Press: Florida, FL, USA, 1981.
14. Branca, C.; Lannace, A.; Di Blasi, C. Devolatilization and Combustion Kinetics of Quercus cerris Bark. *Energy Fuels* **2007**, *21*, 1078–1084. [CrossRef]
15. D' Almeida, A.; Barreto, D.; Calado, V.; d' Almeida, J. Thermal analysis of less common lignocellulose fibers. *J. Therm. Anal. Calorim.* **2008**, *91*, 405–408. [CrossRef]
16. Guerrero, M.; da Silva, M.; Zaragoza, M.; Gutiérrez, J.; Velderrain, V.; Ortiz, A.; Collins, V. Thermogravimetric study on the pyrolysis kinetics of apple pomace as waste biomass. *Int. J. Hydrogen. Energy* **2014**, *39*, 16619–16627. [CrossRef]
17. Hill, J. *For Better Thermal Analysis and Calorimetry*; International Confederation for Thermal Analysis, 1991.
18. Giuntoli, J.; De Jong, W.; Arvelakis, S.; Spliethoff, H.; Verkooijen, A. Quantitative and kinetic TG-FTIR study of biomass residue pyrolysis: Dry distiller's grains with solubles (DDGS) and chicken manure. *J. Anal. Appl. Pyrol.* **2009**, *85*, 301–312. [CrossRef]
19. Zhu, H.; Yan, J.; Jiang, X.; Lai, Y.; Cen, K. Study on pyrolysis of typical medical waste materials by using TG-FTIR analysis. *J. Hazard. Mater.* **2008**, *153*, 670–676. [CrossRef] [PubMed]
20. Koreňová, Z.; Juma, M.; Annus, J.; Markoš, J. Kinetics of pyrolysis and properties of carbon black from a scrap tire. *Chem. Pap.* **2006**, *60*, 422–426. [CrossRef]
21. Sun, J.; Wang, W.; Liu, Z.; Ma, Q.; Zhao, C.; Ma, C. Kinetic study of the pyrolysis of waste printed circuit boards subject to conventional and microwave heating. *Energies* **2012**, *5*, 3295–3306. [CrossRef]

22. White, J.; Catallo, W.; Legendre, B. Biomass pyrolysis kinetics: A comparative critical review with relevant agricultural residue case studies. *J. Anal. Appl. Pyrol.* **2011**, *91*, 1–33. [CrossRef]
23. Singh, S.; Wu, C.; Williams, P. Pyrolysis of waste materials using TGA-MS and TGA-FTIR as complementary characterisation techniques. *J. Anal. App. Pyrol.* **2012**, *94*, 99–107. [CrossRef]
24. Cao, R.; Naya, S.; Artiaga, R.; García, A.; Varela, A. Logistic approach to polymer degradation in dynamic TGA. *Polym. Degrad. Stab.* **2004**, *85*, 667–674. [CrossRef]
25. Brown, M. Steps in a minefield: Some kinetic aspects of thermal analysis. *J. Therm. Anal. Calorim.* **1997**, *49*, 17–32. [CrossRef]
26. Brown, M.; Maciejewski, M.; Vyazovkin, S.; Nomen, R.; Sempere, J.; Burnham, A.; Opfermann, J.; Strey, R.; Anderson, H.; Kemmler, A.; et al. Computational aspects of kinetic analysis: Part A: The ICTAC kinetics project-data, methods and results. *Thermochim. Acta* **2000**, *355*, 125–143. [CrossRef]
27. Friedman, H. Kinetics of thermal degradation of char-forming plastics from thermogravimetry. Application to a phenolic plastic. *J. Polym. Sci. Pol. Sym.* **1964**, *6*, 183–195. [CrossRef]
28. Ozawa, T. A new method of analyzing thermogravimetric data. *B. Chem. Soc. Jpn.* **1965**, *38*, 1881–1886. [CrossRef]
29. Kissinger, H. Reaction kinetics in differential thermal analysis. *Anal. Chem.* **1957**, *29*, 1702–1706. [CrossRef]
30. Akahira, T. Trans. Joint convention of four electrical institutes. *Res. Rep. Chiba Inst. Technol.* **1971**, *16*, 22–31.
31. Pintor, L.; Carrillo, A.; Herrera, R.; López, P.; Rutiaga, J. Physical and chemical properties of timber byproducts from *Pinus leiophylla, P. montezumae and P. pseudostrobus* for a bioenergetic use. *Wood Res.* **2017**, *62*, 849–862.
32. Téllez, C.; Ochoa, H.; Sanjuan, R.; Rutiaga, J. Componentes químicos del duramen de Andira inermis (W. Wright) DC. (Leguminosae). *Rev. Chapingo Ser. Cienc. For. Ambient.* **2010**, *16*, 87–93.
33. UNE-EN 14775. *Solid Biofuels. Method for the Ash Content Determination*; CONFEMADERA, AENOR, Grupo 9: Madrid, España, 2010; 10p. (In Spanish)
34. Vyazovkin, S.; Burnham, A.; Criado, J.; Pérez, L.; Popescu, C.; Sbirrazzuoli, N. ICTAC Kinetics Committee recommendations for performing kinetic computations on thermal analysis data. *Thermochim. Acta* **2011**, *520*, 1–19. [CrossRef]
35. Damartzis, T.; Vamvuka, D.; Sfakiotakis, S.; Zabaniotou, A. Thermal degradation studies and kinetic modeling of cardoon (Cynara cardunculus) pyrolysis using thermogravimetric analysis (TGA). *Bioresour. Technol.* **2011**, *102*, 6230–6238. [CrossRef]
36. Flynn, J. The isoconversional method for determination of energy of activation at constant heating rates: Corrections for the Doyle approximation. *J. Therm. Anal. Calorim.* **1983**, *27*, 95–102. [CrossRef]
37. Sadhukhan, A.; Gupta, P.; Saha, R. Modelling of pyrolysis of large wood particles. *Bioresour. Technol.* **2009**, *100*, 3134–3139. [CrossRef] [PubMed]
38. Min, F.; Zhang, M.; Chen, Q. Non-isothermal kinetics of pyrolysis of three kinds of fresh biomass. *J. China Univ. Min. Technol.* **2007**, *17*, 105–111. [CrossRef]
39. Capart, R.; Khezami, L.; Burnham, A. Assessment of various kinetic models for the pyrolysis of a microgranular cellulose. *Thermochim. Acta* **2004**, *417*, 79–89. [CrossRef]
40. Vlaev, L.; Markovska, I.; Lyubchev, L. Non-isothermal kinetics of pyrolysis of rice husk. *Thermochim. Acta* **2003**, *406*, 1–7. [CrossRef]
41. Galwey, A.; Brown, M. Application of the Arrhenius equation to solid state kinetics: Can this be justified? *Thermochim. Acta* **2002**, *386*, 91–98. [CrossRef]
42. Steinfeld, J.; Francisco, J.; Hase, W. *Chemical Kinetics and Dynamics*, 2nd ed.; Prentice-Hall: Upper Saddle River, NJ, USA, 1999.
43. Atkins, P.W. *Physical Chemistry*, 5th ed.; W.H. Freeman: New York, NY, USA, 1994.
44. Flynn, J. The 'temperature integral'—Its use and abuse. *Thermochim. Acta* **1997**, *300*, 83–92. [CrossRef]
45. Liu, M.; Gao, L.; Zhao, Q.; Wang, Y.; Yang, X.; Cao, S. Thermal degradation process and kinetics of poly (dodecamethyleneisophthalamide). *Chem. J. Internet.* **2003**, *5*, 43.
46. Biagini, E.; Guerrini, L.; Nicolella, C. Development of a variable activation energy model for biomass devolatilization. *Energy Fuels* **2009**, *23*, 3300–3306. [CrossRef]
47. Kantarelis, E.; Yang, W.; Blasiak, W.; Forsgren, C.; Zabaniotou, A. Thermochemical treatment of E-waste from small household appliances using highly pre-heated nitrogen-thermogravimetric investigation and pyrolysis kinetics. *Appl. Energy* **2011**, *88*, 922–929. [CrossRef]

48. Flynn, J.; Wall, L. General treatment of the thermogravimetry of polymers. *J. Res. Nat. Bur. Stand.* **1966**, *70*, 487–523. [CrossRef] [PubMed]
49. Tang, T.; Chaudhri, M. Analysis of dynamic kinetic data from solid-state reactions. *J. Therm. Anal. Calorim.* **1980**, *18*, 247–261. [CrossRef]
50. Doyle, C. Kinetic analysis of thermogravimetric data. *J. Appl. Polym. Sci.* **1961**, *5*, 285–292. [CrossRef]
51. Chao, M.; Li, W.; Wang, X. Influence of antioxidant on the thermal–oxidative degradation behavior and oxidation stability of synthetic ester. *Thermochim. Acta* **2014**, *591*, 16–21. [CrossRef]
52. Coats, A.; Redfern, J. Kinetic parameters from thermogravimetric data. *Nature* **1964**, *201*, 68–69. [CrossRef]
53. Lyon, R. An integral method of nonisothermal kinetic analysis. *Thermochim. Acta* **1997**, *297*, 117–124. [CrossRef]
54. Obernberger, I.; Thek, G. *The Pellet Handbook*, 1st ed.; Earthscan: London, UK; Washington, DC, USA, 2010.
55. Van Lith, S.; Alonso, V.; Jensen, P.; Frandsen, F.; Glarborg, P. Release to the gas phase of inorganic elements during wood combustion. Part 1: Development and evaluation of quantification methods. *Energy Fuels* **2006**, *20*, 964–978. [CrossRef]
56. Werkelin, J.; Lindberg, D.; Boström, D.; Skrifvars, B.; Hupa, M. Ash-forming elements in four Scandinavian wood species part 3: Combustion of five spruce samples. *Biomass Bioenergy* **2011**, *35*, 725–733. [CrossRef]
57. Miles, T. *Alkali Deposits Found in Biomass Power Plants*; NREL Report 443-8142; Vol 1. Sand 96-8225; National Techrucd Momation Service (NTIS): Springfield, VA, USA, 1995.
58. Du, Z.; Sarofim, A.; Longwell, J. Activation energy distribution in temperature-programmed desorption: Modeling and application to the soot oxygen system. *Energy Fuels* **1990**, *4*, 296–302. [CrossRef]
59. Yahiaoui, M.; Hadoun, H.; Toumert, I.; Hassani, A. Determination of kinetic parameters of Phlomis bovei de Noé using thermogravimetric analysis. *Bioresour. Technol.* **2015**, *196*, 441–447. [CrossRef]
60. Vamvuka, D.; Kakaras, E.; Kastanaki, E.; Grammelis, P. Pyrolysis characteristics and kinetics of biomass residuals mixtures with lignite. *Fuel* **2003**, *82*, 1949–1960. [CrossRef]
61. Meesri, C.; Moghtaderi, B. Lack of synergetic effects in the pyrolytic characteristics of woody biomass/coal blends under low and high heating rate regimes. *Biomass Bioenergy* **2002**, *23*, 55–66. [CrossRef]
62. Jeguirim, M.; Trouvé, G. Pyrolysis characteristics and kinetics of Arundo donax using thermogravimetric analysis. *Bioresour. Technol.* **2009**, *100*, 4026–4031. [CrossRef] [PubMed]
63. Gašparovič, L.; Koreňová, Z.; Jelemenský, Ľ. Kinetic study of wood chips decomposition by TGA. *Chem. Pap.* **2010**, *64*, 174–181. [CrossRef]
64. Quan, C.; Li, A.; Gao, N. Thermogravimetric analysis and kinetic study on large particles of printed circuit board wastes. *Waste Manag.* **2009**, *29*, 2353–2360. [CrossRef]
65. Kumar, A.; Wang, L.; Dzenis, Y.; Jones, D.; Hanna, M. Thermogravimetric characterization of corn stover as gasification and pyrolysis feedstock. *Biomass Bioenergy* **2008**, *32*, 460–467. [CrossRef]
66. Wang, G.; Li, W.; Li, B.; Chen, H. TG study on pyrolysis of biomass and its three components under syngas. *Fuel* **2008**, *87*, 552–558. [CrossRef]
67. Mishra, R.; Mohanty, K. Pyrolysis kinetics and thermal behavior of waste sawdust biomass using thermogravimetric analysis. *Bioresour. Technol.* **2018**, *251*, 63–74. [CrossRef]
68. Vyazovkin, S.; Wight, C. Model-free and model-fitting approaches to kinetic analysis of isothermal and nonisothermal data. *Thermochim. Acta* **1999**, *340*, 53–68. [CrossRef]
69. Amutio, M.; Lopez, G.; Aguado, R.; Artetxe, M.; Bilbao, J.; Olazar, M. Kinetic study of lignocellulosic biomass oxidative pyrolysis. *Fuel* **2012**, *95*, 305–311. [CrossRef]
70. Domínguez, J.; Santos, T.; Rigual, V.; Oliet, M.; Alonso, M.; Rodriguez, F. Thermal stability, degradation kinetics, and molecular weight of organosolv lignins from *Pinus radiata*. *Ind. Crop. Prod.* **2018**, *111*, 889–898. [CrossRef]
71. Vyazovkin, S. Computational aspects of kinetic analysis.: Part C. The ICTAC Kinetics Project-the light at the end of the tunnel? *Thermochim. Acta* **2000**, *355*, 155–163. [CrossRef]
72. Aboyade, A.; Hugo, J.; Carrier, M.; Meyer, E.; Stahl, R.; Knoetze, J.; Görgens, J. Non-isothermal kinetic analysis of the devolatilization of corn cobs and sugar cane bagasse in an inert atmosphere. *Thermochim. Acta* **2011**, *517*, 81–89. [CrossRef]
73. Gai, C.; Dong, Y.; Zhang, T. The kinetic analysis of the pyrolysis of agricultural residue under non-isothermal conditions. *Bioresour. Technol.* **2013**, *127*, 298–305. [CrossRef] [PubMed]

74. Huang, Y.; Kuan, W.; Chiueh, P.; Lo, S. A sequential method to analyze the kinetics of biomass pyrolysis. *Bioresour. Technol.* **2011**, *102*, 9241–9246. [CrossRef]
75. Turmanova, S.; Genieva, S.; Dimitrova, A.; Vlaev, L. Non-isothermal degradation kinetics of filled with rise husk ash polypropene composites. *Express Polym. Lett.* **2008**, *2*, 133–146. [CrossRef]
76. Kondo, T. *Hydrogen Bonds in Cellulose and Cellulose. Derivatives, Polysaccharides II—Structural Diversity and Functional Versatility*; Dumitriu, S., Ed.; Marcel Dekker: New York, NY, USA, 2005; Chapter 3.
77. Popescu, C.; Popescu, M.; Vasile, C. Structural analysis of photodegraded lime wood by means of FT-IR and 2D IR correlation spectroscopy. *Int. J. Biol. Macromol.* **2011**, *48*, 667–675. [CrossRef]
78. Baeza, J.; Freer, J. Chemical characterization of wood and its components. In *Wood and Cellulosic Chemistry*; Hon, D.-S., Shiraishi, N., Eds.; Marcel Dekker Inc.: New York, NY, USA, 2001; Chapter 8; pp. 275–384.
79. Popescu, C.; Popescu, M.; Vasile, C. Characterization of fungal degraded lime wood by FT-IR and 2D IR correlation spectroscopy. *Microchem. J.* **2010**, *95*, 377–387. [CrossRef]
80. Xu, C.; Etcheverry, T. Hydro-liquefaction of woody biomass in sub-and super-critical ethanol with iron-based catalysts. *Fuel* **2008**, *87*, 335–345. [CrossRef]
81. Kim, J.; Hwang, H.; Oh, S.; Kim, Y.; Kim, U.; Choi, J. Investigation of structural modification and thermal characteristics of lignin after heat treatment. *Int. J. Biol. Macromol.* **2014**, *66*, 57–65. [CrossRef]
82. Popescu, M.; Froidevaux, J.; Navi, P.; Popescu, C. Structural modifications of Tilia cordata wood during heat treatment investigated by FT-IR and 2D IR correlation spectroscopy. *J. Mol. Struct.* **2013**, *1033*, 176–186. [CrossRef]

© 2020 by the authors. Licensee MDPI, Basel, Switzerland. This article is an open access article distributed under the terms and conditions of the Creative Commons Attribution (CC BY) license (http://creativecommons.org/licenses/by/4.0/).

Article

Evaluating the Potential for Combustion of Biofuels in Grate Furnaces

Małgorzata Wzorek

Department of Process Engineering, Faculty of Mechanical Engineering, Opole University of Technology, ul. Prószkowska 76, 45-758 Opole, Poland; m.wzorek@po.edu.pl; Tel.: +48-77-449-8779

Received: 11 March 2020; Accepted: 10 April 2020; Published: 15 April 2020

Abstract: The paper assesses the impact of combustion of biofuels produced based on municipal sewage sludge in stoker-fired boilers on the amount of pollutant emissions and examines the tendency of ash deposition of biofuels formed during the combustion process. The combustion tests were performed in a laboratory system enabling simulation of a combustion process present in stoker-fired boilers. The study was conducted for three types of biofuels; i.e., fuel from sewage sludge and coal slime (PBS fuel), sewage sludge and meat and bone meal (PBM fuel) and fuel based on sewage sludge and sawdust (PBT) with particle size of 35 mm and 15 mm. This paper describes and compares the combustion process of biofuels with different granulation and composition and presents the results of changes in emission values of NO_x, SO_2, CO, and CO_2. The emission results were compared with the corresponding results obtained during combustion of hard coal. The results showed that biofuels with lower particle sizes were ignited faster and the shortest ignition time is achieved for fuel based on sewage sludge and coal slime-PBS fuel. Also, the highest NO and SO_2 emissions were obtained for PBS fuel. During the combustion of fuel based on sewage sludge and meat and bone meal (PBM), on the other hand, the highest CO_2 emissions were observed for both granulations. Biofuels from sludge show a combustion process that is different compared to the one for hard coal. The problems of ash fouling, slagging, and deposition during biofuels combustion were also identified. The tendency for ash slagging and fouling is observed, especially for fuel from sewage sludge and meat and bone meal (PBM) and fuel based on sewage sludge and sawdust (PBT) ashes which consist of meat and bone meal and sawdust which is typical for biomass combustion.

Keywords: sewage sludge; biofuels; combustion; grate furnace; emission; ash deposition

1. Introduction

Municipal sewage sludge is a product of the water-cleaning process in wastewater treatment plants. The amount of generated sewage sludge depends on many factors, mainly on the content of pollutants in the sludge and on the technology of its treatment. An amount of sewage sludge cannot be prevented and is reduced in line with the requirements regarding the quality of treated sewage.

The problem of sewage sludge disposal has two aspects: quantitative and qualitative aspect, which results from the specific properties of the waste and the legal aspect.

The problem with the disposal of municipal sewage sludge also results from the introduction of new, increasingly stringent legal regulations concerning sewage sludge management, limiting the use of the sludge for agricultural and natural purposes and prohibiting its storage [1,2].

In this situation, processes that are becoming increasingly important are thermal use processes which are among the most radical methods in terms the possibility of a significant reduction in the mass and volume of sewage sludge. They also allow use of the energy contained in the sludge and to reduce CO_2 emissions in accordance with the principles of sustainable development.

The parameter that is very important for the use of waste to generate energy is the stability of the properties which determines the efficiency of the combustion process, and in the case of sewage sludge it is difficult to talk about stability, since the properties of the sludge vary widely and are dependent on many factors.

Another problem in the thermal use of municipal sludge is its high water content. With a dry matter content of 20%–30%, municipal sewage sludge can be incinerated only with the help of additional fuel, and only after partial drying up to 50% can they be burnt autothermally [3]. The total drying of sludge, up to about 10%, allows their use in co-combustion with coal in industrial processes [4,5]. Another way to use sludge for energy purpose is to use it as an ingredient in the production of fuel with fixed composition and properties [6].

Research on the thermal degradation of sludge combustion and co-combustion with coal and other fuels are conducted both on a laboratory and industrial scale.

Thermogravimetric analysis (TGA) is a technique that is most widely and commonly used for this purpose. Many authors emphasize the specificity of behavior of sewage sludge in combustion process. For example, Lin et al. [7] showed that the co-combustion of sewage sludge and oil shale with a proportion of 10% of the sludge gave the best promoter effect. It was observed that the ignition temperature shifted to an earlier temperature when sewage sludge was added.

Chen et al. [8] had studied co-combustion characteristics of sewage sludge and coffee grounds mixtures (mixing ratios of 9:1, 8:2, 7:3 and 6:4) using thermogravimetric analysis coupled to artificial neural networks modeling. The results showed interactions between the components, and with the addition of coffee grounds ignition temperature, maximum mass loss rate, and the reactivity of sewage sludge increased while charring was reduced.

The authors also presented testes of co-combustion sewage sludge with straw [9], olive and animal waste [10], shiitake substrate [11], rice husk [12] and also with water hyacinth in CO_2/O_2 atmosphere [13]. All the above research was carried out for mixtures of fuels in the form of powder, for samples of 10 mg.

Investigations of co-combustion of sewage sludge in a pelletized form, on a small laboratory scale, were conducted, among others by Akdağ et al. [14], who studied co-combustion of sewage sludge with coal (3%, 5%, 10%, 20% and 30%) in a laboratory batch reactor. Kijo-Kleczkowska et al. [15] tested the co-combustion of pelletized (10 mm) sewage sludge with coal and willow Salix viminalis. Junga et al. [16] studied the combustion of sewage sludge-based pellets and agriculture waste in 10 kW understocker boiler.

The literature also reports on the large-scale combustion and co-combustion of coal and sewage sludge in grate furnaces [17,18]. This process is mainly carried out in fluidized and pulverized-fuel boilers [19,20].

The impact of co-combustion of sewage sludge on boiler efficiency, the amount of pollutant emissions and its impact on the environment is widely discussed in the literature.

The grate boiler furnaces are used for co-combustion of hard coal with biomass and waste fuels, including sewage sludge [5,21,22]. Werle [23] presented, among others, an analysis of the possibility of co-combustion of sewage sludge with coal (blends of 0%–20%) in a WR-25 power station. It was found that an increase in the mass of the sewage sludge in the fuel blend causes a significant reduction in CO_2 emissions to the atmosphere.

Nadziakiewicz et al. [18] investigated the changes in emissions of CO, NO_x, and SO_2 during co-combustion of dried sewage sludge with coal in the laboratory stoker-fired boiler. The tests show that the emissions of air pollutants increase with the increase in sludge ratio in the fuel mixture. Houshfar et al. [24], on the other hand, conducted an experimental investigation on the NO_x formation and reduction, among others, for sewage sludge mixtures with straw and wood pellets.

The main issue in burning waste fuels and biomass in power boilers is a different, than that of coal, chemical, and mineral composition of ash. The presence of components with low melting point in the ash from those fuels poses the risk of slag formation and problem of powdered material (pollutants) sedimentation of heated surfaces of heat exchangers.

Ashes in the temperature range between softening and melting point tend to stick (adhere). The transformations of minerals contained in the fuel, taking place under the furnace chamber conditions, often also lead to the formation of compounds (or their combinations) characterized by particularly low melting points [25].

To assess the propensity of fuel to slag and contaminate the heating surface, several different value indicators based on ash oxide analysis [25–27] have been developed, which include:

- B/A ratio—the ratio of the alkaline to acidic oxides in the ash base-to-acid ratio:

$$B/A = \frac{Fe_2O_3 + CaO + MgO + Na_2O + K_2O}{SiO_2 + Al_2O_3 + TiO_2} \qquad (1)$$

  for biomass it also includes P_2O_5 content—B/A_{+P} index
- Slagging index R_S

$$R_s = \left(\frac{B}{A}\right) \cdot A^d \qquad (2)$$

  where: A^d is the percentage of ash in dry fuel.
- Fouling index F_u—index of the probability of heated surfaces fouling

$$F_u = \left(\frac{B}{A}\right) \cdot (Na_2O + K_2O) \qquad (3)$$

- Slag viscosity index S_R

$$S_R = \frac{SiO_2}{SiO_2 + Fe_2O_3 + CaO + MgO} \qquad (4)$$

- Fe_2O_3/CaO ratio informing about the emergence of slag-promoting eutectic.

Numerous studies have been devoted to the mechanism of ash formation and the behavior of mineral matter in the processes of co-combustion of coal with other fuels with low caloric value, in particular biomass, among others with biomass [28–31] and sewage sludge [25,32,33] in large and small scale.

Furthermore, an important issue, while using fuels from waste and biomass is also sufficient mixing of the fuel on the grate, which is expected to prevent from such detrimental effects as local material overheating, leading to slag formation, furnace chamber damage, and grate overburning.

An important parameter of the fuel combusted on a grate is its grain composition. Improper choice of the fuel grain size composition may lead to considerable loss of unburnt carbon in the slag and fly ash, since the content of combustibles in the slag may be even 25–30% of its weight, and in the fly ash 15–20% [34].

In stoker-fired boilers, fuels with grain sizes 0–25 mm are usually burnt, with less than 25% 0–2 mm fractions.

Most of the stoker-fired boilers in use are not equipped in any flue gas treatment systems, except for the simplest dedusting equipment. The need to upgrade the flue gas treatment system, in order to comply with the emission standards and to adjust the flue system to the requirements of waste co-combustion process (conditions related to the minimum flue gas presence and minimum temperature in the furnace chamber) is the main subject of numerous discussions.

The aim of the study is to evaluate the potential of the combustion of biofuels made of municipal sewage sludge and other materials in grate furnaces. Three types of biofuels were tested; i.e., fuel based on sewage sludge and coal slime (PBS), sewage sludge and meat and bone meal (PBM) and sewage sludge and sawdust (PBT) with particle size of 35 mm and 15 mm. The impact of the fuel pellet size on the combustion process and on the emission of pollutants was taken into account. Additionally, the objective of this paper is the evaluation of ash deposits formed during the tests to establish the deposition behavior of biofuel pellets.

2. Materials and Methods

2.1. Materials

The combustion tests were performed on fuels made of municipal sewage sludge and other materials such as coal sludge, meat and bone meal, and sawdust. The technology of the sewage sludge fuel production consists of the initial mixing of the components and the subsequent proportions and simultaneous granulation and drying in the purpose made drum granulator. The drum is equipped with a feeder system ensuring the granulate diameter in the range from 15 to 35 mm [10,35,36]. The combustion tests were performed on biofuels made of:

- 60 wt.% of sewage sludge, 34 wt.% coal sludge and 6 wt.% of quicklime-PBS fuel,
- 75 wt.% of sewage sludge, 24 wt.% of meat and bone meal, 1 wt.% of quicklime-PBM fuel,
- 80 wt.% of sewage sludge, 19 wt.% of sawdust and 1wt.% of quicklime-PBT fuel.

The purpose of mixing sewage sludge with other components is, among others, to reduce its initial moisture of sewage sludge. The mixtures-following mixing and granulation-present a moisture content ranging from 40 to 60%. They are subsequently dried to content of moisture about 10% in a solar drier using solar energy. The method has been patented [37] and recommended because of much lower production costs compared with conventional methods of high-temperature drying.

Energy properties of the fuels are listed in Table 1.

Table 1. Energy properties of fuels from sewage sludge.

Parameter	Unit	PBS	PBM	PBT
Lower Heating Value, LHV	MJ/kg	19.30	14.59	13.23
Moisture	%	8.58	8.67	10.37
Voltaire matter	% d.m.	34.44	55.29	59.87
Ash	% d.m.	27.26	33.72	20.36
Elementary analysis				
Carbon		50.28	36.64	31.42
Hydrogen		3.91	4.12	4.43
Oxygen	% d.m.	15.01	17.95	40.50
Nitrogen		1.72	6.67	2.61
Sulphur		1.16	0.68	0.65
Chlorine		0.06	0.02	0.03

d.m.—dry mass.

Fuels with pellet sizes 35 mm and 15 mm were tested in order to evaluate the effect of grain size on the combustion process and emission of pollutants.

The argument confirming the implementation of the research is the fact that most of the standards and techniques used to determine fuel parameters are based on the use of fuels in a powder form. Conducting research on combustion of sludge fuels under primary grain size analysis and under conditions simulating the real facility in which these fuels can be used, may give a picture of the combustion process that is more accurately resembling the process in industrial installations.

2.2. Combustion Testes

The combustion tests were performed on a laboratory scale at the Silesian University of Technology, Department of Technologies and Installations for Waste Management. This system is used for testing combustion of various types of waste and waste fuels.

The main component of the test stand is a boiler with special construction that enables simulation combustion processes present in water boilers with a stationary and mechanical grate. The boiler comprises two main parts: the bottom one with adjustable heating temperature (up to 1200 °C), and the top one with water jacket.

The test stand is schematically shown in Figure 1. In the tests, the type of fuel fired was adopted as the input value (variable), and based on a series of initial tests and literature data, the following constant values were adopted as the system operating parameters:

- thickness of the bed of fuel being burnt -ca. 75 mm (each time 2.5 kg sample was burnt);
- process duration –3600 s from starting air feeding to the combustion chamber;
- air excess ratio in the furnace chamber $\lambda = 1.8$;
- secondary air stream–5 Nm³/h;
- initial temperature in the combustion chamber –900 °C ± 10K;
- minimum temperature during the combustion process –800 °C.

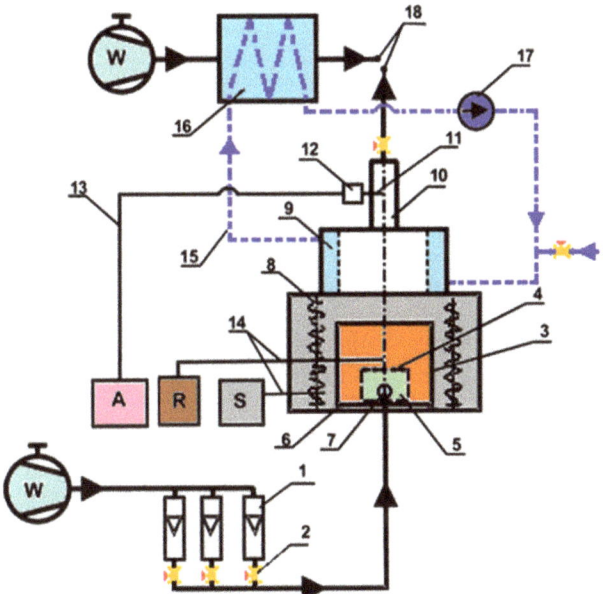

Figure 1. Test stand diagram [38]: A-flue gas analyzer, S-furnace control system, R-recorder, W-ventilator, 1-rotameter, 2-valve, 3-reciprocal bed with a grate, 4-grate, 5-plenum system (ash pit), 6-rail, 7-air nozzle, 8-heating element (electrical), 9-water jacket, 10-stack, 11-measuring probe, 12-probe head, 13-heated hose, 14-thermo elements with compensating cables, 15-cooling water circuit, 16-water/air heat exchanger (cooler), 17-circulation pump, 18 – ambience.

The measurement method which was used based on placing samples in a chamber preheated to the required temperature of 900 °C. To expedite loading the fuel sample into the oven chamber and arranging it on the grate as needed, the grate was mounted on a movable bed which, having been heated to the required temperature, was then slid inside.

While the sample was being heated, but prior to the ignition point, secondary air was supplied to the combustion chamber at a constant rate of 5m³.

The secondary air was supplied after one of the below listed parameters had reached the limit value:

- CO concentration at the measuring point −6000 ppm,
- CO_2 concentration higher than 1%.

Two different streams of air fed to the process took place during measurements (one for the ignition phase and another one for the combustion phase).

Table 2 also shows the times the primary air was first fed to the combustion chamber (the starting point of the combustion process after the sample had been placed in the combustion chamber).

Table 2. Starting time of the combustion process.

Kind of Biofuels	PBS		PBM		PBT	
	35 mm	15 mm	35 mm	15 mm	35 mm	15 mm
Starting Time of the Combustion Process, s	600	420	1800	1500	330	480

During the tests, MGA 5 MRU flue gas analyzer was used, with a heated probe and internal flue gas conditioner. The analyzer allows measuring of flue gas composition with reference methods (measuring CO_2, CO, NO, NO_2, SO_2-NDIR sensor; O_2-electrochemical sensor). During the tests, the concentration of measured gases: CO_2 (0%–21%), CO (0%–5%), NO (0–10000 ppm), NO_2 (0–500 ppm), SO_2 (0–10000 ppm) in the flue gas was continuously measured.

Since the systems with fixed grates operate mainly in quasi-steady state at the most, to determine fuel behavior on the grate, gaseous pollutants were measured from feeding the fuel to the combustion chamber to the process end. Presentation of results for the entire combustion time enables forecasting of system operation in steady conditions (average value) and is a source of knowledge allowing forecasting the degree of emission variations in unstable operating conditions.

2.3. Anlayses of Ash and Slag

The chemical composition of ashes was analyzed by using the ICP method, and the remaining component levels-using the PANalitical XRF method.

Ash behavior and deposition tendencies were predicted through the use of empirical indices according to equations 1 to 4.

The residues after combustion process were analyzed for the presence of combustible matter in the slag and ash according to PN-90-G-04512.

Additionally, slag and ash were tested for leaching hazardous compounds. For that reason, water extracts were created according to PN-EN 12457-4:2006, and in the water eluates, among others, chlorides (according to PN-ISO 9297), sulfides (PN-74/C-04566) and the heavy metal ions were determined with Perkin-Elmer Plasma 400 ICP Emission Spectrometer.

3. Results and Discussion

3.1. Combustion Process of Fuels from Sewage Sludge

The combustion process of fuels from sewage sludge, along with the variations of CO_2, CO, NO, SO_2 emissions, is graphically shown in Figures 2–5.

As the '0' moment, the start of air feeding to the combustion process was assumed. The earlier time is marked with negative values. Fuels were introduced into the combustion chamber in the time marked as −1800 s.

Emissions measured while burning fuels from sewage sludge were compared to the average values for hard coal combustion (particle diameter 15 to 30 mm) [38]. Biofuels were feed to chamber at time marks such as (1800 s). Changes of CO_2 emissions shown in Figure 2 best reflect the variations of combustion intensity. Analyzing the variations of CO_2 emission while combustion of biofuels with the same composition, but different particle size, great similarity can be noted. The most intense burning process was observed between 250 and 1800 s; after that the CO_2 decreased slowly.

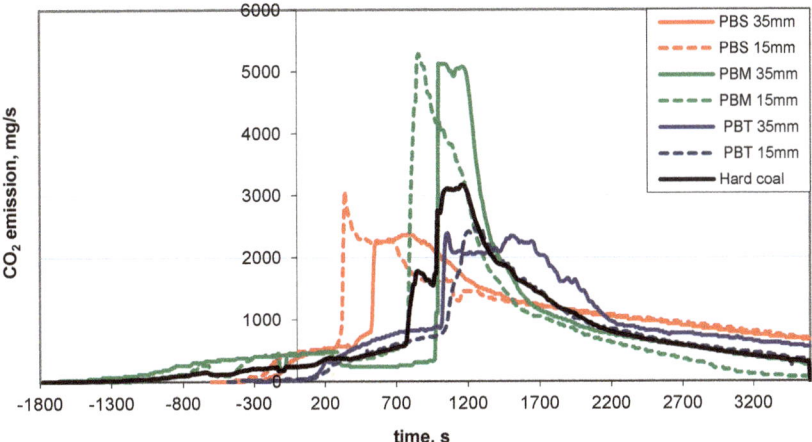

Figure 2. Variations of CO_2 emissions during combustion of fuels.

Maximum values of CO_2 emission were observed while burning the PBM fuel with both particle sizes. It contains meat and bone meal, and animal derived waste, as proved by the research of [34], burn faster than other fuels e.g., lignite. This can explain different behavior of PBM while combustion, as compared to other sludge fuels.

Analyzing Figure 2, it can be noted that for sewage sludge fuels, the fuels with lower particle sizes were ignited faster. The shortest ignition time is achieved for the PBS with particle size 15 mm. Since the ignition time of a fuel depends mainly on the moisture content and the time of thermal decomposition of the organic matter, it can be concluded that the cause for a faster ignition of PBS is the lower moisture content than in other sludge fuels.

Figure 3 presents CO emissions achieved while combustion of the sewage sludge fuels.

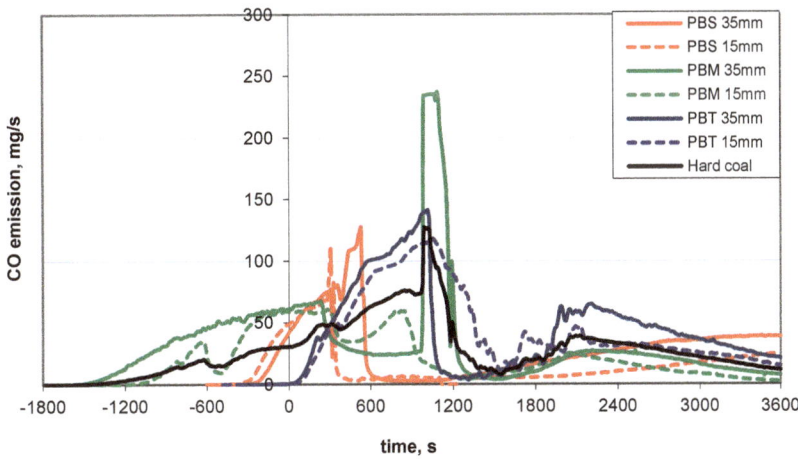

Figure 3. Variations of CO emissions during combustion of fuels.

Since the CO emission was observed to rise at point 0, primary air was supplied. As a result, after a momentary increase, the CO emission would drop to 0. The lower CO emissions were the result of a higher efficiency of the combustion process (and de facto reduced losses from incomplete combustion). Between 500 and 1500 s, after reaching the maximum values for individual fuels, CO emissions to values dropped close to 0. After that, the CO emission values increased, not exceeding 60 mg/s. According to Kozioł [39], such emission drops and increments are typical for burning fuels in grate furnaces. Naturally, adding some oxidizer to intensify the combustion process resulted in afterburning CO to CO_2, as demonstrated by higher CO_2 emissions (Figure 2). The delays observed in peaks (i.e., a large peak appearing after more than ten seconds following addition of the air) may result from the system inertia, reaction delay times, and the delay of the emission analyzer measurement path.

The combustion intensity was due to introducing air into the combustion chamber which in turn led to combustion of the flammable compounds of the fuel. For example, PBS fuel with 15 mm (Figure 3) produced high CO emissions beginning at −600s time which continued until approximately 300s after the air was fed at point "0", only to drop suddenly (almost to 0 mg/s) within roughly 300 s. It means adding the air intensified the fuel combustion process, a fact attested to by the rapid increase of CO_2 emission, i.e., the combustion product. Following this phase, CO_2 emissions stabilized (starting at about 600s to 800s) owing to the complete combustion. CO oxidation occurs most likely in the CO + OH· = CO_2 + H· reaction, and thus it is the amount of hydroxyl radicals OH· in the emissions that most likely determines the chances of overreacting. At the same time, the OH· concentration drops rapidly in sync with the temperature drop while the CO concentration remains higher than an equivalent one by an order of magnitude.

The so-called CO freeze effect may be explained by the higher emission levels of the substance in the latter phase of the combustion process. On the other hand, however, the CO_2 emission reduction results primarily from the fuel being entirely used up (reduced combustion intensity) and, on the other hand, from a higher degree of incomplete combustion, even though its share in this case is rather marginal. One must also keep in mind the primary reasons for the CO emissions is the insufficient time the fuel substance remains inside the grate or its excessive cooling. The above described phenomenon may also be observed with other types of fuel and the time delay difference is attributable to the oxidant penetrating fuel particles, or the ratio of volatile matter to fix carbon in a given particle volume.

While measuring the NO_2 concentration, in neither of the tests performed was it found in the flue gas. It may be due to the stabilization of temperature in the combustion chamber during the tests at the temperature level 900 °C. NO_2 is formed mainly as a result of reactions in temperatures lower than 750 °C [40]. The NO emission changes during combustion are shown in Figure 4. For all sludge fuels, very rapid increase of NO emission was observed in the time interval between 350 and 1500 s.

For particle size 15 mm, this value at 530 s was almost 9 mg/s, and for 35 mm, in the interval (790–830 s) – 7.8 mg/s.

A probable cause of higher NO values, achieved while combustion of sludge fuels than for coal is the higher content of elementary nitrogen in the sludge fuels, which-by the subject literature including Boardman & Smooth [41], Habi et al. [42] and Williams et al. [43]-is referred to as the fuel mechanism, one of three major causes for NO_x formation.

In relation to SO_2 momentary emissions (Figure 5), it can be noted that the highest values were noted while combustion of the PBS fuel (grain size 35 mm), reaching 10 mg/s at 510 s, and for the PBM fuel (grain size 35 mm), for which the maximum was almost 9.90 mg/s at the 1020–1050 s interval. All curves, up to 1200 s, are characterized by rapid fluctuations. The exception is the emission of the PBT fuel (grain size 35 mm), for which the stabilization of SO_2 emission changes occurs only at 2200 s.

The results of measurements carried out on an industrial system of the stoker-fired boiler type ORS-16, in which hard coal was co-combusted with sewage sludge (mass content in the mixture being burnt 10%–30%) presented in publication [44] prove that during combustion an increase of NO_x emission by 10%–60%, and SO_2 by 10%–40% (compared to combustion of pure coal) was observed, depending on the amount of sludge added to coal. Increased emission of those compounds was accompanied by an increased share of sewage sludge in the mixture being burnt.

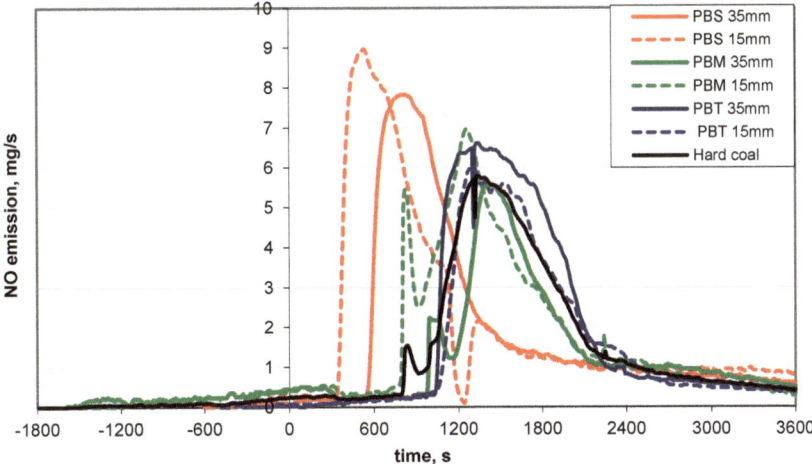

Figure 4. Variations of NO emissions during combustion of fuels.

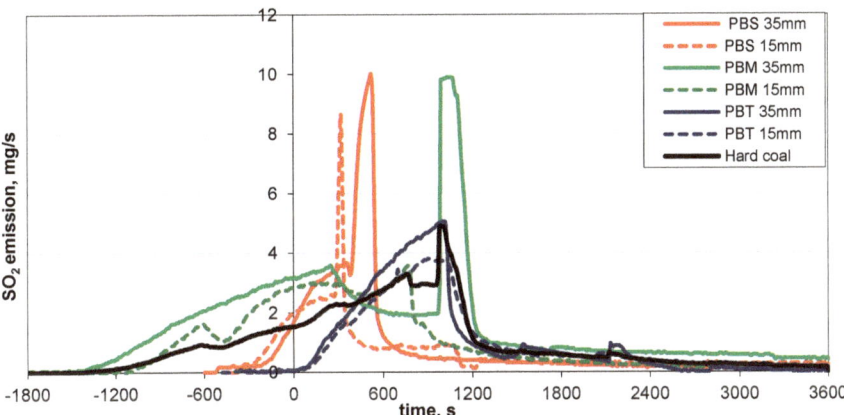

Figure 5. Variations of SO_2 emissions during combustion of fuels.

Similar relationship is also described by Werther & Ogada [3] for fluidized bed boilers, based, among others, on the tests performed by van Doorn et al. [45] as well as research carried out by Vamvuka et al. [46] in which the co-combustion of the sewage sludge was tested with hard coal, lignite, and other biomass materials. In all cases, as the mass content of sludge in the burnt mixture increased, also the NO_x emission was increasing.

However, Morgan & van de Kamp [47] based on their own research, concluded that for mass content of sludge above 50%, the NO_x emission reached its maximum, and after that, reduced with the increased share of sludge in the mixture being burnt.

As the practice shows, co-combustion of dried sludge with coal (with 1% share) does not increase the NO_x concentrations, and the SO_2 concentrations noted are even lower by 12% [44].

3.2. Analysis of Ash and Slag and from Combustion of Sewage Sludge Fuels

Chemical analyses of ashes from sewage sludge fuels compared to hard coal and biomass ashes are presented in Table 3.

Table 3. Chemical composition of ashes.

Parameter *, %	PBS	PBM	PBT	Biomass [48]	Hard Coal [49]
SiO_2	29.20 ± 1.46	17.88 ± 0.89	21.78 ± 1.09	0.12–15.12	50–57
Al_2O_3	17.07 ± 0.85	4.04 ± 0.20	4.92 ± 0.25	0.04–68.18	25–30
Fe_2O_3	8.41 ± 42	7.08 ± 0.35	11.80 ± 0.59	0.13–8.40	3.5–8.0
CaO	25.84 ± 1.29	40.35 + 2.02	38.33 ± 1.92	4.39–83.46	2–4
MgO	0.01 ± 0.00	0.95 ± 0.06	1.15 ± 0.06	1.10–15.12	1.5–3.0
P_2O_5	4.10 + 0.21	21.06 ± 1.05	10.24 ± 0.51	0.45–18.24	n.d.
SO_3	12.10 ± 0.61	0.50 ± 0.27	7.37 ± 0.37	0.36–45.89	0.5–1.2
Mn_3O_4	0.14 ± 0.01	0.16 ± 0.01	0.25 ± 0.52	n.d.	n.d.
TiO_2	0.80 + 0.04	0.36 ± 0.02	0.52 ± 0.03	0.05–28.00	0–1
SrO	0.10 ± 0.01	0.11 ± 0.01	0.15 ± 0.01	n.d.	n.d.
Na_2O	0.70 ± 0.04	1.11 ± 0.06	0.36 ± 0.02	0.14–29.82	0.2–2.0
K_2O	1.94 ± 0.10	1.25 ± 0.06	1.03 ± 0.05	2.19–37.70	2.5–3.0

* extended uncertainty—0.95.

In biomass ash, there is usually a higher content of such components as CaO, MgO, Na_2O, K_2O, P_2O_5 and, at the same time, a lower content of SiO_2, Al_2O_3, TiO_2 in comparison to the ash from coal combustion. Vassilew et al. [50] ranged the oxide composition of fly ash from biomass combustion according to the following rule: SiO_2 > CaO > K_2O > P_2O_5 > Al_2O_3 > MgO > Fe_2O_3 > SO_3 > Na_2O > MnO > TiO_2. In sludge fuels, due to the addition of quicklime serving as a binder, the content of CaO in ashes is high. Particular attention should be paid to the content of approx. 21% of P_2O_5 in ash from fuel produced from sewage sludge and meat and bone meal (PBM). According to Febrero et al. [51] and other authors [52,53] high content of P_2O_5 will have effect on melting phases. Pronobis [25] stated that when ash fraction consists of pentoxide, hemispherical temperature (HT) is 569 °C which enhances of low-melting-point phases in the fly ash. Tests carried out by Li et at. [52] have shown that SiO_2 and Al_2O_3 are all favorable to increase the ash fusion temperature but Al_2O_3 is more effective than SiO_2 in reducing the slagging tendency. The SiO_2 content in fuel ashes is higher than that of Al_2O_3.

Table 4 shows the selected ash deposition indexes calculated according to formulas 1–4 in comparison to the criteria reported in the literature [25,53].

Table 4. Ash deposition indexes and associated criteria.

Parameter	PBS	PBM	PBT	Coal [24]	Biomass [24]	Criteria [25,53]			
						Low	Medium	High	Extremely High
B/A, -	0.781	2.277	1.935	0.556	0.950	<0.5	0.5–0.7	0.7–1	>1.0
						<0.4		>0.7	
B/A_{+P}, -	0.868	3.222	2.311	0.557	0.980				
R_b, %	36.460	52.740	52.670	35.71	48.00	35–55			
S_R, -	46.241	26.985	29.811	62.55	76.34	>72	65–72	≤65	
R_s, -	0.211	0.767	0.394	0.742	0.106	<0.6	0.6–2.0	2–2.6	>2.6
F_u, -	2.063	5.375	2.689	1.975	30.891	<0.6		0.6–40	>40
Fe_2O_3/CaO, -	0.33	0.17	0.31	0.810	0.025			0.3–3.0 *	
SiO_2, %	29.20	17.88	21.78	53.73	50	<20	20–25	>25	>0.5
Cl^r, %	0.06	0.02	0.03	0.5	0.5	<0.2	0.2–0.3	0.3–0.5	>0.5 [a]

* eutectics enhancing slag formation. [a]—limit values for substance leaching according to Annex 3 of the Regulation of the Minister of Economy and Labor on the criteria and procedures for referring waste deposition on neutral waste landfill.

Base-to-acid ratio (B/A) ratio indicates the potential tendencies for slagging and fouling. The highest B/A index of the tested ashes is the one for PBM fuel and PBT fuel ash, which classifies them as extremally high prone to slagging and fouling (according to the criteria specified by Pronobis [25]). PBS fuel shows high tendency to be subject to those phenomena. For the base-to-acid ratios, B/A_{+P} which including P_2O_5 and is more closely to approach for biomass application a similar trend is clearly visible. According to research conducted by García-Maraver et al. [48] B/A ratio values for wood and woody biomass may range from 2.16–64.46, and even as high as 192.62 and 339.69 for paulownia wood and black poplar chips, respectively. For meat and bone meal (MBM), on the other hand, B/A ratio is 38.90, for sewage sludge it is 1.08, and 1.50 according to [25].

Additionally, for PBM ash the fouling index F_U revealed that it had a strong tendency to fouling. Slag viscosity index S_R corresponds to high viscosity and therefore to low slagging inclination [25]. Values obtained for sewage fuel ashes demonstrate a high predisposition to slagging inclination similar to RDF fuel and agriculture biomass [48].

Ash from sludge fuels contains small amounts of chlorine, within 0.02%–0.06%, which may indicate a low susceptibility to chloride corrosion of certain metal elements in combustion installation.

The residues after combustion process were analyzed for the presence of combustible matter in the ash and slag. The results are shown in Table 5. In the slag and ash after combustion process of sludge fuels less than 5% of combustible matter content was determined (condition specified by [54] for slags and ashes from co-combustion installations).

Table 5. Combustible matter content in slag and ash from sewage sludge fuels.

Fuel Type		Combustible Matter Content, %
PBS	35 mm	2.61
	15 mm	1.60
PBM	35 mm	2.80
	15 mm	1.30
PBT	35 mm	1.43
	15 mm	1.49

Unburnt carbon can also be found in the ashes. For example, grate boilers often produce fly ash with 50% or more of unburnt carbon. As Demirbas [55] claims, the fly ash from biomass-fired grate boilers contains also high levels of unburnt carbon. The presence of this carbon indicates inefficient fuel use and can reduce ash stabilization (chemical hardening) and significantly increases ash volume.

The problem of storage of the residue after burning the sewage sludge fuels was also analyzed. One of the criteria determining the possibility to refer the waste to landfill, other than for hazardous waste, are the acceptable values of leaching the pollutants, as defined by [56], which acknowledges the correct choice of the combustion process parameters and its proper performance.

Therefore, the residue after combustion sewage sludge fuels was tested for leaching hazardous compounds. The results of analyses were compared with water extracts from biomass ashes and the criteria for approving waste for deposition on a neutral waste landfill, according to the regulation are listed in Table 6.

The data in Table 5 prove that the leaching of pollutants, in the extracts from sludge fuels residues after their combustion was, in each case, lower or equal (for mercury) the level determined by the regulation, so the ash and slag can be deposited on neutral waste landfill.

It is also possible that the ashes from sewage sludge fuels could be used in different ways. Ashes from combustion of conventional solid fuels have been used in production of building materials for many years and are used predominantly by the cement industry for cement and concrete production [57–61]. Their popularity in the construction engineering results, first and foremost, from its high fineness (close to cement), chemical and phase composition (close to mineral loam resources) and reactivity. The existing standard [62] defines the criteria of application of ashes as an additive to concrete and determines the maximum content of the fly ash from co-combustion materials to be 30%.

Table 6. Analysis of water extracts from the residues after combustion of sewage sludge fuels and biomass.

Parameter	Unit	PBS	PBM	PBT	Biomass [56]	Limit Value [a]
pH	-	9.60	9.80	9.00	12.9–13.3	-
Phosphates	mg PO_4/L	<0.03	<0.03	<0.03	n.d.	-
Chlorides	mg Cl/L	699.5	216	585	n.d.	800
Sulfides	mg SO_4/L	960	553	994	n.d.	1000
As	mg/L	0.10	0.10	0.10	<0.022–0.024	0.5
Cr		0.43	0.28	0.32	0.065–2.85	0.5
Zn		0.10	0.10	0.10	<0.0315	4.0
Cd		0.01	0.01	0.01	<0.0007	0.04
Cu		0.10	0.10	0.10	0.0565	2.0
Pb		0.10	0.10	0.10	<0.00065–0.007	0.5
Hg		0.01	0.01	0.01	n.d.	0.01
Se		0.30	0.30	0.30	0.007–0.135	0.1
Fe		0.01	0.14	0.47	<0.012	-
Mn		0.10	0.10	0.10	0.0055–0.0325	-
Ba		0.33	10.0	14.0	n.d.	20
B		0.30	0.30	0.30	n.d.	-

[a]—limit values for substance leaching according to Annex 3 of the Regulation of the Minister of Economy and Labor on the criteria and procedures for referring waste deposition on neutral waste landfill; n.d.—no data

As has already been mentioned, the chemical composition of biomass ashes includes mostly oxides, such as: SiO_2, CaO, K_2O. High phosphorus content (phosphate ions) can cause significant inhibition of hydration and postponement of the commencement and end of the cement curing time as well as a decline in its early strengths [60].

A prerequisite for use of ashes from new fuels in civil engineering, underground mines and other industries is fulfilment of legal requirements, including preparation of the relevant standards for new applications.

4. Conclusions

Despite the existence of many methods for the disposal of municipal sewage sludge, the problem of its management still exists. The use of sludge as a component of a biofuel with a fixed composition and properties allows the energetic use of sludge with lower calorific values and adjusting the quality of fuels to the requirements of a specific combustion installation.

While combustion of the sewage sludge transformed into pelletized biofuel, typical problems of co-combustion of dried sludge with coal can be avoided. Grain size composition of the sludge fuels is close to that of pea coal and adjusted to the grate firing process.

To sum up, it can be concluded that the tests performed showed some difference of the process of burning sewage sludge fuels as compared to hard coal. Momentary emission of CO_2, NO, and SO_2 while combustion of fuels with the same composition, differing as to the grain size is similar. A noticeable difference between both particle sizes was the reduced ignition time and reduced emission of CO for fuels with smaller particles (15 mm).

Unfortunately, as in the case of co-combustion of dried sludge, the problem of NO_x emission still remains. However, while burning sludge fuels with coal, reducing the emission of nitrogen compounds can be expected, and the NO_x emission (due to the fuel origin of nitrogen in the process discussed) will limit the share of sewage sludge fuels in the mixture being burnt.

The tendency for ash slagging and fouling is also observed, especially for PBM and PBT ashes which consist of MBM and sawdust which is typical for biomass combustion. Relatively high slagging and fouling indices of sludge limits its use in combustion.

It can be concluded that the test performed proved the potential of using sewage sludge fuels in co-combustion processes with coal in grate furnaces.

Funding: This research was funded by the POLISH MINISTRY OF SCIENCE AND HIGHER EDUCATION, project No. R1401601.

Conflicts of Interest: The authors declare no conflict of interest.

References

1. *Council Directive 86/278/EEC of 12 June 1986 on the Protection of the Environment, and in Particular of the Soil, when Sewage Sludge is Used in Agriculture*; European Union: Brussels, Belgium, 2018.
2. *Council Directive 99/31/EC of April 1999 on the Landfill of Waste*; European Union: Brussels, Belgium, 2018.
3. Werther, J.; Ogada, T. Sewage sludge combustion. *Prog. Energy Combust. Sci.* **1999**, *25*, 55–116. [CrossRef]
4. Stasta, P.; Boráň, J.; Bébar, L.; Stehlik, P.; Oral, J. Thermal processing of sewage sludge. *Appl. Therm. Eng.* **2006**, *26*, 1420–1426. [CrossRef]
5. Sahu, S.; Chakraborty, N.; Sarkar, P. Coal–biomass co-combustion: An overview. *Renew. Sustain. Energy Rev.* **2014**, *39*, 575–586. [CrossRef]
6. Wzorek, M.; Troniewski, L. Application of sewage sludge as a component of alternative fuel. In *Environmental Engineering*; Dudzińska, M., Pawłowski, L., Eds.; Taylor & Francis Group: New York, NY, USA, 2007; pp. 311–316.
7. Lin, Y.; Liao, Y.; Yu, Z.; Fang, S.; Ma, X. The investigation of co-combustion of sewage sludge and oil shale using thermogravimetric analysis. *Thermochim. Acta* **2017**, *653*, 71–78. [CrossRef]
8. Chen, J.; Liu, J.; He, Y.; Huang, L.; Sun, S.; Sun, J.; Chang, K.; Kuo, J.; Huang, S.; Ning, X. Investigation of co-combustion characteristics of sewage sludge and coffee grounds mixtures using thermogravimetric analysis coupled to artificial neural networks modeling. *Bioresour. Technol.* **2017**, *225*, 234–245. [CrossRef] [PubMed]
9. Xiao, H.; Ma, X.-Q.; Lai, Z. Isoconversional kinetic analysis of co-combustion of sewage sludge with straw and coal. *Appl. Energy* **2009**, *86*, 1741–1745. [CrossRef]
10. Yilmaz, E.; Wzorek, M.; Akçay, S. Co-pelletization of sewage sludge and agricultural wastes. *J. Environ. Manag.* **2018**, *216*, 169–175. [CrossRef]
11. Chen, G.-B.; Chatelier, S.; Lin, H.-T.; Wu, F.-H.; Lin, T.-H. A Study of Sewage Sludge Co-Combustion with Australian Black Coal and Shiitake Substrate. *Energies* **2018**, *11*, 3436. [CrossRef]
12. Rong, H.; Wang, T.; Zhou, M.; Wang, H.; Hou, H.; Xue, Y. Combustion Characteristics and Slagging during Co-Combustion of Rice Husk and Sewage Sludge Blends. *Energies* **2017**, *10*, 438. [CrossRef]
13. Huang, L.; Liu, J.; He, Y.; Sun, S.; Chen, J.; Sun, J.; Chang, K.; Kuo, J.; Ning, X. Thermodynamics and kinetics parameters of co-combustion between sewage sludge and water hyacinth in CO_2/O_2 atmosphere as biomass to solid biofuel. *Bioresour. Technol.* **2016**, *218*, 631–642. [CrossRef]
14. Akdağ, A.S.; Atak, O.; Atimtay, A.T.; Sanin, F.D. Co-combustion of sewage sludge from different treatment processes and a lignite coal in a laboratory scale combustor. *Energy* **2018**, *158*, 417–426. [CrossRef]
15. Kijo-Kleczkowska, A.; Środa, K.; Kosowska-Golachowska, M.; Musiał, T.; Wolski, K. Experimental research of sewage sludge with coal and biomass co-combustion, in pellet form. *Waste Manag.* **2016**, *53*, 165–181. [CrossRef] [PubMed]
16. Junga, R.; Kaszubska, M.; Wzorek, M. Technical and environmental performance of 10 kW understocker boiler during combustion of biomass and conventional fuels. *E3S Web Conf.* **2017**, *19*, 1009. [CrossRef]
17. Fleck, E.; Scholz, S. Co-combustion of sewage sludge in grate-based combustion plants. In *Waste Management*; Thome-Kozminski, K.J., Pelloni, L., Eds.; TK Verlag Karl-Kozminsky: Neuruppin, Gremany, 2011; pp. 779–798.
18. Nadziakiewicz, J.; Kozioł, M. Co-combustion of sludge with coal as a possible methods of its utilization in Poland. *Appl. Energy* **2003**, *75*, 239–248. [CrossRef]
19. Hroncová, E.; Ladomerský, J.; Musil, J. Problematic issues of air protection during thermal processes related to the energetic uses of sewage sludge and other waste. Case study: Co-combustion in peaking power plant. *Waste Manag.* **2018**, *73*, 574–580. [CrossRef] [PubMed]
20. Jang, H.-N.; Kim, J.-H.; Back, S.-K.; Sung, J.-H.; Yoo, H.-M.; Choi, H.S.; Seo, Y.-C. Combustion characteristics of waste sludge at air and oxy-fuel combustion conditions in a circulating fluidized bed reactor. *Fuel* **2016**, *170*, 92–99. [CrossRef]
21. Beckmann, M.; Pohl, M.; Bernhardt, D.; Gebauer, K. Criteria for solid recovered fuels as a substitute for fossil fuels—A review. *Waste Manag Res.* **2012**, *30*, 354–369. [CrossRef]

22. Yina, C.; Li, S. Advancing grate-firing for greater environmental impacts and efficiency for decentralized biomass/wastes combustion. *Energy Procedia* **2017**, *120*, 373–379. [CrossRef]
23. Werle, S. Multivariate analysis of possibility of co-combustion of sewage sludge in coal fired power boilers. *Arch. Waste Manag. Environ. Prot.* **2011**, *13*, 21–38.
24. Houshfar, E.; Løvås, T. Skreiberg, Øyvind Experimental Investigation on NOx Reduction by Primary Measures in Biomass Combustion: Straw, Peat, Sewage Sludge, Forest Residues and Wood Pellets. *Energies* **2012**, *5*, 270–290. [CrossRef]
25. Pronobis, M. The influence of biomass co-combustion on boiler fouling and efficiency. *Fuel* **2006**, *85*, 474–480. [CrossRef]
26. Masiá, A.T.; Buhre, B.; Gupta, R.; Wall, T. Characterising ash of biomass and waste. *Fuel Process. Technol.* **2007**, *88*, 1071–1081. [CrossRef]
27. Baxter, L.L. Ash deposition during biomass and coal combustion: A mechanistic approach. *Biomass-Bioenergy* **1993**, *4*, 85–102. [CrossRef]
28. Theis, M.; Skrifvars, B.-J.; Hupa, M.; Tran, H. Fouling tendency of ash resulting from burning mixtures of biofuels. Part 1: Deposition rates. *Fuel* **2006**, *85*, 1125–1130. [CrossRef]
29. Jeong, T.-Y.; Sh, L.; Kim, J.-H.; Lee, B.-H.; Jeon, C.-H. Experimental Investigation of Ash Deposit Behavior during Co-Combustion of Bituminous Coal with Wood Pellets and Empty Fruit Bunches. *Energies* **2019**, *12*, 2087. [CrossRef]
30. Lee, J.M.; Kim, D.-W.; Kim, J.-S.; Na, J.-G.; Lee, S.-H. Co-combustion of refuse derived fuel with Korean anthracite in a commercial circulating fluidized bed boiler. *Energy* **2010**, *35*, 2814–2818. [CrossRef]
31. Dunnu, G.; Maier, J.; Scheffknecht, G. Ash fusibility and compositional data of solid recovered fuels. *Fuel* **2010**, *89*, 1534–1540. [CrossRef]
32. Kupka, T.; Mancini, M.; Irmer, M.; Weber, R. Investigation of ash deposit formation during co-firing of coal with sewage sludge, saw-dust and refuse derived fuel. *Fuel* **2008**, *87*, 2824–2837. [CrossRef]
33. Kanchanapiya, P.; Sakano, T.; Kanaoka, C.; Mikuni, T.; Ninomiya, Y.; Zhang, L.; Masui, M.; Masami, F. Characteristics of slag, fly ash and deposited particles during melting of dewatered sewage sludge in a pilot plant. *J. Environ. Manag.* **2006**, *79*, 163–172. [CrossRef]
34. Karcz, H.; Kozakiewicz, A.; Kantorek, M.; Dziugan, P.; Wierzbicki, K. Czy spalanie odpadów komunalnych w kotłach rusztowych jest właściwe. *Instal* **2011**, *10*, 24–30.
35. Wzorek, M. Characterization of the properties of alternative fuels contain sewage sludge. *Fuel Process Technol.* **2012**, *104*, 80–89. [CrossRef]
36. Wzorek, M. Physical and chemical properties of fuel containing animal waste. In *Waste Management and the Environment IV*; Zamorano, M., Ed.; WIT Transactions on Ecology and the Environment: Southampton, UK, 2008; Volume 109, pp. 69–77.
37. Wzorek, M.; Głowacki, T. A device for mixing, especially sewage sludge. Utility model PL W.121829, 31 July 2014.
38. Wzorek, M.; Kozioł, M.; Scierski, W. Emission characteristics of granulated fuel produced from sewage sludge and coal slime. *J. Air Waste Manag. Assoc.* **2010**, *60*, 1487–1493. [CrossRef] [PubMed]
39. Kozioł, M. Ecological and technical aspects of co-combustion of coal with substantial fraction of sludge in the grate-type boilers. *Arch. Comb.* **2010**, *30*, 409–426.
40. Wilk, K. Podstawy Niskoemisyjnego Spalania. Wydawnictwo Gnome PAN: Katowice, Poland, 2000; pp. 100–120.
41. Boardman, R.D.; Smoot, L.D. Pollutant formation and control. In *Fundamentals of Coal Combustion for Clean and Efficient Use*; Boardman, R.D., Smoot, L.D., Eds.; Elsevier Science Ltd.: Amsterdam, The Netherlands, 1993; pp. 433–506.
42. Habib, M.; Elshafei, M.; Dajani, M. Influence of combustion parameters on NOx production in an industrial boiler. *Comput. Fluids* **2008**, *37*, 12–23. [CrossRef]
43. Williams, A.; Pourkashanian, M.; Jones, J.M.; Rowlands, L. A review of NO_x formation and reduction mechanisms in combustion systems with particular reference to coal. *J. Energy Inst.* **1997**, *70*, 102–113.
44. Wandrasz, J.W.; Kozioł, M.; Landrat, M.; Ścierski, W.; Wandrasz, A.J. Możliwości współspalania osadów z oczyszczalni ścieków z węglem w kotłach rusztowych. *Gospod. Paliwami Energia* **2000**, *8*, 10–15.

45. van Doorn, J.; Bruyn, P.; Kos, B.; Hanse, J. Combined combustion of biomass, municipal sewage sludge and coal in a atmospheric fluidised bed installation. In *Proceedings of the 9th European Biomass Conference for Energy and the Environment*; Pergamon: Oxford, UK, 1996; pp. 199–210.
46. Vamvuka, D.; Alexandrakis, S.; Galetakis, M. Combustion Performance of Sludge from a Wastewater Treatment Plant in Fluidized Bed. Factorial Modeling and Optimization of Emissions. *Front. Energy Res.* **2019**, *7*. [CrossRef]
47. Morgan, D.J.; van de Kamp, W.L. The co-firing of biomass and municipal sewage sludge with pulverised coals in utility boilers. In Proceedings of the Second International Conference Combustion and Emissions Control, London, UK, 4–5 December 1995; pp. 159–168.
48. Garcia-Maraver, A.; Mata-Sanchez, J.; Carpio, M.; Perez, J.A. Critical review of predictive coefficients for biomass ash deposition tendency. *J. Energy Inst.* **2017**, *90*, 214–228. [CrossRef]
49. Jarosiński, A. Mineral and chemical composition of fly ashes deriving from co-combustion of biomass with coal and its application. *J. Pol. Min. Eng. Soc.* **2013**, *14*, 141–148.
50. Vassilev, S.V.; Baxter, D.; Andersen, L.K.; Vassileva, C.G. An overview of the composition and application of biomass ash. *Fuel* **2013**, *105*, 19–39. [CrossRef]
51. Febrero, L.; Granada-Álvarez, E.; Regueiro, A.; Míguez, J.L. Influence of Combustion Parameters on Fouling Composition after Wood Pellet Burning in a Lab-Scale Low-Power Boiler. *Energies* **2015**, *8*, 9794–9816. [CrossRef]
52. Li, Q.; Zhang, Y.; Meng, A.; Li, L.; Li, G. Study on ash fusion temperature using original and simulated biomass ashes. *Fuel Process. Technol.* **2013**, *107*, 107–112. [CrossRef]
53. Viana, H.; Vega-Nieva, D.; Torres, L.O.; Lousada, J.; Aranha, J. Fuel characterization and biomass combustion properties of selected native woody shrub species from central Portugal and NW Spain. *Fuel* **2012**, *102*, 737–745. [CrossRef]
54. Ordinance of the Polish Minister of Economy and Labour on the requirements to the process of thermal conversion of waste for slag and ash from co-combustion of waste. *J. Laws* **2002**, *37*, 339.
55. Demirbas, A. Potential applications of renewable energy sources, biomass combustion problems in boiler power systems and combustion related environmental issues. *Prog. Energy Combust. Sci.* **2005**, *31*, 171–192. [CrossRef]
56. Ordinance of the Polish Minister of Economy and Labour on the criteria and procedures for the acceptance of waste for land filling. *J. Laws* **2005**, *186*, 1553.
57. Maresca, A.; Hyks, J.; Astrup, T. Recirculation of biomass ashes onto forest soils: Ash composition, mineralogy and leaching properties. *Waste Manag.* **2017**, *70*, 127–138. [CrossRef] [PubMed]
58. Kurama, H.; Kaya, M. Usage of coal combustion bottom ash in concrete mixture. *Constr. Build. Mater.* **2008**, *22*, 1922–1928. [CrossRef]
59. Król, A. The role of the silica fly ash in sustainable waste management. 1st International Conference on the Sustainable Energy and Environment Development (SEED). *E3S Web Conf.* **2016**, *10*, 00049. [CrossRef]
60. Kuterasińska, J.; Król, A. Mechanical properties of alkali-acivated binders based on copper slag. *Arch. Civil Eng. Environ.* **2015**, *8*, 61–67.
61. Król, A. Binding chromium ions during hydratation of mineral binders. *Przem. Chem.* **2007**, *86*, 971–973.
62. *PN-EN 450-1: 2012 Fly Ash for Concrete—Part 1: Definitions, Specifications and Compliance Criteria*; Polish Committee for Standardization: Warsaw, Poland, 2012.

© 2020 by the author. Licensee MDPI, Basel, Switzerland. This article is an open access article distributed under the terms and conditions of the Creative Commons Attribution (CC BY) license (http://creativecommons.org/licenses/by/4.0/).

Article

Catalytic Fast Pyrolysis of Forestry Wood Waste for Bio-Energy Recovery Using Nano-Catalysts

Cheng Li, Xiaochen Yue, Jun Yang, Yafeng Yang, Haiping Gu * and Wanxi Peng *

School of Forestry, Henan Agricultural University, Zhengzhou 450002, China; lichengzzm@163.com (C.L.); yuexiaochen95@163.com (X.Y.); yangjun940207@163.com (J.Y.); yafengyangzz@163.com (Y.Y.)
* Correspondence: guhaiping.1357@163.com (H.G.); pengwanxi@hau.edu.cn (W.P.)

Received: 12 September 2019; Accepted: 16 October 2019; Published: 18 October 2019

Abstract: Fast pyrolysis is envisioned as a promising technology for the utilization of forestry wood waste (e.g., widely available from tree logging) as resources. In this study, the potential of an innovative approach was explored to convert forestry wood waste of *Vernicia fordii* (VF) into energy products based on fast pyrolysis combined with nano-catalysts. The results from fast pyrolysis using three types of nano-catalysts showed that the distribution and composition of the pyrolytic product were affected greatly by the type of nano-catalyst employed. The use of nano-Fe_2O_3 and nano-NiO resulted in yields of light hydrocarbons (alkanes and olefins) as 38.7% and 33.2%, respectively. Compared to the VF sample, the use of VF-NiO and VF-Fe_2O_3 led to significant increases in the formation of alkanes (e.g., from 14% to 26% and 31%, respectively). In addition, the use of nano-NiO and nano-Fe_2O_3 catalysts was found to promote the formation of acid, aromatics, and phenols that can be used as chemical feedstocks. The NiO catalyst affected the bio-oil composition by promoting lignin decomposition for the formation of aromatics and phenolics, which were increased from 9.52% to 14.40% and from 1.65% to 4.02%, respectively. Accordingly, the combined use of nano-catalysts and fast pyrolysis can be a promising technique for bio-energy applications to allow efficient recovery of fuel products from forestry wood waste.

Keywords: pyrolysis; catalyst; wood; waste; energy

1. Introduction

Due to the extensive exploitation and consumption of non-renewable fossil fuels, environmental pollution, climate change, and ecological damage have become increasingly severe [1–3]. Therefore, the development and extension of alternative energy resources are required [4–6]. Biomass is extremely abundant in nature in various types as representative renewable resource [7]. It is generally present in diverse forms such as agricultural crops, crop residues, woods, forest industry wastes, and aquatic plants [8–10]. For a long time, biomass was directly burned as a fuel to obtain energy, which inflicted severe environmental pollution. Therefore, the valorization of biomass into biofuels or fine chemicals via chemical conversion technologies could lessen the dependence of modernization on fossil resources [11], thereby alleviating the bottleneck associated with the shrinking fossil resource reserves [12,13]. Furthermore, the efficient utilization of carbon-neutral biomass is vital to mitigate the greenhouse effect provoked by the combustion and/or inappropriate handling of biomass. Hence, an alternative technology is required to facilitate effective conversion of biomass into fuels or fine chemicals of good quality [6,14–16].

Forestry waste can be referred to as the residues produced in the process of forestry production and processing such as the residues of tree cutting and wood processing, urban landscaping waste, forest tending and thinning residues, economic forest pruning waste, and waste wood materials. According to the calculation of all the above waste, there were about 454.04 million tons of forestry waste in China in 2014 [17]. The pruning waste from commercial forests was about 141.74 million tons [17]. *Vernicia fordii* is widely planted in China as an important economic tree species with high utilization and economic value. A good amount of pruning waste is generated from *Vernicia fordii* trees each year, which is often discarded or burned. Therefore, how to make full use of such waste has great significance for the high value utilization of biomass waste.

The use of catalysts in fast pyrolysis can potentially improve the quality of bio-oil. These catalysts could lead to an upgrade of the properties of bio-oil and enhance the formation of valuable chemicals [18]. Banks et al. [19] investigated the bio-oils produced from fast pyrolysis of alkali metal (potassium) impregnated biomass. The potassium promoted the pyrolytic decomposition biomass (cellulose and hemicellulose) and the formation of levoglucosan and hydroxymethyl cyclopentene derivatives. Chen et al. [20] studied fast pyrolysis of biomass with metal nitrides (TiN or GaN) for furfural production, whereby direct decomposition of oligosaccharides was catalyzed to yield furfural. Through catalysis, aromatics could also be obtained from lignin depolymerization during biomass pyrolysis [21–24]. NiO and Fe_2O_3 have attracted extensive interest in recent years for their catalytic and magnetic properties [25–27]. Nanometer-sized NiO and Fe_2O_3 have many improved properties compared to their pristine (or bulk) forms. It was found that nano-NiO particles exerted more effective catalytic effects than micro-NiO particles in biomass pyrolysis [24]. Khelfa et al. [28], using Fe_2O_3 as catalyst, studied the catalytic pyrolysis and gasification of *Miscanthus giganteus*. Their results showed that Fe_2O_3 as a catalyst was active in gasification and hydrogen production. In addition, Fe_2O_3 could break down the tar produced and improved the partial oxidation of phenols during the thermal degradation of the biomass. Despite the potential utility of these catalysts, no reports have been made to describe the effects of the nano-NiO and nano-Fe_2O_3 catalysts on the fast pyrolysis of forestry wood waste.

In light of the high economical value and high availability of forestry wood waste (*Vernicia fordii*), an integrated approach is proposed to combine fast pyrolysis with nano-catalysts (NiO and Fe_2O_3) to convert forestry wood waste (*Vernicia fordii*) into energy products. Characterization of the wood waste was first performed followed by pyrolysis using three types of nano-catalysts to investigate their influence on the distribution and composition of the pyrolytic product. Thermogravimetric analysis and pyrolysis gas chromatography–mass spectrometry were used to analyze the chemical components of the bio-oil produced and to compare the catalytic effect of nano-catalysts on fast pyrolysis of *Vernicia fordii* wood waste.

2. Materials and Methods

2.1. Materials

Forestry wood waste of *Vernicia fordii* wood (VF) was collected from Funiu Mountain, China. The VF was ground and screened to a particle size range of 149–177 µm before its preservation at −3 °C under vacuum conditions. Methanol, benzene, and ethanol of chromatographic grade were purchased from Hunan Huihong Reagent Co., Ltd., China. The nano-catalysts (α-Fe_2O_3 (30 nm, spherical, 99.5%) and NiO (60–120 nm, spherical, 99.5%)) used in this work were directly procured from Shanghai Macklin Biochemical Co., Ltd. For the current study, the mass of VF was fixed as 20 g with and without the addition of 1 wt% nano-catalysts either individually or as a mixture (NiO, Fe_2O_3, or NiO/Fe_2O_3 mixture (equal mass of NiO and Fe_2O_3)). Thus, there were four pyrolysis samples investigated, which were designated as VF, VF-NiO, VF-Fe_2O_3, and VF-NiO/Fe_2O_3. The flow chart of forestry wood waste procedure is shown in Figure 1.

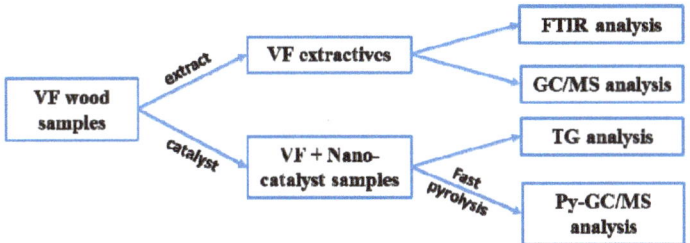

Figure 1. The flow chart of forestry wood waste pyrolysis procedure (Note: VF is the abbreviation of *Vernicia fordii*).

2.2. Characterization of Forestry Wood Waste (Vernicia fordii Wood)

Three pairs of cotton bags and cotton thread were soaked in methanol, ethanol/benzene (1:1), and ethanol/ether (1:1) solution for 12 h, respectively, to remove possible contaminants that possibly arose as interference. About 40 g of *Vernicia fordii* wood (VF) with a particle size range of 149–177 μm was parceled into three cotton bags, which were later tied and sewed with the cotton thread. Henceforth, the extraction of VF was conducted in the Soxhlet extractor for 6 h at 60 °C with 300 mL solvents such as methanol, ethanol/benzene, and ethanol/ether. After extraction, the solvents were removed via rotary evaporation (55 °C, 10–50 Pa) and desiccated with anhydrous sodium sulfate. Prior to any characterization, the resulting VF extracts were stored at −3 °C. Then, the VF extracts were subjected to FTIR and GC/MS for their functional groups and chemical compositions, individually. By using an FTIR spectrophotometer (IR100), the KBr discs containing 1 wt% finely ground sample were scanned with infrared radiation from 4000 to 500 cm^{-1}. The GC/MS analysis was executed with a GC/MS (Agilent 7890B-5977A) equipped with an HP-5MS column (30 m × 25 μm × 0.25 μm). The GC was initiated at 50 °C, heated to 250 °C with a ramping rate of 8 °C/min, and finally heated to 300 °C with a ramping rate of 5 °C/min. Meanwhile, the inlet temperature, column flow, split ratio, and carrier gas were 250 °C, 1.0 mL/min, 20:1, and helium, respectively. For the MS, the electrons of samples were ionized with electron energy of 70 eV from the ion source at 230 °C, while the temperature of the quadrupole was 150 °C. The MS program was capable of detecting compounds in the mass range of 30–600 amu. For qualitative spectrum matching, the Wiley 7n.1 standard spectrum was used [29].

2.3. Catalytic Fast Pyrolysis of Forestry Wood Waste (Vernicia fordii Wood) Using Nano-Catalyst

The fast pyrolysis of forestry wood waste (*Vernicia fordii* wood, VF) using nano-catalysts was investigated by TG and Py-GC/MS analysis. One pristine VF and three amended VF samples (VF-NiO, VF-Fe$_2$O$_3$, and VF-NiO/Fe$_2$O$_3$) were analyzed to scrutinize the effect of nano-catalyst addition. In this study, the catalyst-to-feed-ratio was 1 to 100. For TG analysis, about 5 mg of samples were loaded on the platinum pan inside a thermal gravimetric analyzer (TA Instruments Q50) to examine the thermal decomposition of samples. The non-isothermal TG curves were obtained by heating the samples from ambient temperature to 850 °C under N$_2$ atmosphere with two different heating rates (20 °C/min and 55 °C/min) [30].

In order to conduct in-situ analysis of the bio-oil contents, the samples were further analyzed via Py-GC/MS analysis by using integrated pyrolysis-GC/MS instrumentation (CDS Pyroprobe 5000-Agilent 7890B-5977A). In the pyrolyzer, the samples were subjected to fast pyrolysis via heating under inert helium flow to 850 °C at a high heating rate of 250 °C/s, whereby the maximum pyrolysis temperature was maintained for 15 s. Apart from creating an inert environment, the helium also acted as a carrier gas that delivered the vaporized bio-oil from the pyrolyzer to the GC/MS. The temperature of the pyrolysis product transfer line and injection valve was set to 300 °C to prevent the recondensation of vaporized bio-oil within the instrumentation. The GC was furnished with a capillary column (TR-5MS) and operated in split mode, wherein the split ratio and total flow rate used were 50:1 and

50 mL/min, respectively. The GC oven initial temperature was 40 °C (holding 2 min), heated to 120 °C (ramping rate of 5 °C/min), and then increased to 200 °C (holding 15 min at ramping rate of 10 °C/min). For MS, the temperature of electron ionization and scanning range were fixed as 230 °C and 28–500 amu, respectively.

It is known that the GC/MS technique cannot quantify the chemical compounds. However, there is a considered linear relationship between the chromatographic peak area of a compound and its quantity. Therefore, in this study, the peak area and peak area % values with different catalysts were calculated and used to reveal the different yields for each product [31,32].

3. Results and Discussion

3.1. Characterization of Forestry Wood Waste (Vernicia fordii Wood) Extracts

Through GC/MS analysis, the total ion chromatograph of VF extracts from Soxhlet extraction with different solvents (methanol, ethanol/benzene, or ethanol/ether) were acquired. Supplementary Materials Figures S2–S4 depict the total ion chromatographs while the chemical composition of VF extracts are tabulated in Tables S2–S4. The compositional difference of VF extracts was rendered by the different affinity of solvents towards the extractable components of different polarities in VF. From Figures S2–S4 and Tables S2–S4 in Supplementary Materials, a total of 77 distinct chemical compounds were identified from the GC/MS analysis of VF extracts. For ease of discussion, these compounds were classified in terms of common functional groups (e.g., acids, alcohols, aldehydes, esters, amines, phenols, ketones, aromatics, olefins, and saccharides). In Supplementary Materials, Figure S5 presents the chemical composition of *Vernicia fordii* wood (VF) extracts that had been sorted into the aforementioned functional groups.

The GC/MS analysis confirmed that the VF extracts contained high value chemical constituents, which have widely promising and potential applications. For instance, linoleic acid has several medical applications such as lowering blood lipids, softening blood vessels, lowering blood pressure, and reducing cardiovascular diseases. In a few reports, a high potential of linoleic acid was suggested for the prevention of cancer, inflammation, and arthritis [33]. In addition, the *n*-hexadecanoic acid in VF could serve as a renewable feedstock for the production of soaps, cosmetics, and industrial mold release agents. Sitosterol can lower serum cholesterol while butorphanol can act as a pain reliever [34,35]. The esters of VF can be used as the precursors for emulsifiers, wetting agents, stabilizers, and plasticizers. The aromatics of VF could be utilized for the synthesis of more complex compounds through substitution reactions of simple aromatics. In brief, VF possesses a wide range of useful chemical compounds, which could be used as feedstocks in medical and industrial applications.

3.2. Catalytic Fast Pyrolysis of Forestry Wood Waste (Vernicia fordii Wood)

Lignocellulosic biomass can be divided into three major components: cellulose, hemicellulose, and lignin [36,37]. Cellulose is a polymer formed by the polymerization of glucose through β-1,4-glycosidic bonds. Hemicellulose is a polymer formed by the polymerization of hexose and pentose sugars. Lignin is mainly composed of guaiacol, syringyl, and *para*-hydroxy-phenyl alcohol, which are relatively complex and difficult to depolymerize [38]. As a biomass, the VF is also mainly composed of three lignocellulosic components called cellulose, hemicellulose, and lignin. The thermal decomposition of VF is very complex owing to the different reactivity and stability of these lignocellulosic components as well as the interactions between them. Based on the thermogravimetric (TG) analysis, the thermal decomposition of *Vernicia fordii* wood samples (e.g., VF, VF-NiO, VF-Fe$_2$O$_3$, and VF-NiO/Fe$_2$O$_3$) was elucidated with fast pyrolysis at two specified heating rates (25 °C/min and 55 °C/min). Figure 2 presents TG and the first derivative of thermogravimetric (DTG) curves of VF samples.

All the VF samples went through three stages during pyrolysis (Figure 2). During the first stage, when the temperature increased from room temperature to 35 °C to 200 °C, the weight loss of all samples mainly involved the evaporation of water and small molecular weight components.

The second decomposition stage occurred in the temperature range of 200–400 °C, when all DTG curves (Figure 2) showed that there was a primary peak of weight loss, which could be due to the process of decomposition of cellulose, hemicellulose, and part of lignin [39–41]. According to a report by Yu et al. [24], with increasing temperature, the decomposition of cellulose increased rapidly and was almost completed at 400 °C. According to Biagini's report, the onset temperature of hemicellulose (xylan as model compound) was 253 °C. The cellulose exhibited the maximum weight loss in the range of 200–400 °C. The onset and maximum weight loss were 319 and 354 °C [42]. At around 200 °C, the decomposition of lignin can be attributed to the dehydration reactions. Then, the cleavage of α- and β-aryl–alkyl-ether linkages occurred at around 300 °C. Meanwhile, the aliphatic side chains started to split off from the aromatic ring of lignin [43,44]. These results are consistent with our observation made in this work.

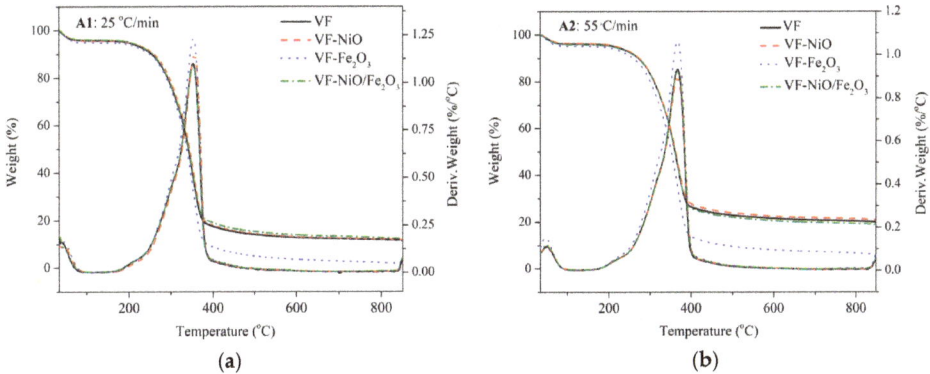

Figure 2. TG and DTG curves of *Vernicia fordii* wood samples at two different heating rates: 25 °C/min (a) and 55 °C/min (b).

Compared to cellulose and lignin, hemicellulose is the most unstable component in *Vernicia fordii* wood. The onset of its decomposition at about 200 °C is assumed to reflect a lower degree of polymerization compared to cellulose and lignin [45]. The primary weight loss occurred in the temperature range of 320–400 °C. In addition, due to the complex stable aromatic rings with various branch structures, the pyrolysis of lignin has been proven to occur continuously through a range of 200–900 °C [46]. At the third stage, above 400 °C, the weight loss was slower and relatively small due to the pyrolysis of lignin and the residues of char [41,47–49].

The thermal decomposition of wood biomass could be influenced by various factors such as temperature, chemical composition, heating rate, etc. [50]. In order to investigate the effect of heating rate on the thermal decomposition of VF and nano-VF samples, the heating rate of 25 °C and 55 °C were recorded, as shown in Figures 2 and 3. It can be seen that the heating rate had significant influence on the thermal decomposition of VF and nano-VF samples. With the increase of heating rate, the peak temperatures of all samples were increased from around 354 °C to around 368 °C. In addition, the weight loss of all the different decomposition stages also obviously was changed as the heating rate increased. Comparing the influence of nano-catalysts, it can be seen that both the TG and DTG curves only experienced slight changes when nano-NiO catalyst and nano-NiO/Fe_2O_3 were added. During the second stage (200–400 °C), only one peak was observed in the DTG curves of *Vernicia fordii* (VF). The peak temperature was almost the same with the addition of nano-Fe_2O_3, while the weight loss rate increased from 0.9%/°C to 1.25%/°C. This indicated that the nano-Fe_2O_3 catalyst can promote the pyrolysis of cellulose and hemicellulose during this range of temperature, causing the production of more acids and ketone compounds. In the third stage, the varying DTG values indicated that the introduction of the nano-Fe_2O_3 catalyst has a significant effect on the decomposition of VF. As seen in

Figure 3c, in the range of 200–400 °C, the weight loss of VF, VF-NiO, VF-Fe$_2$O$_3$, and VF-NiO/Fe$_2$O$_3$ were 69.03%, 68.07%, 80.60%, and 69.31%, respectively. Meanwhile, the VF-Fe$_2$O$_3$ sample had the highest weight loss rate (1.06%/min, Figure 3, SZ-D2). These indicated that the catalyst of nano-Fe$_2$O$_3$ improved the decomposition of cellulose and lignin in VF. At 400–800 °C, the weight loss of VF, VF-NiO, VF-Fe$_2$O$_3$, and VF-NiO/Fe$_2$O$_3$ were 6.18%, 6.41%, 7.39%, and 6.70%, respectively. The VF-Fe$_2$O$_3$ sample had the biggest weight loss compared to the others. This indicated that nano-Fe$_2$O$_3$ promoted the pyrolysis of lignin and the remaining solid residues of cellulose and hemicellulose in the last stage. In addition, we can also see in Figure 3 that the VF-Fe$_2$O$_3$ sample had the lowest residues, which showed that the VF-Fe$_2$O$_3$ sample had the biggest weight loss compared to the others. Compared with nano-NiO and nano-NiO/Fe$_2$O$_3$, the catalyst of nano-Fe$_2$O$_3$ had a significant effect on the whole pyrolysis process of *Vernicia fordii* wood. The main reason might be attributed to the fact that the nano-Fe$_2$O$_3$ can promote the breaking of ether bond in the lignin and lignin derivative structures. Nano-Fe$_2$O$_3$ improved the cleavage of α- and β-aryl–alkyl-ether linkages and the splitting of the aliphatic side chains from the aromatic ring. However, for the catalyst of nano-NiO/Fe$_2$O$_3$, the catalysis of nano-Fe$_2$O$_3$ was restricted by nano-NiO.

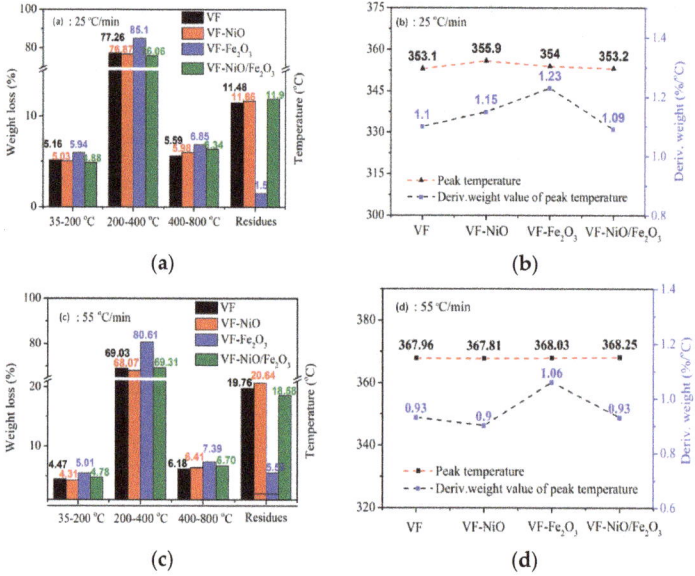

Figure 3. Thermal properties of *Vernicia fordii* wood: (**a**) the weight loss of different decomposition stages at the heating rate of 25 °C/min; (**b**) the peak temperature of DTG curves and the decomposition rate (25 °C/min); (**c**) the weight loss of different decomposition stages at the heating rate of 55 °C/min; (**d**) the peak temperature of DTG curves and the decomposition rate (25 °C/min).

For the utilization of lignocellulosic biomass, fast pyrolysis is the most felicitous technology since it is two to three times more economical than liquefaction and gasification processes. By fast pyrolysis, lignocellulosic biomass could be ameliorated into a liquid product, which is often known as the bio-oil. However, the bio-oil has an extremely complicated composition, with different proportions of ethers, esters, aldehydes, ketones, phenols, organic acids, aromatics, and alcohol compounds. It is believed that these compounds in the bio-oil could serve as precursors of value-added biofuels and fine chemicals [51,52]. Many studies have been concerned with the mechanisms of biomass pyrolysis, especially in relation to single lignocellulosic components such as lignin, cellulose, and hemicellulose [53–59]. Py-GC/MS is a rapid, reliable, and powerful method to scrutinize biomass fast pyrolysis because it facilitates the elucidation of chemical mechanisms by detecting the pyrolysis products [60]. In this study, the bio-oil vapor released from non-catalytic and catalytic pyrolysis VF samples were analyzed in-situ by Py-GC/MS. For all of the VF samples, the total ion chromatograms of their bio-oil vapor are shown in Figure 4, with the product of bio-oil vapor summarized in Supplementary Materials Tables S5–S8. Similar to the VF extracts, a wide range of organic compounds were found in the fast pyrolysis product of VF samples. Likewise, these organic compounds were categorized into common functional groups, viz. acids, alcohols, aldehydes, aromatics, amines, alkanes, esters, furans, ketones, olefins, phenolics, and others. Figure 5 compiles the chemical composition of bio-oil vapor released from the fast pyrolysis of *Vernicia fordii* wood (VF) samples sorted by common functional groups.

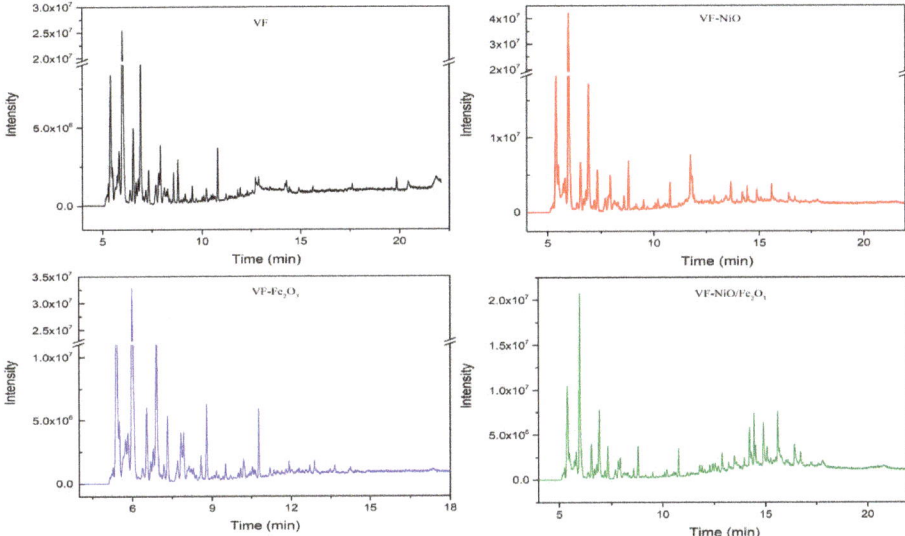

Figure 4. Total ion chromatograms of bio-oil vapor released from the fast pyrolysis of *Vernicia fordii* wood samples (VF, VF-NiO, VF-Fe$_2$O$_3$, and VF-NiO/Fe$_2$O$_3$).

The Py-GC/MS analysis revealed that the bio-oils from VF, VF-NiO, VF-Fe$_2$O$_3$, and VF-NiO/Fe$_2$O$_3$ were composed of 72, 68, 69, and 70 chemical compounds. As observed in Figure 5, non-catalytic pyrolysis of VF and catalytic fast pyrolysis of VF-NiO, VF-Fe$_2$O$_3$, and VF-NiO/Fe$_2$O$_3$ almost produced bio-oil with similar product distribution, although the functional group contents were non-identical. The bio-oil from VF was comprised of acids (0.23%), alcohols (29.72%), aldehydes (3.24%), alkanes (14.43%), amines (7.38%), aromatics (9.52%), esters (1.81%), furans (9.01%), ketones (4.03%), phenolics (1.65%), olefines (14.53%), and others (4.45%). The bio-oil from VF was rich with 2-methyl-3-buten-1-ol (28.00%), ethylcyclopropane (11.88%), 1,3-butadiene (10.60%), 3-methylfuran (4.95%), 3-iodo-1*H*-pyrazole (3.67%), 2-butenal (2.99%), felbamate (2.14%), toluene (2.10%), and benzene (1.83%).

Meanwhile, the bio-oil from VF-NiO was made up of acids (1.17%), alcohols (27.83%), aldehydes (0.66%), alkanes (25.91%), amines (1.92%), aromatics (14.40%), esters (2.33%), furans (6.96%), ketones (4.07%), phenolics (4.02%), olefines (7.32%), and others (3.43%). The bio-oil from VF-NiO primarily contained 2-methyl-3-buten-1-ol (25.07%), methylenecyclopropane (12.12%), ethylcyclopropane (9.67%), dimethylethylborane (4.12%), 4,4'-methylenedianiline (4.02%), 2-methylfuran(3.44%), 1,3-pentadiene (3.30%), toluene (2.65%), 3-iodo-1*H*-pyrazole (2.64%), benzene (2.63%), 1,3-butadien-1-ol (2.15%), 3-hexen-1-yne (1.54%), *L-β*-homoserine (1.17%), and furfural (1.15%).

For VF-Fe$_2$O$_3$, its bio-oil was comprised of acids (3.23%), alcohols (29.01%), aldehydes (0.40%), alkanes (30.78%), amines (2.19%), aromatics (9.56%), esters (0.28%), furans (9.78%), ketones (2.86%), phenolics (0.84%), olefines (7.91%), and others (3.14%). The bio-oil of VF-Fe$_2$O$_3$ was abundant with 2-methyl-3-buten-1-ol (25.77%), methylenecyclopropane (17.05%), ethylcyclopropane (10.58%), 2-methylfuran (4.06%), 1,3-pentadiene (3.90%), benzene (3.28%), toluene (3.21%), 3-iodo-1H-pyrazole (2.87%), acetic acid (2.86%), dimethylethylborane (2.51%), furfural (2.42%), 1,3-butadien-1-ol (2.29%), 2,5-dimethylfuran (1.34%), 1-penten-3-one (1.13%), lidocaine (1.10%), and p-xylene (0.95%), respectively.

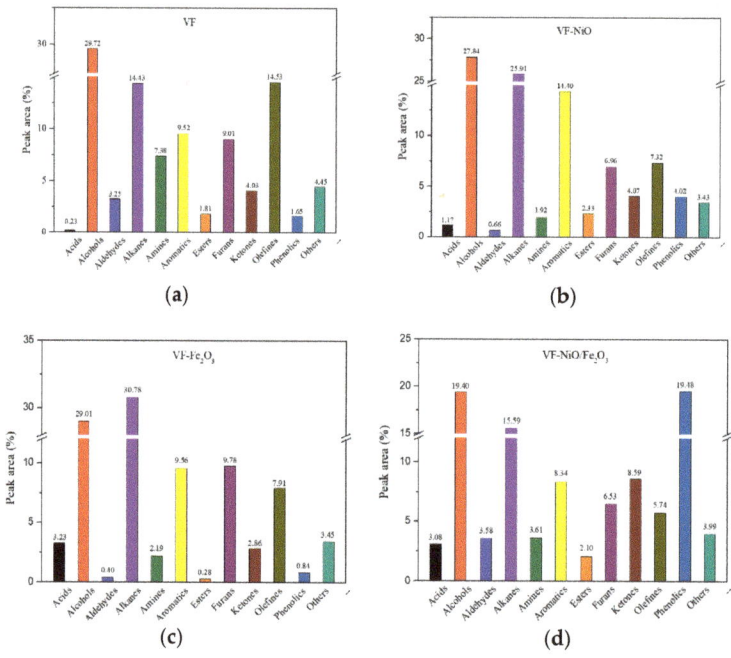

Figure 5. Chemical composition of bio-oil vapor released from fast pyrolysis of *Vernicia fordii* wood (VF) samples that were sorted by common functional groups. (a) VF, (b) VF-NiO, (c) VF-Fe$_2$O$_3$, and (d) VF-NiO/Fe$_2$O$_3$.

The constituents of VF-NiO/Fe$_2$O$_3$ bio-oil were acids (3.08%), alcohols (19.40%), aldehydes (3.58%), alkanes (15.59%), amines (3.61%), aromatics (8.34%), esters (2.10%), furans (6.53%), ketones (8.59%), phenolics (19.48%), olefines (5.74%), and others (3.99%). The bio-oil of VF-NiO/Fe$_2$O$_3$ was rich in 2-methyl-3-buten-1-ol (17.25%), methylenecyclopropane (9.48%), ethylcyclopropane (6.08%), isoeugenol (4.14%), acetoveratrone (3.58%), 4-vinylguaiacol (3.41%), *DL*-2-aminoadipic acid (3.06%), syringol (2.92%), 2-methylfuran (2.87%), 4-allyl-2,6-dimethoxyphenol (2.67%), 3-iodo-1H-pyrazole (2.56%), benzene (2.34%), 1,3-pentadiene (2.33%), toluene (1.94%), vanillin (1.64%), furfural (1.44%), piperonal (1.40%), guaiacol (1.40%), creosol (1.38%), 3-hexen-1-yne (1.16%), and anandamide (1.08%).

When the NiO catalyst was introduced, the aldehyde, amine, furan, and olefin products in the VF bio-oil decreased significantly, but aromatic, alkane, and phenolic compounds increased remarkably (Figure 5). In the presence of NiO, the peak area proportion of aromatics and phenolics increased from 9.52 to 14.40% and from 1.65 to 4.02%, respectively. Thus, the NiO catalyst affected the bio-oil composition by promoting lignin decomposition for the formation of aromatics and phenolics. Meanwhile, the VF bio-oil produced in the presence of Fe$_2$O$_3$ had a greater proportion of alkanes and furans than that of non-catalytic pyrolysis at the expense of lower productions of aldehydes, amines, esters, ketones, and olefins. Seemingly, the bio-oil of VF and VF-Fe$_2$O$_3$ had almost similar peak area proportions of aromatics (9.52% and 9.56%, respectively); nonetheless, the distribution of the aromatics was different. In the presence of Fe_2O_3, the slight increment of furans and drastic decrement of aldehydes implied the inhibition of cellulose pyrolysis by the Fe_2O_3 catalyst. Furthermore, the lower proportion of phenols and aromatics in the bio-oil of VF-Fe_2O_3 than that of VF eventually corroborated the catalytic effect of Fe_2O_3 on lignin pyrolysis. The Fe_2O_3 catalyst was also effective to produce VF bio-oil with high hydrocarbon yield.

Moreover, the NiO/Fe_2O_3 mixture had a poor performance in forming olefins, alkanes, and alcohols. Hydrocarbons have high value in the fuel industry [61]. The NiO/Fe_2O_3 completely converted acids, aldehydes, and sugars besides significantly reduced furans. The synergistic effect between NiO and Fe_2O_3 caused substantial enhancement of ketones and phenols, whereby phenols and their alkylated derivatives are useful chemical precursors. Due to the high heating value, olefines and alkanes have a high value for fuel applications. The catalytic pyrolysis of VF over NiO and Fe_2O_3 considerably increased the formation of alkanes in the VF bio-oil from 14.43% to 25.91% and 30.78% peak areas, respectively (Figure 5). There was a disparity in the bio-oils between VF-NiO/Fe_2O_3 and other samples since the employment of NiO/Fe_2O_3 gave rise to the highest yield of ketones and phenolics.

4. Conclusions

In this study, the solvent extracts were analyzed by FTIR and GC-MS methods, which revealed that the *Vernicia fordii* wood contained a large number and diversity of chemical compounds. These natural product active molecules of the *Vernicia fordii* wood could be used as drug and biomedical active ingredients, further indicating that *Vernicia fordii* wood extractives have broad application prospects as raw materials in industrial and medical fields. The Py-GC-MS analysis indicated that the catalyst type significantly influenced the compositions of the pyrolysis of the *Vernicia fordii* wood. The results revealed that the nano-NiO and nano-Fe_2O_3 catalysts influenced the formation of acid, aromatics, phenols, and alkanes compounds, and inhibited the formation of olefins and amines. In the presence of nano-NiO, the formation of aromatics and phenolics was increased from 9.52% to 14.40% and from 1.65% to 4.02%, respectively. In addition, the NiO/Fe_2O_3 mixture had a poor performance in forming olefins, alkanes, and alcohols.

Supplementary Materials: The following are available online at http://www.mdpi.com/1996-1073/12/20/3972/s1, Figure S1: FTIR spectra of the Vernicia fordii wood (VF) extracts from Soxhlet extraction with either methanol, ethanol/benzene (1:1), or ethanol/ether (1:1), Figure S2: Total ion chromatogram of Vernicia fordii wood (VF) extract from methanol extraction, Figure S3: Total ion chromatogram of Vernicia fordii wood (VF) extract from ethanol/benzene extraction, Figure S4: Total ion chromatogram of Vernicia fordii wood (VF) extract from ethanol/ether extraction, Figure S5: Chemical composition of Vernicia fordii wood (VF) extracts sorted by common functional groups, Table S1: The classification of functional groups and compounds obtained from the extracts of Vernicia fordii wood (VF) from different solvent based on FTIR method, Table S2: Chemical composition of Vernicia fordii wood (VF) extract from methanol extraction, Table S3: Chemical composition of Vernicia fordii wood (VF) extract from ethanol/benzene extraction, Table S4: Chemical composition of Vernicia fordii wood (VF) extract from ethanol/ether extraction, Table S5: Chemical composition of bio-oil vapor released from fast pyrolysis of VF, Table S6: Chemical composition of bio-oil vapor released from fast pyrolysis of VF-NiO, Table S7: Chemical composition of bio-oil vapor released from fast pyrolysis of VF-Fe2O3, Table S8: Chemical composition of bio-oil vapor released from fast pyrolysis of VF-NiO/Fe_2O_3.

Author Contributions: Writing—original draft preparation, C.L.; investigation, X.Y., J.Y., and Y.Y.; data curation, X.Y.; writing—review and editing, C.L. and H.G.; supervision, W.P.; funding acquisition, W.P. and H.G.

Funding: This research was funded by the National Natural Science Foundation of China, grant number 41,701,360.

Acknowledgments: The authors acknowledge financial support by the National Natural Science Foundation of China (41701360). Yiyang Li has contribution on this research. Thanks for her excellent work.

Conflicts of Interest: The authors declare no conflict of interest.

References

1. Morales, M.; Ataman, M.; Badr, S.; Linster, S.; Kourlimpinis, I.; Papadokonstantakis, S.; Hatzimanikatis, V.; Hungerbühler, K. Sustainability assessment of succinic acid production technologies from biomass using metabolic engineering. *Energy Environ. Sci.* **2016**, *9*, 2794–2805. [CrossRef]
2. Grams, J.; Niewiadomski, M.; Ruppert, A.M.; Kwapiński, W. Influence of Ni catalyst support on the product distribution of cellulose fast pyrolysis vapors upgrading. *J. Anal. Appl. Pyrolysis* **2015**, *113*, 557–563. [CrossRef]
3. Laurent, A.; Espinosa, N. Environmental impacts of electricity generation at global, regional and national scales in 1980–2011: What can we learn for future energy planning? *Energy Environ. Sci.* **2015**, *8*, 689–701. [CrossRef]
4. Morales, M.; Dapsens, P.Y.; Giovinazzo, I.; Witte, J.; Mondelli, C.; Papadokonstantakis, S.; Hungerbühler, K.; Pérez-Ramírez, J. Environmental and economic assessment of lactic acid production from glycerol using cascade bio- and chemocatalysis. *Energy Environ. Sci.* **2015**, *8*, 558–567. [CrossRef]
5. Nel, W.P.; Cooper, C.J. Implications of fossil fuel constraints on economic growth and global warming. *Energy Policy* **2009**, *37*, 166–180. [CrossRef]
6. Op de Beeck, B.; Dusselier, M.; Geboers, J.; Holsbeek, J.; Morré, E.; Oswald, S.; Giebeler, L.; Sels, B.F. Direct catalytic conversion of cellulose to liquid straight-chain alkanes. *Energy Environ. Sci.* **2015**, *8*, 230–240. [CrossRef]
7. Vardon, D.R.; Franden, M.A.; Johnson, C.W.; Karp, E.M.; Guarnieri, M.T.; Linger, J.G.; Salm, M.J.; Strathmann, T.J.; Beckham, G.T. Adipic acid production from lignin. *Energy Environ. Sci.* **2015**, *8*, 617–628. [CrossRef]
8. Cai, H.; Wang, J.; Feng, Y.; Wang, M.; Qin, Z.; Dunn, J.B. Consideration of land use change-induced surface albedo effects in life-cycle analysis of biofuels. *Energy Environ. Sci.* **2016**, *9*, 2855–2867. [CrossRef]
9. Ding, J.; Wang, H.; Li, Z.; Cui, K.; Karpuzov, D.; Tan, X.; Kohandehghan, A.; Mitlin, D. Peanut shell hybrid sodium ion capacitor with extreme energy–power rivals lithium ion capacitors. *Energy Environ. Sci.* **2014**, *8*, 941–955. [CrossRef]
10. Mason, P.M.; Glover, K.; Smith, J.A.C.; Willis, K.J.; Woods, J.; Thompson, I.P. The potential of CAM crops as a globally significant bioenergy resource: Moving from 'fuel or food' to 'fuel and more food'. *Energy Environ. Sci.* **2015**, *8*, 2320–2329. [CrossRef]
11. Jung, J.I.; Risch, M.; Park, S.; Kim, M.G.; Nam, G.; Jeong, H.Y.; Shao-Horn, Y.; Cho, J. Optimizing nanoparticle perovskite for bifunctional oxygen electrocatalysis. *Energy Environ. Sci.* **2016**, *9*, 176–183. [CrossRef]
12. Liu, B.; Zhang, Z. Catalytic Conversion of Biomass into Chemicals and Fuels over Magnetic Catalysts. *ACS. Catal.* **2015**, *6*, 326–338. [CrossRef]
13. Shi, M.; Zhang, P.; Fan, M.; Jiang, P.; Dong, Y. Influence of crystal of Fe_2O_3 in magnetism and activity of nanoparticle CaO/Fe_2O_3 for biodiesel production. *Fuel* **2017**, *197*, 343–347. [CrossRef]
14. Bridgwater, A.V. Production of high grade fuels and chemicals from catalytic pyrolysis of biomass. *Catal. Today* **1996**, *29*, 285–295. [CrossRef]
15. Kunkes, E.L.; Simonetti, D.A.; West, R.M.; Serrano-Ruiz, J.C.; Gärtner, C.A.; Dumesic, J.A. Catalytic conversion of biomass to monofunctional hydrocarbons and targeted liquid-fuel classes. *Science* **2008**, *322*, 417–421. [CrossRef]
16. Hara, M. Biomass conversion by a solid acid catalyst. *Energy Environ. Sci.* **2010**, *3*, 601–607. [CrossRef]
17. Duan, X.F.; Zhou, Z.F.; Xu, J.M.; Tian, Y.; Wang, R. Utilization situation and suggestion of forestry residues resources in China. *China Wood Panel* **2017**, *24*, 1–5.
18. Lu, Q.; Zhang, Z.F.; Dong, C.Q.; Zhu, X.F. Catalytic Upgrading of Biomass Fast Pyrolysis Vapors with Nano Metal Oxides: An Analytical Py-GC/MS Study. *Energies* **2010**, *3*, 1805–1820. [CrossRef]
19. Banks, S.W.; Nowakowski, D.J.; Bridgwater, A.V. Impact of Potassium and Phosphorus in Biomass on the Properties of Fast Pyrolysis Bio-oil. *Energy Fuels* **2016**, *30*, 8009–8018. [CrossRef]
20. Chen, X.; Yang, H.; Chen, Y.; Chen, W.; Lei, T.; Zhang, W.; Chen, H. Catalytic fast pyrolysis of biomass to produce furfural using heterogeneous catalysts. *J. Anal. Appl. Pyrolysis* **2017**, *127*, 292–298. [CrossRef]
21. Jae, J.; Coolman, R.; Mountziaris, T.J.; Huber, G.W. Catalytic fast pyrolysis of lignocellulosic biomass in a process development unit with continual catalyst addition and removal. *Chem. Eng. Sci.* **2014**, *108*, 33–46. [CrossRef]

22. Liu, Q.; Zhong, Z.; Wang, S.; Luo, Z. Interactions of biomass components during pyrolysis: A TG-FTIR study. *J. Anal. Appl. Pyrolysis* **2011**, *20*, 213–218. [CrossRef]
23. Wang, K.; Kim, K.H.; Brown, R.C. Catalytic pyrolysis of individual components of lignocellulosic biomass. *Green Chem.* **2014**, *16*, 727–735. [CrossRef]
24. Yu, J.; Paterson, N.; Blamey, J.; Millan, M. Cellulose, xylan and lignin interactions during pyrolysis of lignocellulosic biomass. *Fuel* **2017**, *191*, 140–149. [CrossRef]
25. Li, J.; Yan, R.; Xiao, B.; Liang, D.T.; Du, L. Development of nano-NiO/Al2O3 catalyst to be used for tar removal in biomass gasification. *Environ. Sci. Technol.* **2008**, *42*, 6224–6229. [CrossRef] [PubMed]
26. Zou, J.; Oladipo, J.; Fu, S.L.; Al-Rahbib, A.; Yang, H.P.; Wu, C.F.; Cai, N.; Williams, P.; Chen, H.P. Hydrogen production from cellulose catalytic gasification on CeO_2/Fe_2O_3 catalyst. *Energy Conves. Mang.* **2018**, *171*, 241–248. [CrossRef]
27. Xie, Y.; Ge, S.; Jiang, S.; Liu, Z.; Chen, L.; Wang, L.; Chen, J.; Qin, L.; Peng, W. Study on biomolecules in extractives of Camellia oleifera fruit shell by GC-MS. *Saudi J. Biol. Sci.* **2018**, *25*, 234–236. [CrossRef]
28. Khelfa, A.; Sharypov, V.; Finqueneisel, G.; Weber, J.V. Catalytic pyrolysis and gasification of Miscanthus Giganteus: Haematite (Fe_2O_3) a versatile catalyst. *J. Anal. Appl. Pyrolysis* **2009**, *84*, 84–88. [CrossRef]
29. Peng, W.; Lin, Z.; Wang, L.; Chang, J.; Gu, F.; Zhu, X. Molecular characteristics of *Illicium verum* extractives to activate acquired immune response. *Saudi J. Biol. Sci.* **2016**, *23*, 348–352. [CrossRef]
30. Li, C.; Zhang, J.; Yi, Z.; Yang, H.; Zhao, B.; Zhang, W.; Li, J. Preparation and characterization of a novel environmentally friendly phenol—Formaldehyde adhesive modified with tannin and urea. *Int. J. Adhes. Adhes.* **2016**, *66*, 26–32. [CrossRef]
31. Lu, Q.; Zhang, Z.B.; Yang, X.C.; Dong, C.Q.; Zhu, X.F. Catalytic fast pyrolysis of biomass impregnated with K3PO4 to produce phenolic compounds: Analytical Py-GC/MS. *J. Anal. Appl. Pyrolysis* **2013**, *104*, 139–145. [CrossRef]
32. Sun, L.; Zhang, X.; Chen, L.; Zhao, B.; Yang, S.; Xie, X. Comparision of catalytic fast pyrolysis of biomass to aromatic hydrocarbons over ZSM-5 and Fe/ZSM-5 catalysts. *J. Anal. Appl. Pyrolysis* **2016**, *121*, 342–346. [CrossRef]
33. Peyrat-Maillard, M.N.; Cuvelier, M.E.; Berset, C. Antioxidant activity of phenolic compounds in 2, 2-azobis (2-amidinopropane) dihydrochloride (AAPH)-induced oxidation Synergistic and antagonistic effects. *JOACS* **2003**, *80*, 1007–1012. [CrossRef]
34. Ikeda, I.; Tanabe, Y.; Sugano, M. Effects of Sitosterol and Sitostanol on Micellar solubility of Cholesterol. *J. Nutr. Sci. Vitaminol.* **1989**, *35*, 361–369. [CrossRef] [PubMed]
35. Salen, G.; Shore, V.; Tint, G.S.; Forte, T.; Shefer, S.; Horak, I.; Horak, E.; Dayal, B.; Nguyen, L.; Batta, A.K. Increased sitosterol absorption, decreased removal, and expanded body pools compensate for reduced cholesterol synthesis in sitosterolemia with xanthomatosis. *J. Lip. Res.* **1989**, *30*, 1319–1330.
36. Luterbacher, J.S.; Azarpira, A.; Motagamwala, A.H.; Lu, F.; Ralph, J.; Dumesic, J.A. Lignin monomer production integrated into the γ-valerolactone sugar platform. *Energy Environ. Sci.* **2015**, *8*, 2657–2663. [CrossRef]
37. Rinaldi, R.; Schüth, F. Design of solid catalysts for the conversion of biomass. *Energy Environ. Sci.* **2009**, *2*, 610. [CrossRef]
38. Li, D.L.; Wu, J.Q.; Peng, W.X.; Xiao, W.F.; Wu, J.G.; Zhuo, J.Y.; Yuan, T.Q.; Sun, R.C. Effect of lignin on bamboo biomass self-bonding during hot-pressing lignin structure and characterization. *Bioresources* **2015**, *10*, 6769–6782. [CrossRef]
39. Burhenne, L.; Messmer, J.; Aicher, T.; Laborie, M.P. The effect of the biomass components lignin, cellulose and hemicellulose on TGA and fixed bed pyrolysis. *J. Anal. Appl. Pyrolysis* **2013**, *101*, 177–184. [CrossRef]
40. Chen, Z.; Hu, M.; Zhu, X.; Guo, D.; Liu, S.; Hu, Z.; Xiao, B.; Wang, J.; Laghari, M. Characteristics and kinetic study on pyrolysis of five lignocellulosic biomass via thermogravimetric analysis. *Bioresour. Technol.* **2015**, *192*, 441–450. [CrossRef]
41. Ondro, T.; Vitázek, I.; Húlan, T.K.; Lawson, M.; Csáki, Š. Non-isothermal kinetic analysis of the thermal decomposition of spruce wood in air atmosphere. *Res. Agric. Eng.* **2018**, *64*, 41–46.
42. Biagini, E.; Barontini, F.; Tognotti, L. Devolatilization of biomss fuels and biomass components stutied by TG/FTIR technique. *Ind. Eng. Chem. Res.* **2006**, *45*, 4486–4493. [CrossRef]
43. Balat, M. Mechanisms of thermochemical biomass conversion processes. Part 1: Reactions of pyrolysis. *Energy Sour. Part A* **2008**, *30*, 620–635. [CrossRef]

44. Demirbas, A. Biofuels securing the plant's future energy needs. *Energ. Convers. Manag.* **2009**, *50*, 2239–2249. [CrossRef]
45. Zhao, C.; Jiang, E.; Chen, A. Volatile production from pyrolysis of cellulose, hemicellulose and lignin. *J. Energy Inst.* **2017**, *90*, 902–913. [CrossRef]
46. Xing, S.; Yuan, H.; Huhetaoli Qi, Y.; Lv, P.; Yuan, Z.; Chen, Y. Characterization of the decomposition behaviors of catalytic pyrolysis of wood using copper and potassium over thermogravimetric and Py-GC/MS analysis. *Energy* **2016**, *114*, 634–646. [CrossRef]
47. Svenson, J.; Pettersson, J.B.C.; Davidsson, K.O. Fast Pyrolysis of the Main Components of Birch Wood. *Combust. Sci. Technol.* **2004**, *176*, 977–990. [CrossRef]
48. Zhang, X.; Yang, W.; Blasiak, W. Thermal decomposition mechanism of levoglucosan during cellulose pyrolysis. *J. Anal. Appl. Pyrolysis* **2012**, *96*, 110–119. [CrossRef]
49. Zhang, X.; Yang, W.; Dong, C. Levoglucosan formation mechanisms during cellulose pyrolysis. *J. Anal. Appl. Pyrolysis* **2013**, *104*, 19–27. [CrossRef]
50. Vitázek, I.; Tkáč, Z. Isothermal kinetic analysis of thermal decomposition of woody biomass: The thermogravimetric study. *AIP Conf. Proc.* **2019**, *2118*. [CrossRef]
51. Wang, H.; Lee, S.J.; Olarte, M.V.; Zacher, A.H. Bio-oil Stabilization by Hydrogenation over Reduced Metal Catalysts at Low Temperatures. *ACS Sustain. Chem. Eng.* **2016**, *4*, 5533–5545. [CrossRef]
52. Papari, S.; Hawboldt, K.; Helleur, R. Production and Characterization of Pyrolysis Oil from Sawmill Residues in an Auger Reactor. *Ind. Eng. Chem. Res.* **2017**, *56*, 1920–1925. [CrossRef]
53. Fushimi, C.; Katayama, S.; Tsutsumi, A. Elucidation of interaction among cellulose, lignin and xylan during tar and gas evolution in steam gasification. *J. Anal. Appl. Pyrolysis* **2009**, *86*, 82–89. [CrossRef]
54. Gani, A.; Naruse, I. Effect of cellulose and lignin content on pyrolysis and combustion characteristics for several types of biomass. *Renew. Energy* **2007**, *32*, 649–661. [CrossRef]
55. Giudicianni, P.; Cardone, G.; Ragucci, R. Cellulose, hemicellulose and lignin slow steam pyrolysis: Thermal decomposition of biomass components mixtures. *J. Anal. Appl. Pyrolysis* **2015**, *100*, 213–222. [CrossRef]
56. Hosoya, T.; Kawamoto, H.; Saka, S. Cellulose—Hemicellulose and cellulose—Lignin interactions in wood pyrolysis at gasification temperature. *J. Anal. Appl. Pyrolysis* **2007**, *80*, 118–125. [CrossRef]
57. Caballero, J.A.; Font, R.; Marcilla, A. Comparative study of the pyrolysis of almond shells and their fractions, holocellulose and lignin: Product yields and kinetics. *Thermochim. Acta* **1996**, *276*, 57–77. [CrossRef]
58. Lu, Q.; Dong, C.Q.; Zhang, X.M.; Tian, H.Y.; Yang, Y.P.; Zhu, X.F. Selective fast pyrolysis of biomass impregnated with $ZnCl_2$ to produce furfural: Analytical Py-GC/MS study. *J. Anal. Appl. Pyrolysis* **2011**, *90*, 204–212. [CrossRef]
59. Mettler, M.S.; Mushrif, S.H.; Paulsen, A.D.; Javadekar, A.D.; Vlachos, D.G.; Dauenhauer, P.J. Revealing pyrolysis chemistry for biofuels production: Conversion of cellulose to furans and small oxygenates. *Energy Environ. Sci.* **2012**, *5*, 5414–5424. [CrossRef]
60. Zhang, B.; Zhong, Z.; Ding, K.; Cao, Y.; Liu, Z. Catalytic Upgrading of Corn Stalk Fast Pyrolysis Vapors with Fresh and Hydrothermally Treated HZSM-5 Catalysts Using Py-GC/MS. *Ind. Eng. Chem. Res.* **2014**, *53*, 9979–9984. [CrossRef]
61. Imrana, A.; Bramer, E.A.; Seshanb, K.; Brem, G. Catalytic flash pyrolysis of oil-impregnated-wood and jatropha cake using sodium based catalysts. *J. Anal. Appl. Pyrolysis* **2016**, *117*, 236–246. [CrossRef]

© 2019 by the authors. Licensee MDPI, Basel, Switzerland. This article is an open access article distributed under the terms and conditions of the Creative Commons Attribution (CC BY) license (http://creativecommons.org/licenses/by/4.0/).

Article

Proof-of-Concept of High-Pressure Torrefaction for Improvement of Pelletized Biomass Fuel Properties and Process Cost Reduction

Bartosz Matyjewicz [1], Kacper Świechowski [1,*], Jacek A. Koziel [2] and Andrzej Białowiec [1,2]

[1] Faculty of Life Sciences and Technology, Institute of Agricultural Engineering, Wrocław University of Environmental and Life Sciences, 51-630 Wrocław, Poland; 110829@student.upwr.edu.pl (B.M.); andrzej.bialowiec@upwr.edu.pl (A.B.)

[2] Department of Agricultural and Biosystems Engineering, Iowa State University, Ames, IA 50011, USA; koziel@iastate.edu

* Correspondence: kacper.swiechowski@upwr.edu.pl

Received: 18 August 2020; Accepted: 11 September 2020; Published: 14 September 2020

Abstract: This paper provides a comprehensive description of the new approach to biomass torrefaction under high-pressure conditions. A new type of laboratory-scale high-pressure reactor was designed and built. The aim of the study was to compare the high-pressure torrefaction with conventional near atmospheric pressure torrefaction. Specifically, we investigated the torrefaction process influence on the fuel properties of wooden-pellet for two different pressure regimes up to 15 bar. All torrefaction processes were conducted at 300 °C, at 30 min of residence time. The initial analysis of the increased pressure impact on the torrefaction parameters: mass yields, energy densification ratio, energy yield, process energy consumption, the proximate analysis, high heating value, and energy needed to grind torrefied pellets was completed. The results show that high-pressure torrefaction needed up to six percent less energy, whereas energy densification in the pellet was ~12% higher compared to conventional torrefaction. The presence of pressure during torrefaction did not have an impact on the energy required for pellet grinding ($p < 0.05$).

Keywords: pressure torrefaction; pellet; renewable energy sources; energy consumption; grinding; thermogravimetric analysis; proximate analysis; high heating value; torrefied biomass; biochar

1. Introduction

1.1. Background

Rapid economic development resulted in a significant increase in demand for energy. The importance of renewable energy sources (RES) has been growing in recent decades. One of the most abundant RES is biomass that can be obtained from energy crops and agricultural waste. Moreover, biomass is an inexhaustible, controlled, and flexible energy source [1,2], and for these reasons, it plays an essential role in the energy supply chain in the European Union (EU) and around the world. An underestimated potential for energy production lies in residual biowaste, which cannot be easily recycled, and whose mass is increasing from a growing population and industrial production.

The main drawback of raw biomass and residual biowaste is its high moisture content, which results in low calorific value and low energy density. The consequence of it is that these materials have to be processed before energetic use. One of the ways to process these materials before energetic use is torrefaction, a process that increases carbon and calorific values, and converts the biomass to a stable and hydrophobic material.

The growing problem with (bio)waste management and energy supplies prompts society to look for new solutions. One of the many ways to manage biowaste is the concept of 'Waste-to-Carbon' with the torrefaction process [3]. In recent times, more research-grade installations for torrefaction are built [4]. The products of the torrefaction process can be used as fuel or as an additive to processes [5] and soils [6]. The use of torrefaction to produce fuel is noticeable worldwide. For example, in Portugal, a 720 kg·h^{-1} pilot industrial-scale plant for torrefaction and torrefied wood pelletization was built [7]. In Burkina Faso, small units for valorizing cashew nut shells were set up [8]. In Steiermark, Austria, a 1 Mg·h^{-1} pilot plant was built in 2011 [9]. Industrial installation for biomass torrefaction in eastern Oregon (USA) is under construction [10].

1.2. Torrefaction for Organic Material Valorization

The torrefaction process (also known as biomass 'roasting') is a type of thermochemical treatment of the organic matter, consisting of a slow heating rate <50 °C·min^{-1} [11] to a temperature above 200 °C, usually 280–320 °C, at a pressure close to atmospheric and in the absence of oxygen [12]. The residence time varies from several minutes up to several hours. The torrefaction is assumed to be suitable for processing material with moisture content under 15% [13]. Torrefaction takes place in five steps. At first, the treated material is (I) preheated, followed by (II) pre-drying, where some of the water is evaporated. The next stages are (III) drying and (IV) post-drying and intermediate heating where remaining water is removed. When water is removed, the proper (V) torrefaction process takes place. Two products are formed. Solid fraction (torrefied material) and a gas fraction (torrgas), wherein the liquid fraction may be separated from the torrgas, dividing it between the condensable fraction (water, oils, tars, and other compounds) and the non-condensable fraction (CO, CO_2, CH_4, and other gases). The solid fraction can be used as fuel or as an additive to industry. Torrgas can be used for further processing or for supplying the heat for the torrefaction process [14]. The resulting solid product is a uniform hydrophobic material with lower humidity, higher calorific value, and improved milling properties compared to raw material. The hydrophobicity guarantees the stability of the fuel in varying storage conditions, protects against the bio-decomposition, the development of mold, and microorganisms' growth [12].

A relatively new approach to torrefaction is hydrothermal carbonization (HTC), which resolves problems of wet organic materials (with moisture >15%). The HTC material is processed in subcritical water at 180–250 (300) °C and pressure >1 MPa. The process residence time varies in a wide range but is usually shorter than in the case of the conventional (atmospheric pressure) torrefaction and is ranging from a few minutes to several hours [15]. An additional technological parameter of HTC is solids loading, which ranges from 7 to 25% [16].

The proposed novel concept lies in the selection of the process conditions that are between conventional torrefaction and HTC. A high-pressure torrefaction is proposed herein, where the material is torrefied in temperature and time consistent with traditional torrefaction but under elevated pressure.

Two strategies of process performance may be derived: (1) under steady pressure starting from the beginning and maintained for the whole process, and (2) released gasses in the closed reactor vessel continue to increase the pressure during the process. In this second scenario (during high-pressure torrefaction), the heated material begins to degas, which causes an increase in pressure inside the reactor, which causes an increase in the fixed carbon content, and therefore, the possibility of obtaining better quality material. Wannapeera and Worasuwannarak [17] examined that high-pressure torrefaction allows obtaining a material with a higher calorific value (HHV). Also, high-pressure torrefaction changes the structure of the crosslinking of the material and causes an increase in charring performance compared to the torrefaction under atmospheric pressure [17].

High-pressure torrefaction creates a new pathway for biochar production. The trapped gases increase the pressure, which causes a faster temperature increase compared to conventional torrefaction; as a result, less energy is consumed. Moreover, the biochar from pressured torrefaction has a higher calorific value. For this reason, a high-pressure unit on an industrial scale could allow achieving higher efficiency with lower energy expenditure for a process, and better fuel quality compared with conventional technologies.

1.3. The Pellet Role in Energy Chain Supply

Pelletization is used to concentrate the energy in the material, resulting in lower transport costs, easier storage (takes less place, emits less dust and organic compounds), and facilitates dosing into a household and industrial energy device [18]. For example, wood pellet is used in the Turów power plant in Poland [19], where pellets are ground and mixed with lignite coal at eight percent share. Recently, a combined technology of pelletization and torrefaction at different (downstream and upstream) configurations are considered to increase energy densification and cost reduction. It has been shown that in the EU, the torrefied pellet can be less expensive than the conventional one (4.7 €·GJ^{-1} vs. 5.8 €·GJ^{-1}, respectively) [3,20].

1.4. The Importance and the Aim of the Study

The increased energy demand and residual biowaste overproduction are important problems to solve. One of the solutions is the torrefaction process that is capable of converting residual biomass and residual biowaste to solid fuel for the powerplant. Despite the constant development of torrefaction technology and small industrial-scale plant, there is a need to optimize the torrefaction technology. New solutions, leading to a decrease in the energy demand for the process, and increasing the efficiency should be developed.

In this work, the new type of batch lab-scale reactor for high-pressure torrefaction was built and tested. The aim of the study was to compare the high-pressure torrefaction with conventional atmospheric pressure torrefaction. Specifically, we investigated the torrefaction process influence on the fuel properties wooden-pellet for two different pressure regimes. Additionally, the main operational observations of a new type of reactor were described.

2. Materials and Methods

The experiment of high-pressure torrefaction was conducted in a designed reactor. The reactor design allowed us to perform conventional torrefaction (at atmospheric pressure) and high-pressure torrefaction. The full experiment setup is presented in Figure 1. Before the torrefaction process, the wooden-pellets were dried in a laboratory oven. Then, dry wooden-pellets were placed in the reactor, where torrefaction tests took place (the torrefaction at atmospheric pressure and at high-pressure). During each test, the energy demand for heating the reactor, the temperature inside the reactor, pressure in the reactor, and fractions mass yields were measured. Next, raw pellets and torrefied ones were subjected to the proximate analysis and grinding test. The thermogravimetric analysis of the raw pellet was conducted to better understand the influence of the reactor temperature and pressure changes during the process. Finally, the obtained data analysis and comparisons of the effects of pressure regimes were completed.

2.1. Materials

A commercially available softwood pellet made from sawdust was chosen for the torrefaction experiment (Figure 1). Wooden pellets (instead of woodchips) were selected for several reasons: (i) a higher homogeneity of pellets, (ii) to study the impact of pressure on pellets, (iii) practical aspect—is easier to integrate a torrefaction as a downstream operation in a pellet plant compared to torrefaction as upstream operation. Pellet parameters are: moisture content 6.6–8.3%, ash content 0.8–1.1%, sulfur content 0.01%, low heating value 16 MJ·kg^{-1}, pellet's diameter 6 mm. The pellets were bought in a DIY store located in Wroclaw, Poland. The conducted analysis of pellets showed that moisture content, organic matter content, combustible parts, and ash content were 9.94, 99.63, 99.74, and 0.26%, respectively.

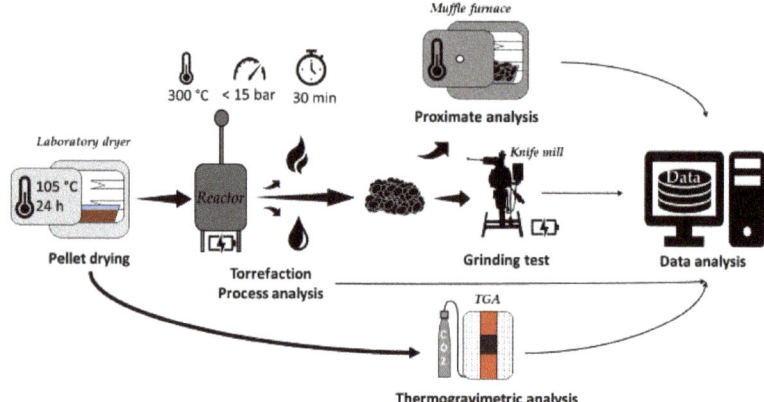

Figure 1. Experiment setup to determine fuel properties of torrefied pellets produced at atmospheric and high-pressure conditions.

2.2. Methods

2.2.1. Reactor Design

The reactor used in this experiment was a prototype dedicated to torrefaction (Figure 2). The reactor can operate at a maximum pressure of 15 bar and a temperature of 300 °C. When the pressure exceeds 15 bar, the safety valve opens (3); it can also be done manually at any time. The safety valve protected the device from bursting. The measuring equipment includes a manometer (4) and a temperature sensor located inside the reactor (6). Heating takes place in an indirect system, where heat is supplied through the walls of the reactor (heat-resistant steel, thickness 4 mm) by two electric heating mats (Conbest, SKU 189-11-3, Kraków, Poland), the total power of 1560 W, controlled by temperature regulator (RKT, REX-C100, China). The heating mats were thermally isolated from the environment by fiberglass insulation tape (thickness~7.5 cm, λ~0,05 $W \cdot m^{-1} \cdot K^{-1}$). The biomass was placed inside of the reactor on the special grille (Figure 3a). The grille was placed at 1/3 of the height of the chamber of the reactor. The total reactor volume was 22.3 dm^3.

2.2.2. Torrefaction Process Procedure

The torrefaction process was conducted in the above-mentioned reactor. Before each torrefaction test, the pellet was dried in a laboratory dryer (WAMED, KBC-65W, Warsaw, Poland) for 24 h at 105 °C. A 600 ± 1 g piece of dry pellet sample was used for each torrefaction test. First, samples were placed into the bottom part of the reactor, on the grille (Figure 3a) (to prevent/limit the adsorption of condensed fraction into a solid fraction after the end of the process when temperature decreased). Next, the reactor was filled with CO_2 inert gas and then was sealed.

Three scenarios were tested:

a. In the case of torrefaction at atmospheric pressure, the upper valve was open, and the end of the rubber pipe (exhaust pipe) was placed into the bottle half-filled with water (acting as a water seal to prevent the oxygen entering into the reactor). The rubber pipe was tightly placed into the bottle (To limit water evaporation resulting from the infusion of high-temperature torrgas at ~70 °C. The torrgas was allowed to escape from the bottle by small holes in the top part of the bottle) (Figure 3b). Then the heating mats were turned on, and the reactor has been heated from room temperature ~20 °C to a setpoint temperature of 300 °C with an average heating rate of 2.6 $°C \cdot min^{-1}$. After the reactor temperature reached the setpoint, the process residence time of 30 min was counted. Finally, the heating mats were turned off, and the upper valve was closed (to stop the water suction from the bottle, which resulted from cooled down gasses in the reactor), and the reactor was left to cool down. The samples from this process are named 'ap1'–'ap3' (atmospheric pressure, numbers represent individual repetitions);

b. In the case of torrefaction at high-pressure, all valves were closed. The heating mats were turned on, and the reactor has been heated from room temperature ~20 °C to a setpoint temperature of 300 °C with an average heating rate of 2.9 °C·min^{-1}. As the temperature rose, the pressure increased (as a result of temperature rise and pellet degassing). In four repetitions for this variant, the pressure was not controlled, which led to the opening of the safety valve. As a result, the pressure decrease occurred. After gas release, the upper valve was closed, and the reactor was left to cool down. The samples for which the safety valve opened were labeled as 'hpd1'–'hpd4' (high-pressure-decrease, numbers represent individual repetitions);

c. During the next four tests, the pressure increase was controlled not to exceed the upper-pressure threshold value (15 bar); therefore, the high-pressure conditions were maintained for the whole process. After the reactor temperature reached the setpoint, the residence time of 30 min was counted, and (if needed) pressure was relieved manually to keep it at 14 ± 1 bar. Finally, the heating mats were turned off, and the upper valve was opened to release pressure from the reactor. The samples where pressure was kept at one steady level have names 'hps1'–'hps4' (high-pressure-steady, numbers represent individual repetitions).

Each experiment ended when the torrefied pellets were retrieved from the reactor, and a condensed fraction trapped on the bottom of the reactor was drained by the lower valve.

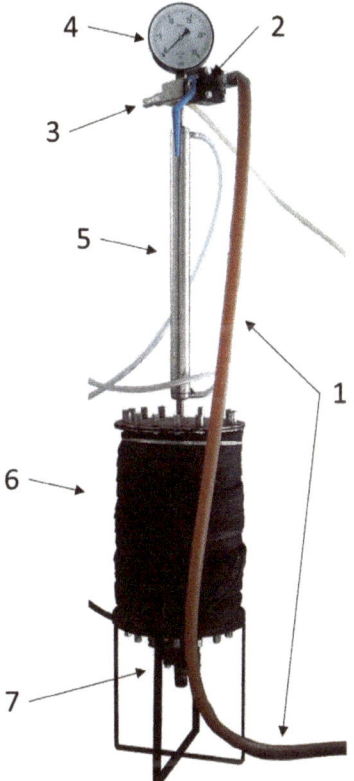

Figure 2. Reactor design. (1) exhaust torrgas pipe, (2) upper valve, (3) safety valve, (4) manometer, (5) cooler, (6) reactor chamber wrapped by heating mats and insulation, (7) lower valve.

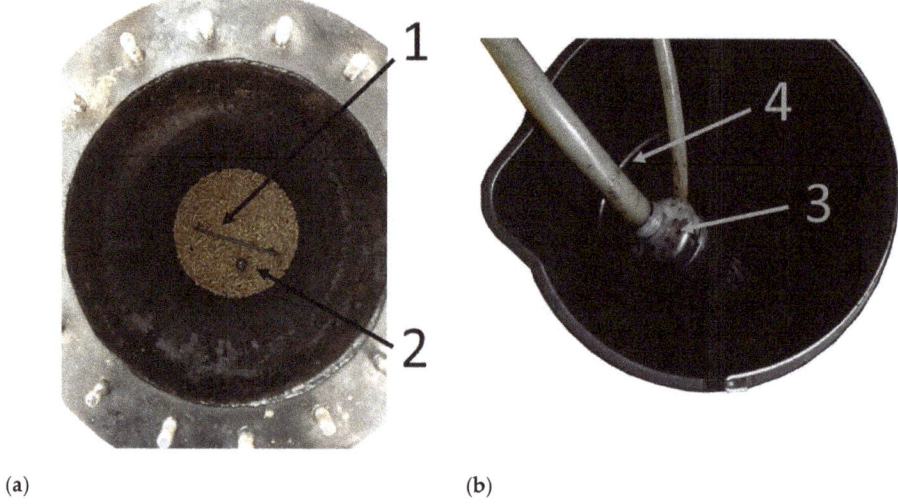

Figure 3. (a) Loading of pellets into the reactor, (b) Protection against air ingress into the reactor. Legend: (1) grille handle, (2) thermocouple pocket, (3) bottle with water and holes in the cover, (4) the terminal of exhaust torrgas pipe.

2.2.3. Torrefaction Process Analysis

For each torrefaction experiment, the energy demand was measured using a power network analyzer (LUMEL, ND40, Zielona Góra, Poland). After the torrefaction at the atmospheric pressure, the solid fraction was measured as solid parts that were found in the reactor (over and under the sieve, grille). Some parts of the pellets crumbled and fell under the sieve of the grille, but did not contact the liquid fraction (all liquid fraction was found inside the bottom valve). In the case of high-pressure torrefaction, the solid fraction was measured separately for the over- and under-side of the grille. Due to the valve closure after gas release at the process end, residual gases and gases that were produced during reactor cooling time, after the cooling down, condensed at the bottom part of the reactor, and were mixed with pellets under the grille.

After the torrefaction at the atmospheric pressure, the condensable fraction was measured as the mass of condensed liquid in a rubber pipe, bottle with water, and bottom of the reactor. After the torrefaction at high-pressure, the condensable fraction was measured as the mass of condensed liquid in a rubber pipe and at the bottom of the reactor.

The mass of the non-condensable fraction was measured by subtraction. The mass yield (MY) of particular fractions were calculated as follow:

Mass yield of the solid fraction was calculated according to Equation (1).

$$MY_{SF} = \frac{mass\ of\ torrefied\ pellet}{initial\ mass\ of\ pellet} \cdot 100 \tag{1}$$

Mass yield of the condensable fraction was determined according to Equation (2).

$$MY_{CF} = \frac{mass\ of\ condensed\ fraction}{initial\ mass\ of\ pellet} \cdot 100 \tag{2}$$

Mass yield of the non-condensable fraction was estimated according to Equation (3).

$$MY_{NCF} = 100\% - MY_{SF} - MY_{CF} \tag{3}$$

For solid fraction, the energy densification ratio (*EDr*), and the energy yield (*EY*) were calculated in accordance with the following Equations (4) and (5) [21]:

$$EDr = \frac{\text{the high heating value of torrefied pellet}}{\text{high heating value of pellet}} \quad (4)$$

$$EY = MY_{SF} \cdot EDr \quad (5)$$

2.2.4. Thermogravimetric Analysis

The pellet was subjected to the thermogravimetric analysis (TGA) to better understand the reactions that occurred during the torrefaction. In the case of torrefaction under atmospheric pressure, the reactor heating rate was slowing down at around 225–250 °C to accelerate again above 250 °C. Interestingly, for high-pressure torrefaction, that temperature phenomenon was not observed.

The thermogravimetric analysis was performed using the stand-mounted tubular furnace (Czylok, RST 40 x 200/100, Jastrzębie-Zdrój, Poland) coupled with the laboratory balance (RADWAG, PS 750.3Y, Radom, Poland). The CO_2 gas was subjected to the tubular furnace to provide the inert condition during analysis. The ~4 g pellet samples were heated from room temperature ~25 °C to 850 °C at three different heating rates: 2.5, 5.0, and 7.5 °C·min^{-1}. The heating rates were chosen to be consistent with the heating rate of the torrefaction reactor. The pellet mass changes were reported with intervals of 1 s. The mass measurements were done within an accuracy of 0.001 g. Obtained raw TGA data were smoothed before further processing. For smoothing, the locally estimated scatterplot smoothing (LOESS) method was used. The calculation of LOESS with parameter *Span* = 0.1 was done using OrginPro 2019b software (OriginLab, OrginPro 2019b, Northampton, MA, USA). Then, from the smoothed TG curve, the derivative curve (DTG) was calculated. DTG is defined as dTG/dT where dTG—mass change, %, dT—temperature change, °C [22].

2.2.5. Proximate Analysis

For pellet and torrefied pellets, the proximate analysis was conducted. The analysis contained moisture content (MC), organic matter content, also known as loss on ignition (OM), ash content (ash), combustible part content (CP), and high heating value (HHV). The used devices and standards for particular variable analysis are given in Table 1. Each analysis was performed in 6 replications.

Table 1. Proximate analysis list and standard methods.

Variable	Analysis Device (Manufacturer, Model, City, Country)	Analysis Standard	Reference
MC	Laboratory dryer (WAMED, KBC-65W, Warsaw, Poland)	PN-EN 14346:2011	[23]
OM	Muffle furnace (SNOL, 8.1/1100, Utena, Lithuania)	PN-EN 15169:2011	[24]
CP	Muffle furnace (SNOL, 8.1/1100, Utena, Lithuania)	PN-Z-15008-04:1993	[25]
Ash	Muffle furnace (SNOL, 8.1/1100, Utena, Lithuania)	PN-Z-15008-04:1993	[25]
HHV	Calorimeter (IKA® Werke GmbH, C200, Staufen, Germany)	PN-G-04513:1981	[26]

2.2.6. Pellet Grinding Test

The samples of the raw pellets (with natural moisture content), dried pellets, and torrefied pellets were subjected to the grinding analysis to measure the influence of high-pressure and conventional torrefaction on the energy consumption of the grinding process. The grinding test was performed with the laboratory knife mill (Testchem, LMN-100, Pszów, Poland). The 2.2 kW knife mill was operated at 2800 rpm. For the test, a 1-mm screen was used. The samples of raw pellets (with the residual moisture in storage) and torrefied pellets were subjected to the grinding to measure the influence on grinding energy consumption of high-pressure torrefaction in comparison to the conventional one. The energy consumed for grinding was measured using a power network analyzer (LUMEL, ND40, Zielona Góra, Poland). The power measurements were taken every 0.2 s. First, the sample (100 ± 1 g) was placed in a knife mill chute, next power analyzer and knife mill were turned on, then after a while (when the knife mill reached an optimal speed), a sample was pushed onto the blades from the top. For each material, three replications were made.

Power data was used to calculate the total specific energy (Es) and the total effective energy (Ee). The Es was calculated by integration of the power vs. time and divided by the sample mass (to show a total consumed energy per pellet mass). The Ee was calculated by subtracting the specific idle energy from Es [27]. The Es and Ee units were recalculated to $Wh \cdot kg^{-1}$.

2.2.7. Statistical Data Analysis

The data obtained from the proximate analysis and pellet grinding test were subjected to the statistical analysis to determine if different process conditions had a statistically significant impact on examined properties. The level of statistical significance was assumed at $\alpha = 0.05$. The one-way analysis of variance (ANOVA) was used. The post-hoc Tukey's HSD test was used when the ANOVA showed that there were significant differences between groups ($p < 0.05$). The statistical analysis was performed using software StatSoft, Statistica 13.3 (TIBCO Software Inc., Palo Alto, CA, USA).

The results of the post-hoc HSD Tukey's test are presented in particular figures in the Results section. The results of the post-hoc test are marked by letters A, B, C. If the letters were the same, there was no statistical difference ($p < 0.05$) between groups. Letters are valid only for a particular plot. At each plot, the value and standard deviation were presented.

3. Results and Discussion

Raw data from the experiments are given in the Supplementary Material.xlsx. Supplementary Material has "Read me" sheet, which serves as a guide on how to find and interpret the data.

3.1. Torrefaction Process

Figures 4–6 show examples of temperature and pressure patterns during torrefaction. Figure 4 presents the results for atmospheric pressure (ap) torrefaction. Figure 5—high-pressure torrefaction with decreasing pressure (hpd), and Figure 6—high-pressure torrefaction with steady pressure (hps). The average heating rate from 25–300 °C was ~2.6, ~2.90, and ~4.95 °C·min^{-1} for ap, hpd, and hps, respectively (Figure 7d). The heating rate for each torrefaction process was the greatest in the range from 50–225 °C (Figure 7a–c). The heating rate for a range of 50–250 °C for ap, hpd, and hps was ~4.73, ~4.56, and ~4.69 °C·min^{-1}, respectively (Figure 7c).

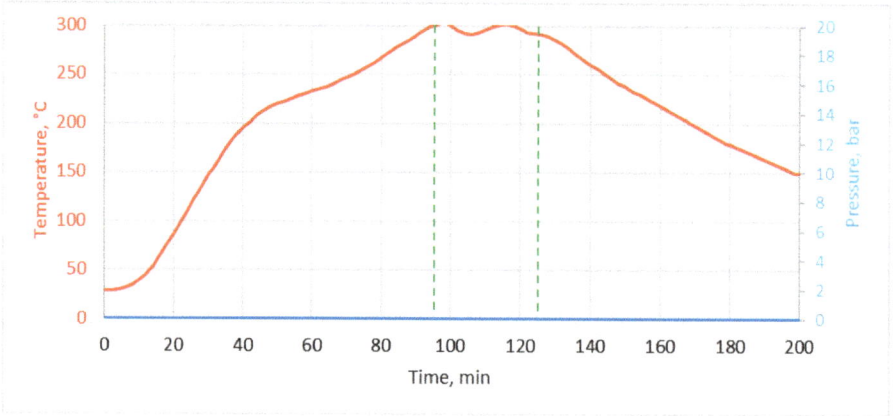

Figure 4. An example of temperature patterns during the atmospheric torrefaction (ap), green dashed line stands for start and end of the torrefaction process (30 min).

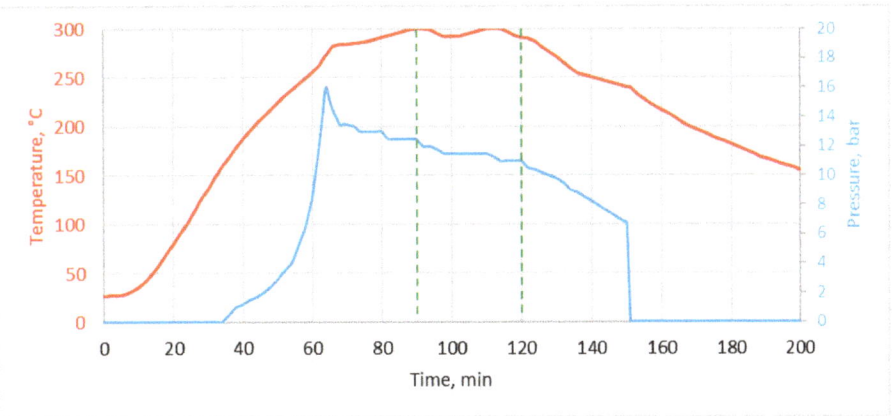

Figure 5. An example of temperature/pressure patterns during the high-pressure torrefaction with decreasing pressure (hpd) torrefaction, green dashed line stands for start and end of the torrefaction process (30 min).

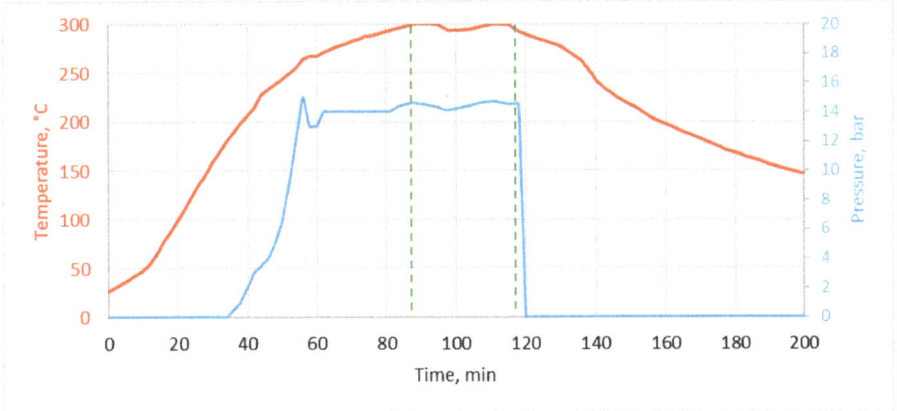

Figure 6. An example of temperature/pressure patterns during the high-pressure torrefaction with steady pressure (hps) torrefaction, green dashed line stand for start and end of the torrefaction process (30 min).

The observed differences in the torrefaction processes for particular heating rates were not significant ($p < 0.05$) (Figure 7a–c). In all cases above ~225–250 °C, the heating rate slowed down. The greatest slowdown was measured for the ap variant (Figure 4, temperature range ~200–250 °C). The heating rate decrease was most likely associated with hot torrgas escape from the reactor—which is illustrated in the results of a TGA test. In the case of hpd and hps, the heating rate decrease was likely a result of insufficient insulation and heat loss. The side of the reactor was wrapped by fiberglass insulation tape (λ~0,05 W·m^{-1}·K^{-1}, thickness~7.5 cm), whereas the top and bottom of the reactor were not insulated. The endothermic reaction of the torrefied pellet was excluded as the reason for the drop in the heating rate, based on the differential scanning calorimetry (DSC) analysis from previous research [28]. There was no additional endothermic phenomenon for wood under this temperature range.

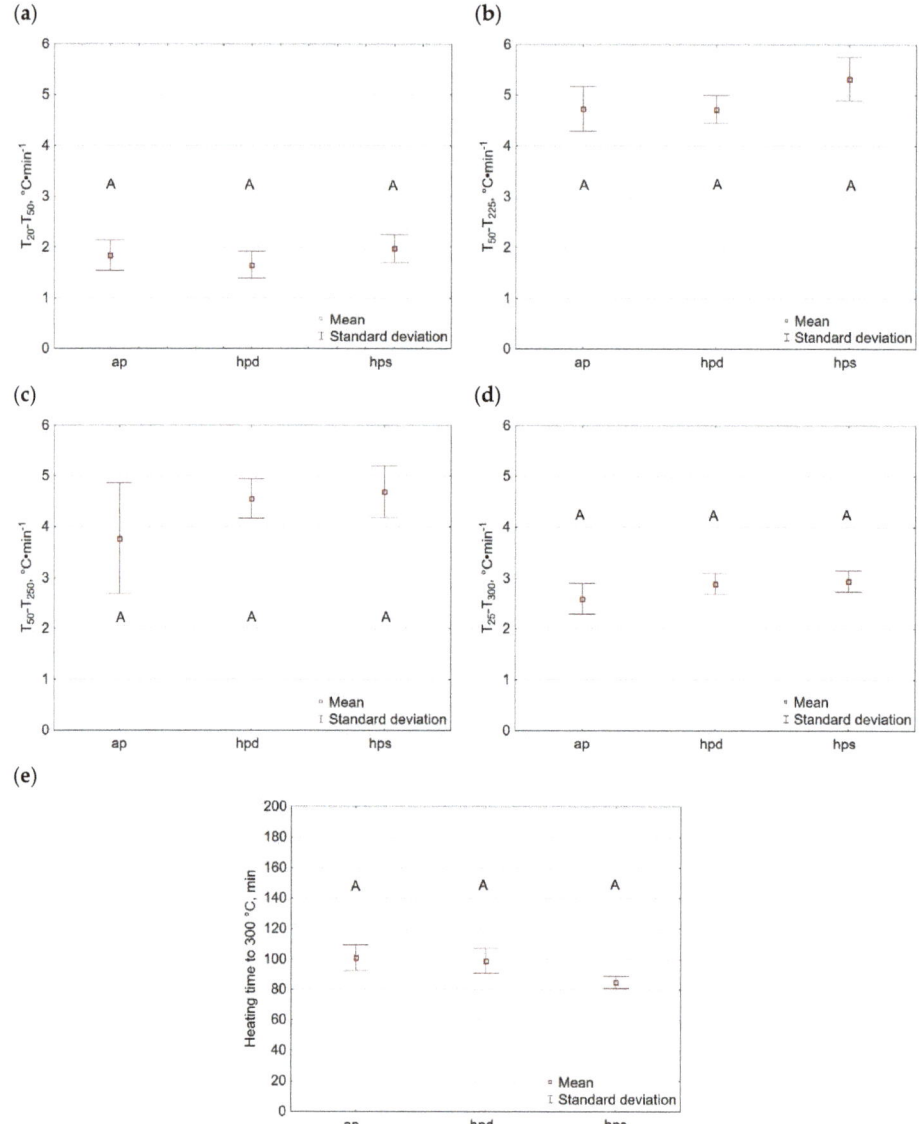

Figure 7. The heating rate of the reactor and heating time to setpoint temperature for a particular torrefaction process, (**a**) heating rate from 20 to 50 °C, (**b**) heating rate from 50 to 225 °C, (**c**) heating rate from 50 to 250 °C, (**d**) heating rate from 25 to 300 °C, (**e**) heating time up to 300 °C.

Different heating rates resulted in the starting point of the torrefaction process on the timeline scale. In the case of ap, the torrefaction started in ~101 min of the experiment, while in the case of hpd and hps in ~99, and ~85 min, respectively, the starting points did not differ significantly ($p < 0.05$) (Figure 7e).

It was observed that after the temperature reached 168–211 °C in hpd, and 183–223 °C in hps, the pressure started to increase, at 36 and 44 min (hpd and hps) (Figure A2). The pressure increase rate was not steady. The pressure started to increase after 34–42 process time (hpd and hps) (Figure A2).

Initially, the pressure increases up to 4 bar was relatively slow (pressure rate of ~0.29 and ~0.42 bar·min^{-1} for hpd and hps, respectively, was observed) (Figure 8a). After that, the pressure rate accelerated up to ~1.09 and ~1.03 bar·min^{-1} for hpd and hps, respectively (Figure 8c). The time needed to reach 4 bar was similar to the time needed for pressure increase from 4 bar up to the maximum of 14 ± 1 bar (Figure 8b,d). The pressure increase rates were similar for hpd and hps. However, for hps, the pressure increase rate was significantly higher ($p < 0.05$). In each process, pressure increase was linked with temperature increase, which is illustrated in the dT and dp diagrams presented in Appendix A, Figure A2.

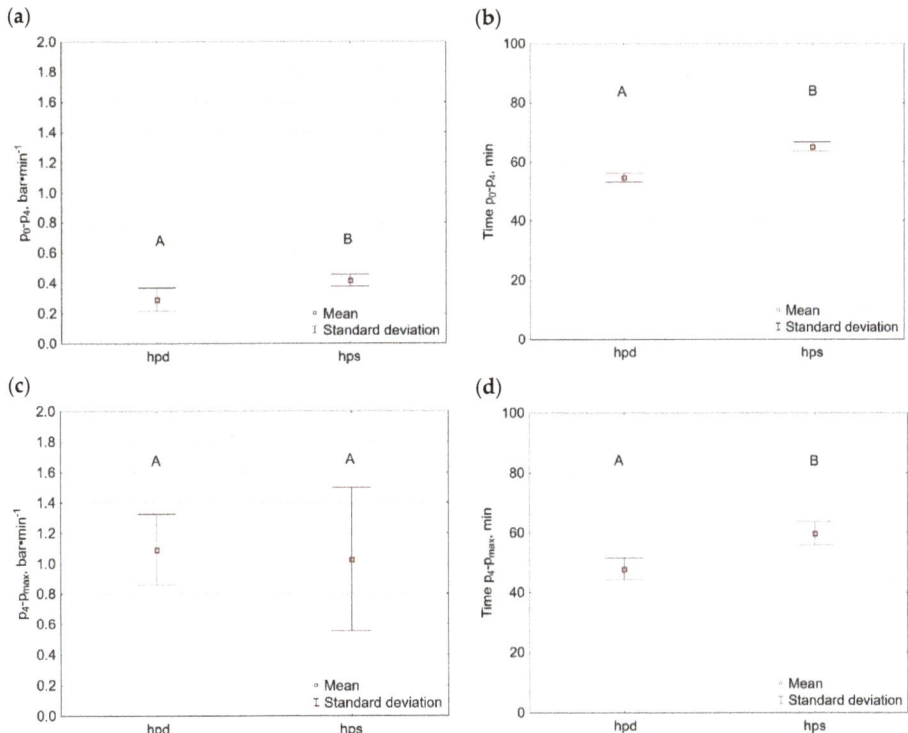

Figure 8. The pressure-increase rate of the reactor and time to a particular pressure point, (**a**) pressure rate for 0 to 4 bar, (**b**) time from pressure increase start to pressure 4 bar, (**c**) pressure rate from 4 bar to max process pressure, (**d**) time from pressure 4 bar to max pressure.

The energy demand for the torrefaction process is presented in Figure 9. The graphic presents the cumulative energy demand. The mean energy demand for torrefaction types was 2892, 2823, and 2705 Wh, for ap, hpd, and hps, respectively. The energy demand for hpd and hps to ap was decreased by around 2 and 6%, respectively. These differences are not statistically significant ($p < 0.05$).

The energy needed for the torrefaction of pellets is presented in Figure 10. The En_{raw} is the energy needed to process 1 kg of the raw pellet into a torrefied pellet (Figure 10a), whereas the $En_{torrefeid}$ is the energy needed to produce 1 kg of the torrefied pellet with account for the weight loss of biomass (Figure 10b). The results show that the average energy demand for processing a pellet by conventional torrefaction was 1.2 Wh·kg^{-1}, and high-pressure torrefaction led to decreasing of this energy demand up to 1.18 (hpd) and 1.13 (hps), respectively. The results also show that the energy needed to produce 1 kg of the torrefied pellet was the lowest for hpd ($En_{torrefied}$ = 2.06 Wh·kg^{-1}) and the highest for hps ($En_{torrefied}$ = 2.13 Wh·kg^{-1}), respectively. For ap, the $En_{torrefied}$ was 2.03 Wh·kg^{-1}. There were no statistical differences in energy needs ($p < 0.05$) (Figure 10).

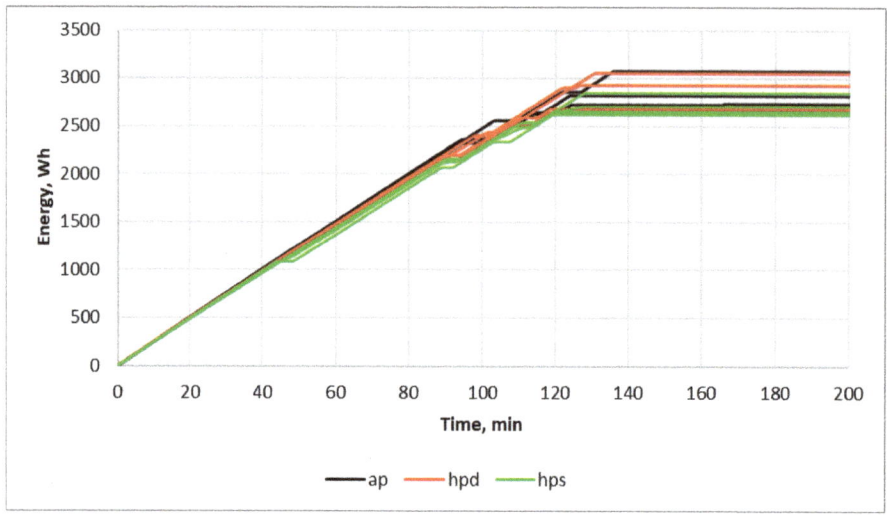

Figure 9. Cumulative energy demand for the torrefaction process.

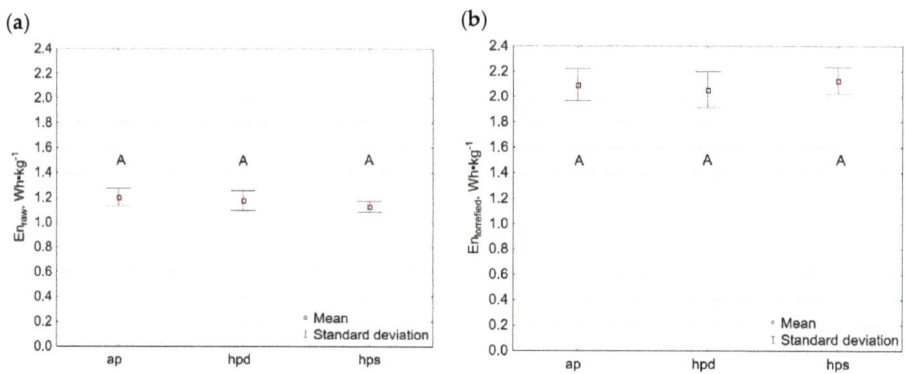

Figure 10. Energy demand to torrefied pellet production, (a) energy used to convert 1 kg of the pellet, (b) energy needed to produce 1 kg of the torrefied pellet.

Figure 11 illustrates pellets after torrefaction at different conditions. The pellet size decrease after torrefaction is apparent. Figure 11 presents a solid fraction from above the grille. Pellets torrefied at atmospheric pressure had partially disintegrated, and part of them fell under the grille. After each ap torrefaction, a part of the pellet that was on the center of the grille partially disintegrated and was not completely torrefied (Figure 12a). Such a problem did not occur for any of high-pressure torrefaction (Figure 12b). This observation may be explained by the inferior heat transfer from the reactor wall to the pellet located inside of the reactor. The used pellets were dried before torrefaction, and this likely resulted in lower thermal conductivity of pellets, and as a result, some pellets did not heat to the setpoint temperature. Whereas, in the case of high-pressure torrefaction, the heat could be better transferred due to the presence of steam that was held inside the reactor.

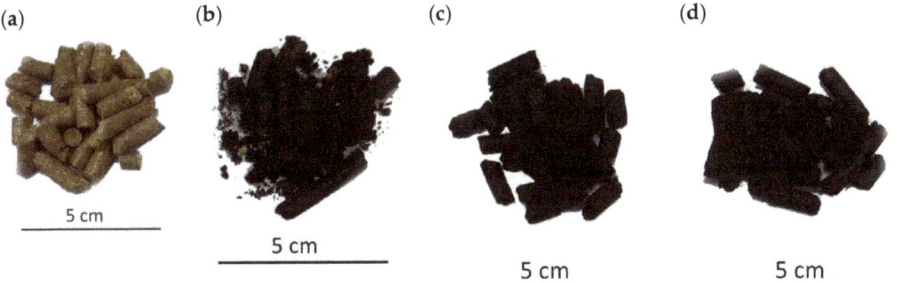

Figure 11. Samples of pellets before and after torrefaction (**a**) raw pellet, (**b**) ap, (**c**) hpd, (**d**) hps.

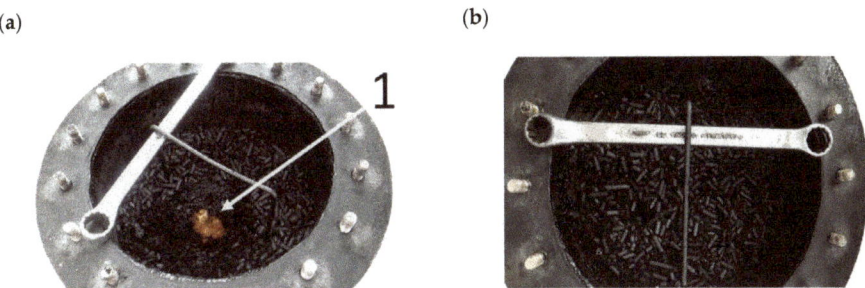

Figure 12. Torrefied pellet at (**a**) atmospheric pressure, (**b**) over atmospheric pressure. 1—pellet that was partially disintegrated and not fully torrefied during the ap process.

Figure 13 shows the fractional mass yields from the torrefaction process. Mean values of the solid fraction at atmospheric torrefaction (ap), at high-pressure torrefaction with decreasing pressure (hpd), and at high-pressure torrefaction with steady pressure (hps) were as follows: 58.1, 51.1, and 46.0%. The results show that, in most cases, the more condensable fraction was collected from the torrgas exhaust pipe (Figure 13c,d). The mean value of total condensable fractions was 23.2, 16.4, and 21.0% for ap, hpd, and hps, respectively. The mean value of non-condensable fraction was similar in the case of hpd and hps and was 26.3 and 26.1%, respectively, while for ap, the non-condensable fraction was 18.6%. There were no statistical differences in solid mass yield between processes ($p < 0.05$). The significant differences occurred for condensable fraction where hpd showed lower mass yield than ap, and hps ($p < 0.05$). On the other hand, the mass yield of the non-condensable fraction was the lowest for the ap process ($p < 0.05$).

Figure 14 shows the energy densification ratio (*EDr*) and energy yield (*EY*) for the solid fraction. In general, all torrefaction processes caused an increase in *EDr* and a decrease of *EY*. The mean value of *EDr* for over grille fractions was ~1.32, 1.38, and 1.44 for ap, hpd, and hps, respectively. In the case of the under-grille fractions, the *EDr* increased to 1.4 and 1.38 for hpd and hps, respectively. The *EY* for pellets from over grille were ~77, 70, and 66% for ap, hpd, and hps, respectively. In the case of under grille fraction, *EY* was 9 and 10% for hpd and hps, respectively. The total *EY* of solid fraction in each process was similar and was 76.8, 78.8, and 75.9% for ap, hpd, and hps, respectively. Observed differences were not statistically different ($p < 0.05$).

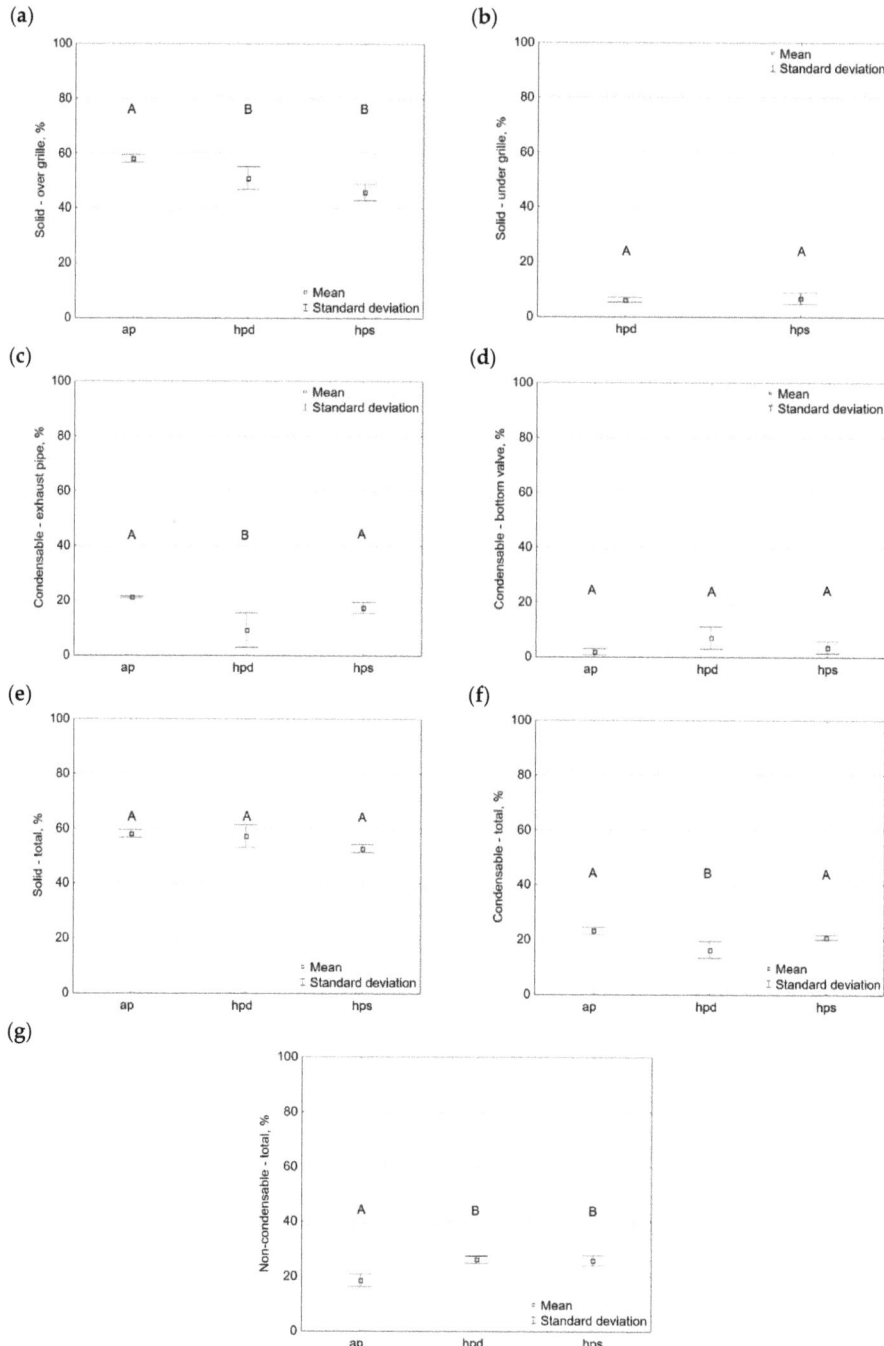

Figure 13. Torrefaction process fractions mass yields, (**a**) mass yield of solid fraction from over grille, (**b**) mass yield of solid fraction from under grille, (**c**) mass yield of condensable fraction from exhaust fraction, (**d**) mass yield of condensable fraction from the bottom valve, (**e**) total mass yield of solid fraction, (**f**) total mass yield of condensable fraction, (**g**) total mass yield of the non-condensable fraction.

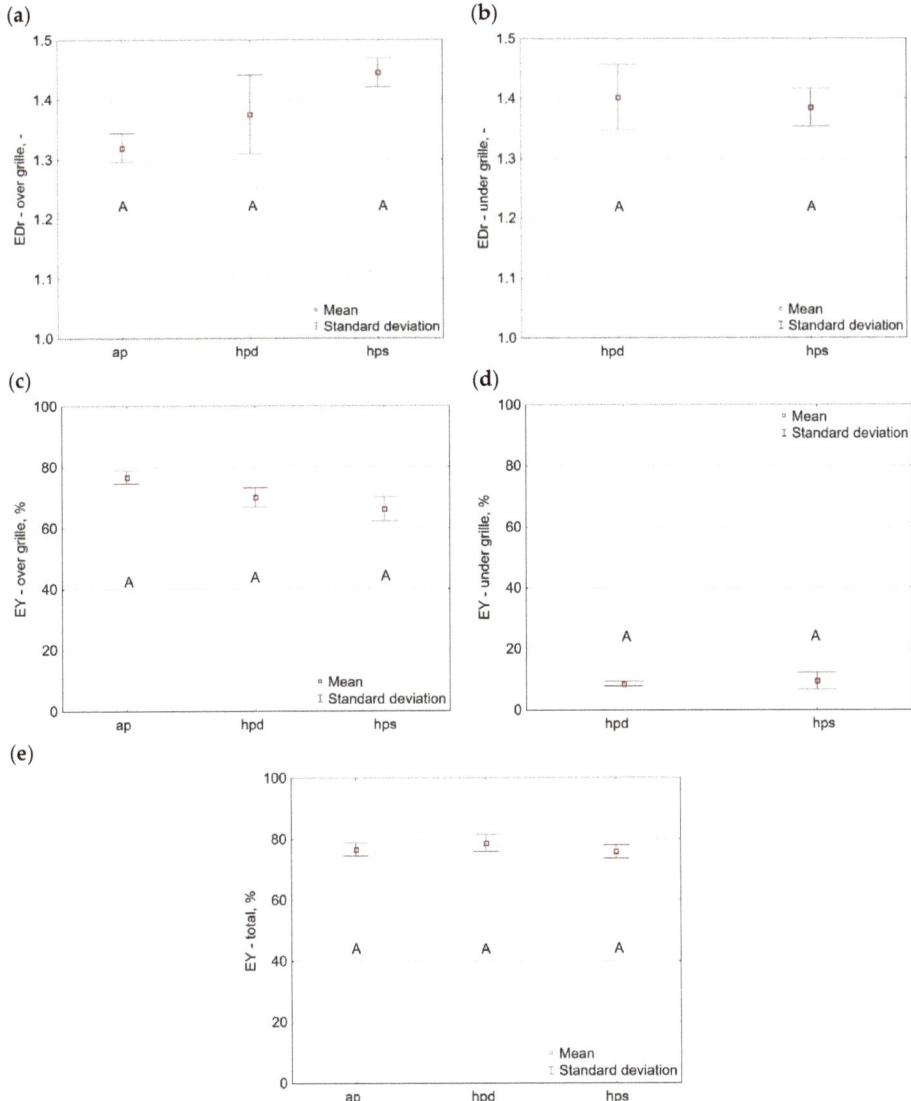

Figure 14. Energy densification ratio (*EDr*) and energy yield (*EY*) of the solid fraction, (**a**) *EDr* of over grille fraction, (**b**) *EDr* of under grille fraction, (**c**) *EY* of over grille fraction, (**d**) *EY* of under grille fraction, (**e**) *EY* of the total solid fraction.

In the work of Manouchehrinejad and Mani [29], the softwood pellet was torrefied at a laboratory scale reactor at 290 °C and 30 min. The *MY* and *EY* of the solid fraction were ~59% and ~82%, respectively. The *MY* of the non-condensable fraction was ~18%, and *MY* of the condensable fraction was 23%. In the work of Arriola et al. [30], a pellet made of forest residues was subjected to the torrefaction experiment in a horizontal tube furnace in at 300 °C and 30 min. The *MY*, *EDr*, and *EY* of solid fraction were 50.5, 1.33, and 67.2%. These results are very similar to those obtained in the present work for the ap torrefaction.

Due to a novelty of the high-pressure torrefaction, research on high-pressure pellet torrefaction was not found. Nevertheless, a Tong et al. [31] conducted a pressure torrefaction of pine sawdust in a batch autoclave reactor (sample mass 15 g, heating rate 10 °C·min^{-1}, pressure 25–50 bar, residence time 15 min, temperature 300 °C) and compared with atmospheric pressure torrefaction. The results showed that MY of pressure torrefaction for solid, liquid, and gas fraction were ~42, ~39, and ~19%, respectively, and for the traditional torrefaction, these values were ~58, ~31, ~11%, respectively [31]. The obtained results (Figure 13e–g) and [31] show that pressure during torrefaction leads to the solid mass yield decrease in favor of liquid and gas fraction. On the other hand, Qin et al. [32] showed that during pressure pyrolysis (1–20 bar), pressure favored the increase of fixed carbon instead of volatile matter, which was due to the fact that a vapor pressure can limit the escape of volatile matter from a substrate. Nevertheless, despite the difference in that fact, this and previous research [31,32] showed the pressure increases HHV of the solid fraction.

The impact of pressure on an increase of energy densification ratio (HHV enhancement factor) of the solid fraction was also studied by Tong et al. [33]. Tong et al. [33] showed that the increase of EDr could also be facilitated by secondary reactions that take place between volatiles and biomass, but with an increase of process residence time, the pressure has more impact.

3.2. Thermogravimetric Analysis

The wooden pellet was subjected to the TGA/DTG analysis at the three-heating rates: 2.5, 5.0, and 7.5 °C·min^{-1}. The results of the analysis are presented in Figure 15. The TGA/DTG analysis also revealed that a tested pellet has a typical composition consisted of hemicellulose (peak 2), cellulose (peak 3), and lignin (peak 4) determined on the base of the degradation temperature peak. At each heating rate, the pellet started losing mass before the main torrefaction process took place <200 °C. This mass loss is probably associated with bounded water release (peak 1) [34].

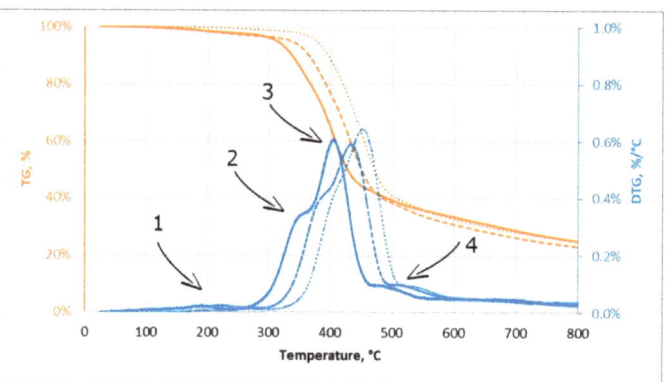

Figure 15. Results of thermogravimetric analysis (TGA)/derivative curve (DTG) analysis Heating rates are presented by different lines. Solid line—2.5 °C·min^{-1}, dashed line—5.0 °C·min^{-1}, dotted line—7.5 °C·min^{-1}. 1—bound water, 2—hemicellulose, 3—cellulose, 4—lignin.

These results match well with the observation of pressure increase in the reactor that started when temperature increased over ~150 °C (Figures 5 and 6). At the 2.5 C·min^{-1} heating rate, a mass loss associated with water evaporation started at ~150 °C and ended at ~250 °C. At 5.0 C·min^{-1}, the mass loss started earlier at 100 °C and ended at 250 °C with the greatest peak at around ~225 °C. At 7.5 C·min^{-1}, the mass losses associated with water evaporation were less visible, but the highest mass loss took place at ~250 °C (likely at this temperature, other volatile compounds were also degassed simultaneously). It is worth mentioning that at this torrefaction temperature, hemicellulose is already degraded (220–315 °C) [35].

The temperature range of 200–225 (250) °C (Figure A2) is also the place in which the greatest slowdown of the reactor heating rate was observed in atmospheric torrefaction, which is likely associated with hot gases released from the reactor. It shows that within this range, more energy is required to heat the reactor. For the high-pressure torrefaction, this phenomenon has a positive impact on the process and leads to faster heating of the reactor and, as a result, leads to energy savings.

3.3. Proximate Analysis

Figure 16 shows the results of the post-hoc HSD Tukey's test for MC, OM, CP, ash, and HHV of raw and torrefied pellets. At each plot, the median value, lower and upper quartile, and the minimum and maximum value are presented. Raw data from the proximate analysis is available in the Supplementary Materials.

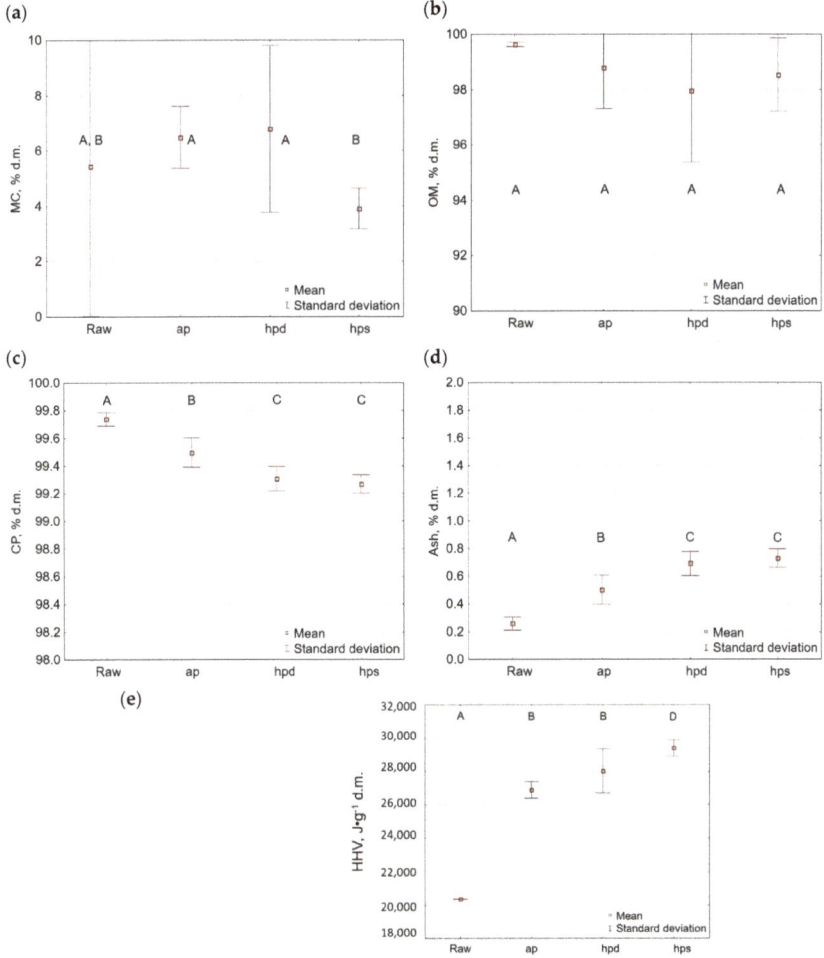

Figure 16. The results of the post-hoc HSD Tukey's test for proximate analysis of torrefied pellet (**a**) moisture content (MC), (**b**) organic matter content (OM), (**c**) combustible part (CP), (**d**) ash, (**e**) high heating value (HHV).

The mean values of moisture content were 5.44, 6.48, 6.79, and 3.90% for raw pellet, ap, hpd, and hps, respectively (Supplementary Materials). The moisture of raw pellet varied, which likely was a result of variations in the storage and time between separating each repetition. The post-hoc test showed that MC in hps is statistically ($p < 0.05$) lower than in the case of ap and hpd torrefaction (Figure 16a). The torrefaction had no impact on OM content (Figure 16b). The mean values of OM were 96.63, 98.79, 97.95, and 98.53% for raw pellet, ap, hpd, and hps, respectively (Supplementary Materials). The post-hoc test for the combustible part (Figure 16c) and ash content (Figure 16d) also showed that high-pressure torrefaction (hpd, hps) had a significant ($p < 0.05$) impact on it. The mean values of CP were 99.74, 99.50, 99.31, and 99.27%, for raw pellet, ap, hpd, and hps, respectively. In the case of ash, the mean values were 0.26, 0.50, 0.69, 0.73% (Supplementary Materials). These results show that high-pressure torrefaction leads to a higher reduction of CP and a higher increase in ash content compared with conventional torrefaction.

The greatest impact of high-pressure pellet torrefaction was seen on the HHV. The mean value of HHV for raw pellet, ap, hpd, and hps process was 20,371 J·g^{-1}, 26,901 J·g^{-1}, 28,028 J·g^{-1}, 29,398 J·g^{-1}, respectively (Supplementary Materials). This means that the conventional torrefaction increased a HHV by ~32%, and high-pressure torrefaction increased it by ~37% (hpd) and ~44% (hpd), and these changes were statistically significant ($p < 0.05$) (Figure 16e).

The post-hoc test was conducted for HHV results of under the grille fraction. We hypothesized that under grille fraction would have a higher value of HHV because this fraction mixed with the oil, which condensed after the process end and drained down into the bottom part of the reactor. In the case of hpd, there were no statistical differences in HHV between over and under grille fraction. For hps, the post-hoc test showed that under grille fraction has lower HHV than over grille fraction, which shows that the initial hypothesis was wrong.

3.4. Pellet Grinding Test

Figure 17a,b show the results of the post-hoc test for the grinding test. The plots present the total specific energy (Es), and effective energy (Ee), respectively. The mean value of the required energy needed to grind raw pellets was Es = 7.35 Wh·kg^{-1}, and Ee = 4.98 Wh·kg^{-1} (Supplementary Materials). The torrefaction in all cases decreased significantly ($p < 0.05$), the energy requirements for pellet grinding (Figure 17). The mean value of the total specific energy and effective energy for the ap was Es = 2.58 Wh·kg^{-1}, and Ee = 1.58 Wh·kg^{-1}, while for the hpd it was Es = 2.45 Wh·kg^{-1} and Ee = 1.29 Wh·kg^{-1}, and for hps was Es = 2.58 Wh·kg^{-1} and Ee = 1.36 Wh·kg^{-1}. No improvements to grinding were observed for high-pressure torrefaction in comparison to the conventional one. In general, energy for pellet grinding after torrefaction decreased by ~65.5%.

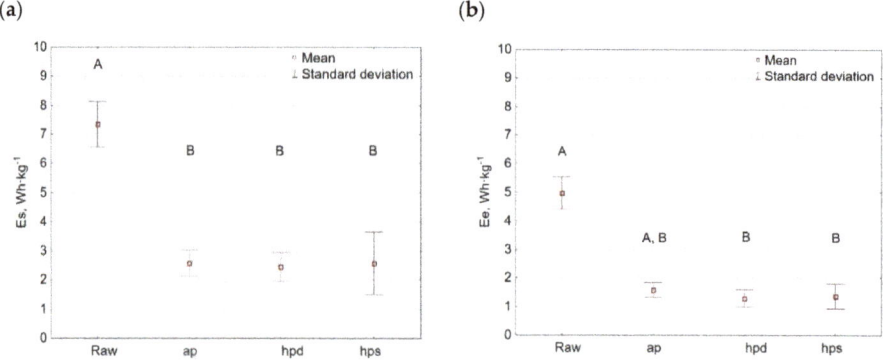

Figure 17. The results of the post-hoc HSD Tukey's test for the grinding test, (**a**) total specific energy (Es), (**b**) specific effective energy (Ee).

The fact that the energy requirement for biomass grinding after the torrefaction is decreasing is well known [36]. Due to a wide range of grinding equipment types (also known as planetary ball mill, bond ball mill, ring roller mill, knife mill) with different energy consumption and optimal capacity [37], the obtained grinding results in this work cannot be directly compared to another research. Nevertheless, based on this limited dataset, it is valid to conclude that wood pellets torrefied at atmospheric pressure and under high-pressure required 2.5 to 3.9 times less energy than the dry pellets. Additionally, the pressure did not have a significant impact on the change of E_a and E_s value ($p < 0.05$) when compared to atmospheric torrefaction (Figure 17).

4. Conclusions

This paper provides a comprehensive description of the new approach to biomass torrefaction under high-pressure conditions. A new type of laboratory pressure reactor was described and tested along with the experience gained during its operation. The initial results of the pressure impact on the key parameters of torrefaction were described, namely: energy demand for torrefaction process, mass and energy yields of particular fractions, proximate analysis with HHV of solid fraction, the energy demand to raw and torrefied pellet grinding. Based on the results, the following main conclusions arise from this research:

- High-pressure torrefaction requires up to six percent less energy than a conventional one;
- High-pressure torrefaction causes less disintegration of pellet compared to the conventional one;
- High-pressure torrefaction leads to higher energy densification in pellets of up to 44% compared to the conventional one up to 32%;
- The presence of high-pressure during torrefaction has no impact on torrefied pellet grinding energy demand in comparison to the conventional one; therefore, this factor appears to be less relevant in the future technology development process.

Pressure-aided torrefaction presents new opportunities to consider designing more efficient waste treatment installations. Such installations may be more expensive to build, but the additional benefits of pressure torrefaction, such as lower energy needs for the process and better quality of biochar, warrant future studies on these types of reactors.

In general, the study indicated that there is a potential to produce better quality solid fuels at a lower cost in comparison to conventional torrefaction. Further study is warranted due to the importance of the topic and gaps in knowledge.

Supplementary Materials: The following are available online at http://www.mdpi.com/1996-1073/13/18/4790/s1, The supplementary materials file contains data on the Initial tests and analysis of high-pressure torrefaction on wood pellets. The spreadsheet "T&p vs t" contains data of temperature and pressure changes of the reactor during torrefaction. The spreadsheet "Torrefaction energy demand" contains data of the power consumption for each torrefaction process. The spreadsheet "Torrefaction process analysis" contains data of fractions mass yield. The spreadsheet "TGA" contains data from the thermogravimetric analysis of the pellet. The spreadsheet "Proximate analysis" contains data of proximate analysis of the raw and torrefied pellets. The spreadsheet "HHV" contains data of the high heating value of raw and torrefied pellets. The spreadsheet "Grinding energy demand)" contains data of the energy used to raw and torrefied pellet grinding.

Author Contributions: Conceptualization, A.B.; methodology, K.Ś., B.M. and A.B.; validation, K.Ś.; formal analysis, B.M. and K.Ś.; investigation, B.M. and K.Ś.; resources, A.B.; data curation, K.Ś.; writing—original draft preparation, K.Ś. and B.M.; writing—review and editing, K.Ś., A.B., B.M. and J.A.K.; visualization, K.Ś. and B.M.; supervision, A.B. and J.A.K. All authors have read and agreed to the published version of the manuscript.

Funding: Partial support came from the Iowa Agriculture and Home Economics Experiment Station: project number IOW05556 (Future Challenges in Animal Production Systems: Seeking Solutions through Focused Facilitation, sponsored by Hatch Act and State of Iowa funds).

Acknowledgments: The presented article results were obtained as part of the activity of the leading research team Waste and Biomass Valorization Group (WBVG), https://www.upwr.edu.pl/research/50121/waste_and_biomass_valorization_group_wbvg.html.

Conflicts of Interest: The authors declare no conflict of interest. The funders had no role in the design of the study; in the collection, analyses, or interpretation of data; in the writing of the manuscript, or in the decision to publish the results.

Appendix A

ap 1

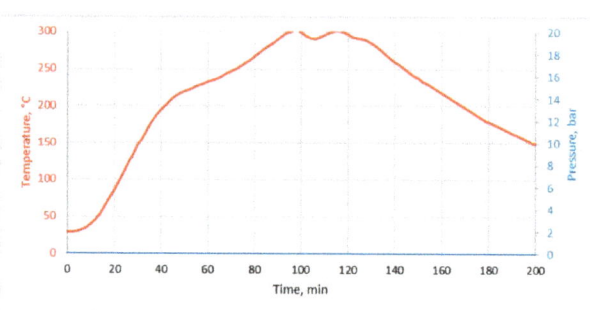

ap 2

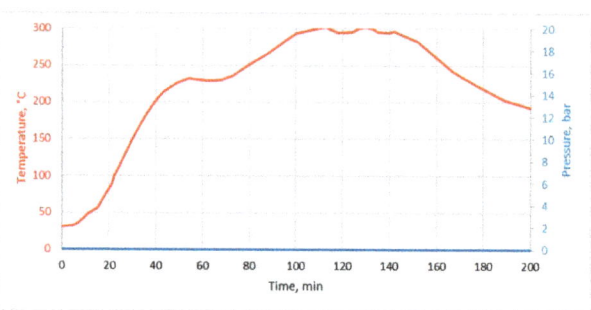

Figure A1. *Cont.*

ap 3

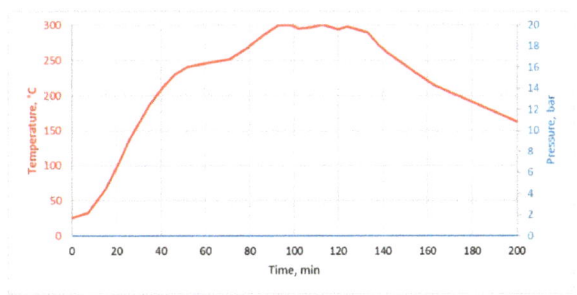

hpd 1

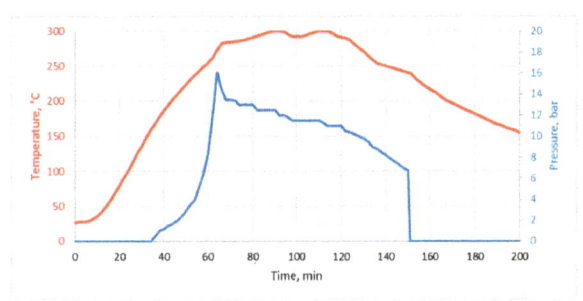

hpd 2

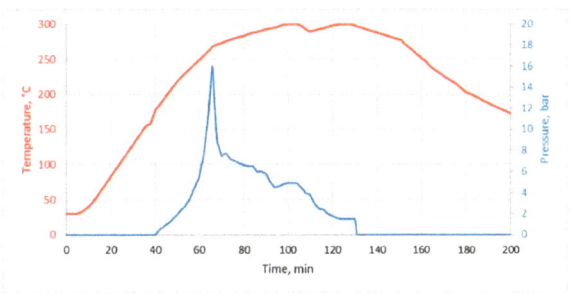

hpd 3

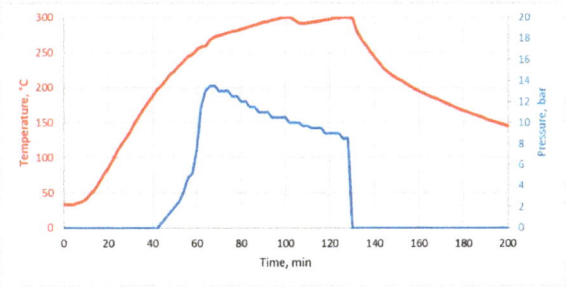

Figure A1. *Cont.*

hpd 4

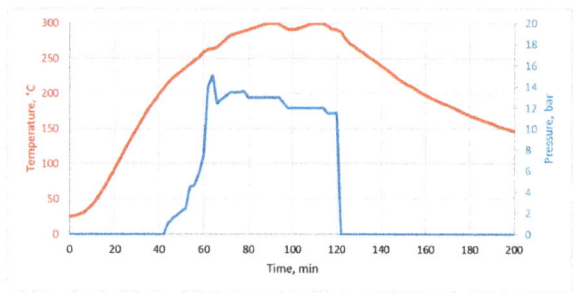

hps 1

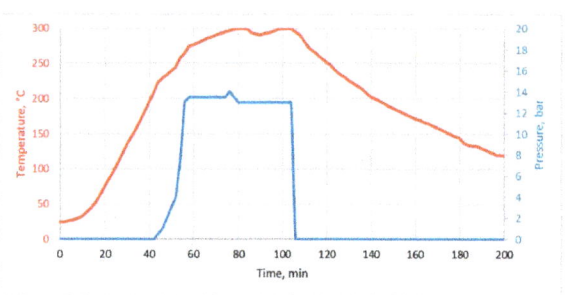

hps 2

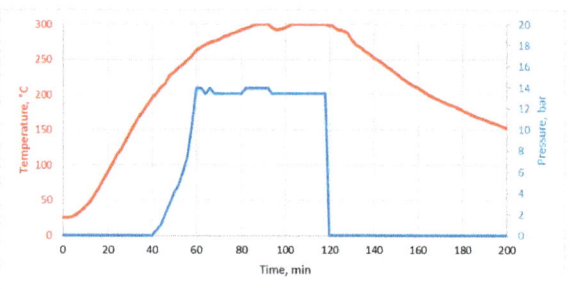

hps 3

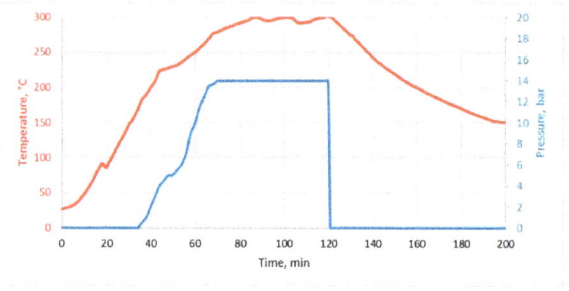

Figure A1. *Cont.*

hps 4

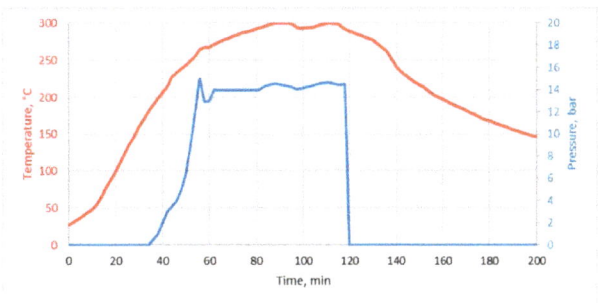

Figure A1. Temperature/pressure patterns during the torrefaction of all experiments.

hpd 1

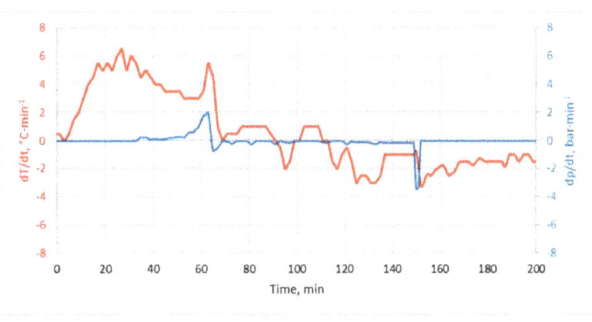

hpd 2

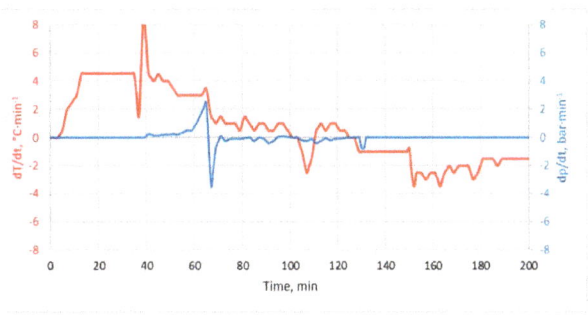

Figure A2. *Cont.*

hpd 3

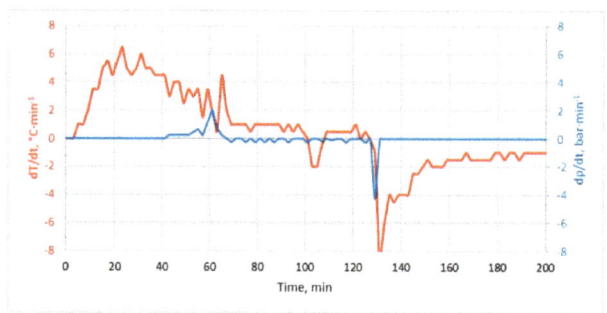

hpd 4

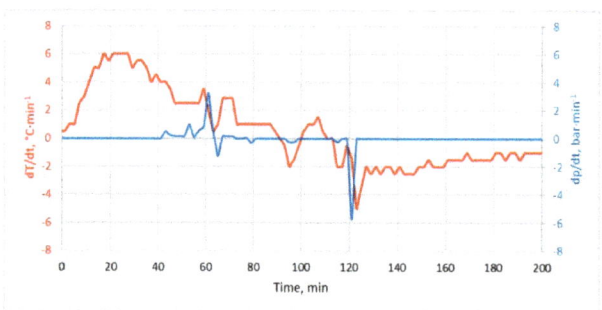

hps 1

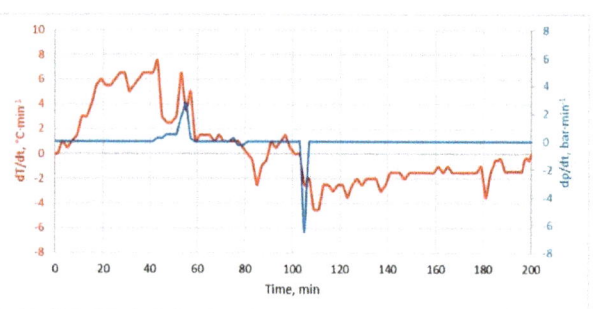

Figure A2. *Cont.*

hps 2

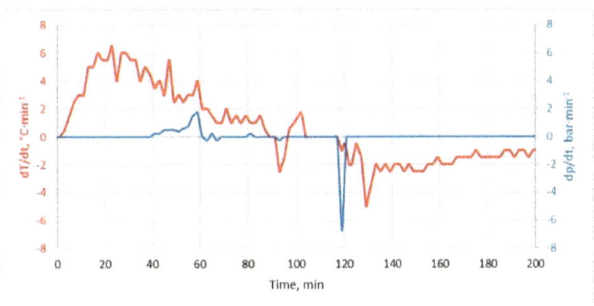

hps 3

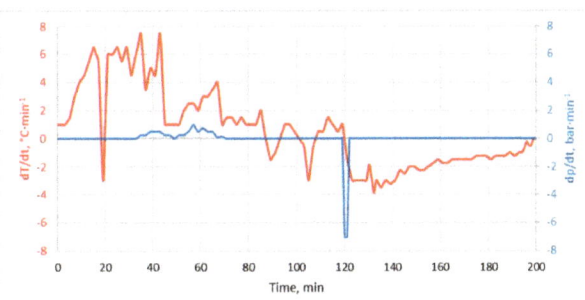

hps 4

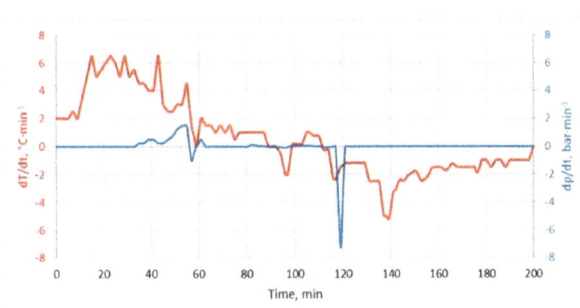

Figure A2. Derivatives of temperature/pressure patterns during the torrefaction of all experiments.

References

1. Stanek, W.; Czarnowska, L.; Gazda, W.; Simla, T. Thermo-ecological cost of electricity from renewable energy sources. *Renew. Energy* **2018**, *115*, 87–96. [CrossRef]
2. Stanek, W.; Simla, T.; Gazda, W. Exergetic and thermo-ecological assessment of heat pump supported by electricity from renewable sources. *Renew. Energy* **2019**, *131*, 404–412. [CrossRef]
3. Białowiec, A.; Micuda, M.; Koziel, J.A. Waste to carbon: Densification of torrefied refuse-derived fuel. *Energies* **2018**, *11*, 3233. [CrossRef]
4. Peng, J.H.; Bi, X.T.; Sokhansanj, S.; Lim, C.J. Torrefaction and densification of different species of softwood residues. *Fuel* **2013**, *111*, 411–421. [CrossRef]

5. Dudek, M.; Świechowski, K.; Manczarski, P.; Koziel, J.A.; Białowiec, A. The effect of biochar addition on the biogas production kinetics from the anaerobic digestion of brewers' spent grain. *Energies* **2019**, *12*, 1518. [CrossRef]
6. Medyńska-Juraszek, A. Biochar as a soil amendment. *Soil Sci. Ann.* **2016**, *67*, 151–157. [CrossRef]
7. Nunes, L.J.R. A case study about biomass torrefaction on an industrial scale: Solutions to problems related to self-heating, difficulties in pelletizing, and excessive wear of production equipment. *Appl. Sci.* **2020**, *10*, 2546. [CrossRef]
8. Ábrego, J.; Plaza, D.; Luño, F.; Atienza-Martínez, M.; Gea, G. Pyrolysis of cashew nutshells: Characterization of products and energy balance. *Energy* **2018**, *158*, 72–80. [CrossRef]
9. Energy Innovation Austria, Current Austrian Developments and Examples of Sustainable Energy Technologies. Available online: https://www.energy-innovation-austria.at/article/torrefizierung/?lang=en (accessed on 25 July 2020).
10. Banse, T. Your Word of the Day is "Torrefaction." First-of-Its Kind Plant to Open in Eastern Oregon. Available online: https://www.kuow.org/stories/your-word-of-the-day-is-torrefaction-first-of-its-kind-plant-to-open-in-eastern-oregon (accessed on 25 July 2020).
11. Kanwal, S.; Chaudhry, N.; Munir, S.; Sana, H. Effect of Torrefaction Conditions on the Physicochemical Characterization of Agricultural Waste (Sugarcane Bagasse). *Waste Manag.* **2019**, *88*, 280–290. [CrossRef] [PubMed]
12. Wei, W.; Mellin, P.; Yang, W.; Wang, C.; Hultgren, A.; Salman, H. *Utilization of Biomass for Blast Furnace in Sweden-Report I: Biomass Availability and Upgrading Technologies*; Technical Report for KTH Royal Institute of Technology: Stockholm, Sweden, December 2013. [CrossRef]
13. Babinszki, B.; Jakab, E.; Sebestyén, Z.; Blazsó, M.; Berényi, B.; Kumar, J.; Krishna, B.B.; Bhaskar, T.; Czégény, Z. Comparison of hydrothermal carbonization and torrefaction of azolla biomass: Analysis of the solid products. *J. Anal. Appl. Pyroly* **2020**, *149*, 104844. [CrossRef]
14. Bergman, P.C.A.; Boersma, A.R.; Zwart, R.W.R.; Kiel, J.H.A. Torrefaction for Biomass Co-Firing in Existing Coal-Fired Power Stations. BIOCOAL. The Netherlands. 2005; p. 71. Available online: https://www.osti.gov/etdeweb/biblio/20670903 (accessed on 25 July 2020).
15. Surup, G.R.; Leahy, J.J.; Timko, M.T.; Trubetskaya, A. Hydrothermal carbonization of olive wastes to produce renewable, binder-free pellets for use as metallurgical reducing agents. *Renew. Energy* **2020**, *155*, 347–357. [CrossRef]
16. Ahmad, F.; Silva, E.L.; Varesche, M.B.A. Hydrothermal processing of biomass for anaerobic digestion–A review. *Renew. Sustain. Energy Rev.* **2018**, *98*, 108–124. [CrossRef]
17. Wannapeera, J.; Worasuwannarak, N. Upgrading of woody biomass by torrefaction under pressure. *J. Anal. Appl. Pyroly.* **2012**, *96*, 173–180. [CrossRef]
18. Malik, B.; Pirzadah, T.B.; Islam, S.T.; Tahir, I.; Kumar, M.; ul Rehman, R. Biomass pellet technology: A green approach for sustainable development. *Agric. Biomass Based Potent. Mater.* **2015**, 403–433. [CrossRef]
19. Kułażyński, M.; Kaczmarczyk, J.; Świątek, Ł.; Pstrowska, K. Technological problems occurring during the implementation process of co-firing brown coal with biomass. *Logistyka* **2015**, 277–282. Available online: http://yadda.icm.edu.pl/yadda/element/bwmeta1.element.baztech-2416d3f1-2fe1-489a-b3a5-b968e2d2280d (accessed on 25 July 2020).
20. Chen, W.H.; Peng, J.; Bi, X.T. A state-of-the-art review of biomass torrefaction, densification and applications. *Renew. Sustain. Energ. Rev.* **2015**, *44*, 847–866. [CrossRef]
21. Świechowski, K.; Liszewski, M.; Babelewski, P.; Koziel, J.A.; Białowiec, A. Fuel properties of torrefied biomass from pruning of oxytree. *Data* **2019**, *4*, 55. [CrossRef]
22. Stępień, P.; Świechowski, K.; Hnat, M.; Kugler, S.; Stegenta-Dąbrowska, S.; Koziel, J.A.; Manczarski, P.; Białowiec, A. Waste to carbon: Biocoal from elephant dung as new cooking fuel. *Energies* **2019**, *12*, 4344. [CrossRef]
23. PN-EN 14346:2011 Standard. Waste Characteristics. Calculation of Dry Mass Based on Dry Residue or Water Content. Available online: https://infostore.saiglobal.com/en-au/Standards/PN-EN-14346-2011-932471_SAIG_PKN_PKN_2197939 (accessed on 25 July 2020).
24. PN-EN 15169:2011 Standard. Waste Characteristics. Determination of Organic Matter Content for Waste, Slurry and Sludge. Available online: http://sklep.pkn.pl/pn-en-15169-2011p.html (accessed on 25 July 2020).

25. PN-Z-15008-04:1993 Standard. Municipal Solid Waste. Analysis of Combustible and Non-Combustible Content. Available online: http://sklep.pkn.pl/pn-z-15008-04-1993p.html (accessed on 25 July 2020).
26. PN-G-04513:1981 Standard. Solid fuels. Determination of the Higher Heating Value and the Lower Heating Value. Available online: http://sklep.pkn.pl/pn-g-04513-1981p.html (accessed on 25 July 2020).
27. Williams, O.; Lester, E.; Kingman, S.; Giddings, D.; Lorimor, S.; Eastwick, C. Benefits of dry comminution of biomass pellets in a knife mill. *Biosyst. Eng.* **2017**, *160*, 42–54. [CrossRef]
28. Świechowski, K.; Syguła, E.; Koziel, J.A.; Stępień, P.; Kugler, S.; Manczarski, P.; Białowiec, A. Low-temperature pyrolysis of municipal solid waste components and refuse-derived fuel—process efficiency and fuel properties of carbonized solid fuel. *Data* **2020**, *5*, 48. [CrossRef]
29. Manouchehrinejad, M.; Mani, S. Torrefaction after pelletization (TAP): Analysis of torrefied pellet quality and co-products. *Biomass Bioenerg.* **2018**, *118*, 93–104. [CrossRef]
30. Arriola, E.; Chen, W.H.; Chih, Y.K.; De Luna, M.D.; Show, P.L. Impact of post-torrefaction process on biochar formation from wood pellets and self-heating phenomena for production safety. *Energy* **2020**, *207*, 1–13. [CrossRef]
31. Tong, S.; Xiao, L.; Li, X.; Zhu, X.; Liu, H.; Luo, G.; Worasuwannarak, N.; Kerdsuwan, S.; Fungtammasan, B.; Yao, H. A gas-pressurized torrefaction method for biomass wastes. *Energ. Convers. Manag.* **2018**, *173*, 29–36. [CrossRef]
32. Qin, L.; Wu, Y.; Hou, Z.; Jiang, E. Influence of biomass components, temperature and pressure on the pyrolysis behavior and biochar properties of pine nut shells. *Bioresour. Technol.* **2020**, *313*, 123682. [CrossRef]
33. Tong, S.; Sun, Y.; Li, X.; Hu, Z.; Dacres, O.D.; Worasuwannarak, N.; Luo, G.; Liu, H.; Hu, Y.; Yao, H. Gas-pressurized torrefaction of biomass wastes: Roles of pressure and secondary reactions. *Bioresour. Technol.* **2020**, *313*, 1–8. [CrossRef] [PubMed]
34. Grycova, B.; Pryszcz, A.; Krzack, S.; Klinger, M.; Lestinsky, P. Torrefaction of biomass pellets using the thermogravimetric analyser. *Biomass Convers. Biorefinery* **2020**. [CrossRef]
35. Yang, H.; Yan, R.; Chen, H.; Lee, D.H.; Zheng, C. Characteristics of hemicellulose, cellulose and lignin pyrolysis. *Fuel* **2007**, *86*, 1781–1788. [CrossRef]
36. Repellin, V.; Govin, A.; Rolland, M.; Guyonnet, R. Energy requirement for fine grinding of torrefied wood. *Biomass Bioenerg.* **2010**, *34*, 923–930. [CrossRef]
37. Williams, O.; Newbolt, G.; Eastwick, C.; Kingman, S.; Giddings, D.; Lorimor, S.; Lester, E. Influence of mill type on densified biomass comminution. *Appl. Energ.* **2016**, *182*, 219–231. [CrossRef]

© 2020 by the authors. Licensee MDPI, Basel, Switzerland. This article is an open access article distributed under the terms and conditions of the Creative Commons Attribution (CC BY) license (http://creativecommons.org/licenses/by/4.0/).

MDPI\
St. Alban-Anlage 66\
4052 Basel\
Switzerland\
Tel. +41 61 683 77 34\
Fax +41 61 302 89 18\
www.mdpi.com

Energies Editorial Office\
E-mail: energies@mdpi.com\
www.mdpi.com/journal/energies

www.ingramcontent.com/pod-product-compliance
Lightning Source LLC
LaVergne TN
LVHW070434100526
838202LV00014B/1596